NUTRITIONAL
PATHOLOGY

THE BIOCHEMISTRY OF DISEASE

A Molecular Approach to Cell Pathology

A Series of Monographs

SERIES EDITORS

Emmanuel Farber

Department of Pathology
University of Toronto
Toronto, Ontario, Canada

Henry C. Pitot

McArdle Laboratory for Cancer Research
University of Wisconsin
Madison, Wisconsin

Other Volumes in Preparation

NUTRITIONAL PATHOLOGY

Pathobiochemistry of Dietary Imbalances

edited by
HERSCHEL SIDRANSKY

Department of Pathology
George Washington University Medical Center
Washington, D.C.

CRC Press
Taylor & Francis Group
Boca Raton London New York

CRC Press is an imprint of the
Taylor & Francis Group, an **informa** business

First published 1985 by Marcel Dekker, Inc.

Published 2019 by CRC Press
Taylor & Francis Group
6000 Broken Sound Parkway NW, Suite 300
Boca Raton, FL 33487-2742

First issued in paperback 2019

No claim to original U.S. Government works

ISBN-13: 978-0-367-45172-1 (pbk)
ISBN-13: 978-0-8247-7303-8 (hbk)

Visit the Taylor & Francis Web site at
http://www.taylorandfrancis.com

and the CRC Press Web site at
http://www.crcpress.com

Library of Congress Cataloging in Publication Data
Main entry under title:

Nutritional Pathology.

   (Biochemistry of disease ; v. 10)
   Includes bibliographies and index.
   1. Nutritionally induced diseases. I. Sidransky,
Herschel, [date] . II. Series. [DNLM: 1. Disease--
etiology. 2. Nutrition Disorders--complications.
W1 BI64F v.10 / QZ 105 N9763]
RC622.N8935 1985   616.3'9   85-4399
ISBN 0-8247-7303-9

I dedicate this volume to my former
colleague, Challakonda N. Murty, Ph.D.,
who died on March 22, 1984.

PREFACE

Nutrition is vital for all living things. Delineation of the requirements for adequate nutrition in living organisms, particularly in mammals, has been developing over many years. Today guidelines are available that enumerate the nutrients and the amounts necessary to promote satisfactory growth and well being. These requirements are constantly being updated and refined with special attention directed toward certain age groups, such as the pediatric and geriatric groups, looking for needs possibly different or altered from those of the norm.

The scope of nutrition has continuously been expanding. Decades ago, the science of nutrition focused primarily on nutritional deficiency states that were explored in much detail. These studies have contributed greatly to our current knowledge about the basic essential components in the diet and to the consequences of individual deficiencies. More recently investigators have become cognizant of many complexities relating to nutrition that occur, such as imbalances, toxicities, altered requirements, fads, endogenous metabolic abnormalities, etc. The boundaries of nutrition thus have expanded to include all forms of nutritional disturbances that affect or influence the organism directly or indirectly. The role that altered or disturbed nutrition plays in the pathogenesis of disease falls within the realm of nutritional pathology.

Man is aware that proper nutrition is of considerable influence to his survival. Indeed, nutrition is now considered to be one of the most important of all environmental factors. Manipulations in nutrition, exogenous or endogenous, may be involved in the pathogenesis of many disease states, ranging from metabolic diseases, such as diabetes and obesity, to the prime killer diseases, such as cancer and cardiovascular disease. The recent resurgence of interest and awareness of nutrition in relation to disease states by scientists as well as by the general public has led to the expansion of nutritionally oriented research, which has been directed into new and important areas and directions. The purpose of this book is to present to the reader a number of exciting recent developments in the field of nutritional investigation. Although only several important topics will be covered, each should serve to illustrate a fundamental aspect in the various approaches and directions whereby research in nutritional pathology is currently taking. Today's state of development indicates that the future of research in nutritional pathology is promising. Many vistas need exploration. The editor feels that the chapters of this book will be informative and hopes that they will encourage others to become interested in the search for further knowledge relating to the relationships between nutrition and diseases, the concern of nutritional pathology.

Herschel Sidransky

CONTRIBUTORS

Sikandar L. Katyal, Ph.D. Department of Pathology, University of Pittsburgh School of Medicine, Pittsburgh, Pennsylvania

David Kritchevsky, Ph.D. The Wistar Institute of Anatomy and Biology, Philadelphia, Pennsylvania

Daniel S. Longnecker, M.D. Department of Pathology, Dartmouth Medical School, Hanover, New Hampshire

Rosemary E. McDanell, B.Sc. Department of Clinical Pharmacology, University College London, London, England

André E. M. McLean, B.M., Ph.D., FRCPath. Department of Clinical Pharmacology, University College London, London, England

Brian L. G. Morgan, B.Sc., M.Sc., Ph.D. Institute of Human Nutrition, Columbia University College of Physicians & Surgeons, New York, New York

Challakonda N. Murty, Ph.D.[†] Department of Pathology, George Washington University Medical Center, Washington, D.C.

Daphne A. Roe, M.D. Division of Nutritional Sciences, Cornell University, Ithaca, New York

Hisashi Shinozuka, M.D., Ph.D. Department of Pathology, University of Pittsburgh School of Medicine, Pittsburgh, Pennsylvania

Herschel Sidransky, M.D. Department of Pathology, George Washington University Medical Center, Washington, D.C.

[†]Deceased.

George V. Vahouny, Ph.D. Department of Biochemistry, George Washington University Medical Center, Washington, D.C.

Myron Winick, M.D. Institute of Human Nutrition, Columbia University College of Physicians & Surgeons, New York, New York

CONTENTS

NUTRITIONAL PATHOLOGY

1

Tryptophan
Unique Action by an Essential Amino Acid

Herschel Sidransky

George Washington University Medical Center
Washington, D.C.

I. INTRODUCTION

Tryptophan has been acknowledged as an essential amino acid for many years, but only in recent years unique biological and metabolic properties have been attributed to this amino acid. Indeed, tryptophan is now known to be involved in regulatory control mechanisms of the central nervous system and of the liver. Thus, in addition to being an essential building block in proteins, tryptophan has other important effects that influence and regulate vital biological mechanisms. In-depth understanding of how one essential nutrient, such as tryptophan, acts in normal as well as in disease states is of importance. Such information should offer an opportunity whereby specific nutritional manipulation of a dietary component may be used in a rational way in the possible prevention and/or treatment of certain disease states.

II. HISTORICAL DATA AND FINDINGS IN DEFICIENCIES

Historically, our knowledge of tryptophan began some 85 years ago. In 1901 Hopkins and Cole (1) isolated tryptophan from a pancreatic digest of casein. Its structure was established by Ellinger and Flamand in 1907 (2). In related nutritional studies, Willcock and Hopkins (3) in 1906 observed that mice failed to grow and even died if their sole source of dietary protein was zein. When tryptophan was added to the ration, the lives of the animals were prolonged. A few years later, Osborne

and Mendel (4) demonstrated that zein plus tryptophan and lysine pro-
moted normal growth in rats and thus established that these two amino
acids were essential nutrients. Subsequently, in the 1950s Rose et al.
(5) demonstrated that L-tryptophan was an essential dietary component
for humans.

Early experiments with mice (3) and with rats (4) showed that tryptophan deficiency leads to a disturbance in growth. This amino acid
was also necessary for the maintenance of nitrogen equilibrium in mature rats (6), mice (7), pigs (8), and dogs (9). A variety of pathological changes in experimental animals have been ascribed to tryptophan deficiency. Cataracts (10,11) and corneal vascularization (12)
have been reported in animals subjected to tryptophan deficiency. Indeed, the only authenticated and reproducible example of experimental
cataract caused by dietary deficiencies was that produced in guinea pigs
and rats by feeding a diet deficient or devoid in tryptophan (13,14).
Hematological manifestations of anemia (15), reduction in plasma proteins (16), fatty liver (16-22), and pancreatic atrophy (19) have been
reported in tryptophan-deficient rats. Scoliosis has been reported
after feeding fish a tryptophan-deficient diet (23,24).

The number of investigations into the biochemical changes in tissues
and organs of experimental animals fed tryptophan-deficient diets has
been limited. Most of these studies were concerned with changes in
the liver. When tryptophan-deficient diets were fed either ad libitum
or by force-feeding, differences in hepatic metabolism were observed
(20). Also, when animals were fed other single essential amino acid-
deficient diets, differences in the pathological and biochemical changes
in the livers that were dependent upon the route of feeding—ad libitum
or force-feeding—have been reported by others (25). In general, more
marked changes were observed after force-feeding than after ad libitum-
feeding of the deficient diets.

Samuels et al. (19) suggested that the fatty livers in rats force-fed
a tryptophan-devoid diet were due to increased synthesis of hepatic
lipid from carbon chains of amino acids that were not being used normally for protein synthesis. However other experimental studies did
not support this suggestion (25). Patrick and Bennington (22) reported that the incorporation of labeled amino acids into liver lipids
was greater in rats force-fed a tryptophan-devoid diet for 5 days compared with those force-fed a complete diet. Since fatty livers develop
after force-feeding other single essential amino acid-devoid diets, other
than a troptophan-devoid diet (25), the mechanism responsible may be
similar regardless of the type of single deficiency. Several studies
have suggested that stimulation of hepatic lipid synthesis may be responsible for the genesis of the fatty liver with single essential amino
acid deficiency (26,27). However because a different mechanism, impairment of lipoprotein released (28,29), is considered to be involved
in most cases of experimental fatty liver, further experimentation is

necessary to clarify the mechanism with respect to amino acid deficiency. Conceivably, impairment of the synthesis of apolipoprotein resulting from a single amino acid deficiency could cause accumulation of triglycerides because of inadequate conversion of triglycerides to lipoprotein.

Hepatic protein synthesis has been evaluated in rats force-fed a tryptophan-devoid diet for 1 to 5 days by viewing the status of hepatic polyribosomes and by measuring labeled amino acid incorporation into hepatic and plasma proteins in vivo (30-32). The results revealed evidence of enhanced hepatic protein synthesis. Also, Patrick and Bennington (22) reported that in vivo hepatic protein synthesis was markedly increased in rats force-fed a tryptophan-devoid diet, but hepatic protein synthesis was only slightly increased in rats fed the deficient diet ad libitum when compared with rats fed comparable control diets. A relationship between the increased protein synthesis in the liver and the decreased protein synthesis in skeletal muscle has been described in rats force-fed a tryptophan-devoid diet (30,31) as well as with other single amino acid deficiencies (25).

In paired-feeding experiments Naito and Kandatsu (32,33) reported that when rats were fed a tryptophan-deficient diet for 2 to 22 days $[^{35}S]$ methionine incorporation into hepatic proteins in vivo was increased over that in control rats. Nimni and Bavetta (34) also reported earlier, using ad libitum fed rats, that the incorporation of $[^{14}C]$ glycine was enhanced in the livers of tryptophan-deficient rats after 13 days. However, Bocker et al. (35) reported that rats pair-fed a complete and a tryptophan-devoid diet for 5 to 15 days showed a diminished in vivo incorporation of a $[^{3}H]$ amino acid mixture into the hepatic proteins of the experimental animals when expressed as a percentage of uptake compared with controls. Thus the results of the majority of the experimental studies suggest that feeding a tryptophan-deficient diet, particularly under force-feeding conditions for 1 or more days, induces enhanced hepatic protein synthesis. These results are similar to those reported with rats force-fed other single essential amino acid-devoid diets (25).

III. OCCURRENCE IN NATURE AND IN MAMMALS

Tryptophan is probably the indole derivative most widely distributed in nature. It is converted into many other substances of important biological significance. The many materials biogenetically related to tryptophan include nicotinic acid (a vitamin), serotonin (a neurohormone), indoleacetic acid (a phytohormone), some pigments found in the eyes of insects, and a number of alkaloids.

Although tryptophan is present in many proteins, it is present in only small amounts (much lower than most of the other amino acids) in

mammalian hepatic proteins (36). Usually it is the least abundant
amino acid in proteins. Indeed, a number of foodstuffs have been
found to be deficient or limited in tryptophan, e.g., corn. On the
other hand, since it is present in low concentration within most proteins, the requirement of tryptophan in the diet is low compared with
that of the other amino acids, particularly the other essential amino
acids. The knowledge that tryptophan is present in low levels in the
diet and in the proteins of tissues and organs, as well as in their free
amino acid pools including that in the blood, has been used in attempting to explain some of the effects of the amino acid. One prominent
belief has been that tryptophan is an important, rate-limiting amino
acid for various metabolic functions.

In 1950, Schurr et al. (37) determined the amino acid concentrations
in various tissues of the rat. If these data are used to calculate tissue/
plasma ratios, it becomes apparent that the relative availability of
plasma tryptophan to tissues is much less than that of the other amino
acids. In 1957, McMenemy et al. (38) described a unique property of
tryptophan: it was the only amino acid in human plasma that was
largely bound to protein. This attribute, specifically the ratio of free
to bound tryptophan in the blood, has much physiological significance.
For example, only the small free fraction of plasma tryptophan has
access to the brain. Factors that influence the equilibrium between
free and bound tryptophan in the plasma have been considered to alter
the availability of tryptophan to the brain, where it has special importance as a precursor of the neurotransmitter 5-hydroxytryptamine
(serotonin) (39-41). Tryptophan differs from other amino acids in that
its concentration in plasma of rats increases (30-40%) after fasting,
after insulin administration, or after consuming a carbohydrate meal
(42).

IV. CHEMICAL PATHWAYS

The degradation of tryptophan in animals occurs mainly in two pathways
(Figure 1). One major pathway, initiated by the action of tryptophan
dioxygenase, involves oxidation of tryptophan to N-formylkynurenine
and the formation of a series of intermediates and by-products, most
of which appear in the urine in varying amounts, the sum of which
accounts for approximately the total metabolism of tryptophan. The
second pathway involves hydroxylation of tryptophan to 5-hydroxy-
tryptophan and decarboxylation of this compound to 5-hydroxytrypt-
amine (serotonin), a potent vasoconstrictor found particularly in the
brain, intestinal tissues, blood platelets, and mast cells. Other minor
pathways also exist in animal tissues. A small percentage (3%) of
dietary tryptophan is metabolized via the pathway to indoleacetic acid.
Tryptophan appears to be converted to a larger number of metabolites
than any of the other amino acids.

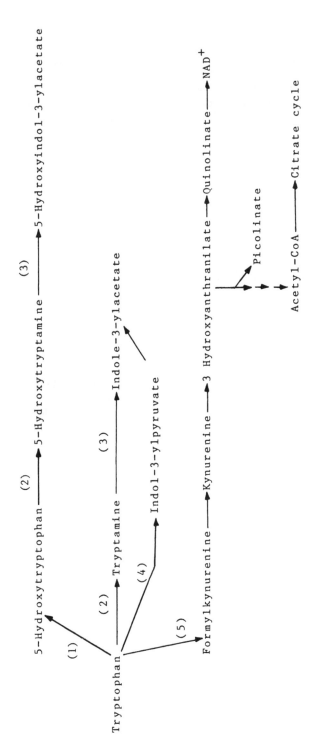

Figure 1 Pathways of tryptophan metabolism in the rat. Enzymes: (1) tryptophan 5-monooxygenase (EC 1.14.16.1); (2) aromatic L-amino acid decarboxylase (EC 4.1.1.28); (3) monoamine oxidase (EC 1.4.3.4); (4) tryptophan aminotransferase (EC 2.6.1.27); (5) tryptophan dioxygenase (EC 1.13.11.11).

Among its physiological and biochemical functions, tryptophan is a precursor of NAD (Figure 1) (43). Ingested tryptophan, via the portal vein, affects many processes in the liver, RNA synthesis, gluconeogenesis and the induction of many enzymes. These aspects of tryptophan's action will be covered in other sections.

V. EFFECT ON THE CENTRAL NERVOUS SYSTEM

Brain tryptophan has special importance as a precursor of the neurotransmitter 5-hydroxytryptamine (serotonin). The brain obtains tryptophan from the blood. However brain and plasma total tryptophan levels (concentrations) do not necessarily correlate, as do the plasma and brain concentrations of other amino acids. This anomaly is related, in part, to a special property of tryptophan, enabling it to bind to plasma albumin (38), which limits its availability to the brain. Only a small free fraction of plasma tryptophan is considered to be directly available to the brain (39-41). Alterations in the equilibrium between bound and free tryptophan in the plasma can influence the availability of tryptophan to the brain. Unesterified fatty acids in the plasma decrease the binding of tryptophan to plasma albumin (44). A number of drugs, e.g., heparin, isoprenoline, aminophylline, and dopa, increase plasma unesterified fatty acid levels by various mechanisms and thereby increase free tryptophan levels in the plasma and in the brain (41). Indeed, pharmacologically, many groups have demonstrated that the administration of drugs that cause sizeable displacement of tryptophan from albumin is associated with increases in brain tryptophan levels (39-41). However Fernstrom (45) believes that under physiological conditions the status of tryptophan in the blood in the free or bound form does not appear to have any significant impact on the availability of tryptophan to the brain. Fernstrom stresses that physiologically two processes, each related to food ingestion, are of importance in their ability to influence tryptophan access to the brain. They are the ingestion of proteins and the insulin secretion that follows the ingestion of almost any food. Each modifies brain tryptophan levels by changing the uptake of the amino acid into the brain. Tryptophan uptake occurs via a competitive transport "carrier" it shares with other large neutral amino acids. The plasma ratio of the total tryptophan to the sum of tyrosine, phenylalanine, leucine, isoleucine, and valine is thought to predict brain tryptophan levels (46,47). Thus it is currently believed that the concentration of brain tryptophan is positively correlated with that of free serum tryptophan and negatively to that of other amino acids competing with tryptophan for the same transport from blood to brain (48).

Research, focusing on the possible effects of food and nutrients upon human behavior, has attracted serious attention in recent times.

In laboratory animals there is good biochemical evidence that alterations in diet can change the amount of various neurotransmitters
synthesized in the brain and can thereby alter behavior. Preliminary
evidence suggests that the same phenomena occur in humans. The
neurotransmitters are precursor-dependent. The rate at which brain
enzymes synthesize the transmitter is limited by the availability of
precursors that derive from food and are transported into the brain.
The most often cited example is that of serotonin and its dependency
on tryptophan. Serotonin is synthesized directly from tryptophan,
and the body's sole source of this amino acid is dietary protein. A
number of reviews have described the important role that serotonin,
a derivative of tryptophan, plays within the central nervous system
(49-51). Serotonin neurons participate in a wide range of behaviors,
including sleep, feeding, aggression, locomotor activity, and pain
sensitivity (52). Dietary manipulations that alter brain tryptophan
levels can, in animals and in humans, affect many of these behaviors.
A few examples of these effects will be cited.

Administration of L-tryptophan to humans has been reported to reduce sleep latency and waking time (53-55). Similar results in reduction of sleep latency also have been observed in rats (56). Since
tryptophan is a precursor of serotonin, which has been implicated in
the mediation of sleep (57), it was assumed that tryptophan produces
these effects by increasing the availability of serotonin at sites where
serotonin naturally occurs in the brain. However, in addition to its
action on brain serotonin, it was suggested that tryptophan may produce its hypnotic effects via a nonserotonergic mechanism (53). The
reduction in concentration of both dopamine or norepinephrine induced in various brain regions by the administration of L-tryptophan
also correlates with reduction in sleep latency (58).

Experimentally, it has been demonstrated that injections of L-
tryptophan can have an effect upon the feeding behavior of rats.
In freely feeding rats injections of L-tryptophan brought about a
significant diminution in the 24-hr food intake and significantly reduced meal size (59). Also, in food-deprived rats tryptophan significantly reduced the size of the first large meal taken after the deprivation period and markedly extended the duration of the postmeal
interval (59). The latter findings differ from those of other investigators (60,61) who reported that tryptophan administered to 18- to
24-hr food-deprived rats failed to demonstrate a significant depression
of food consumption during the first 2 hr after injection. The differences have been attributed to methodological differences in experimental designs (59). Although the action of tryptophan is generally considered to be via its effect on brain serotonin, the mechanisms by
which the short-term administration of tryptophan leads to subtle
adjustments in meal parameters are still unclear. Relative to the
effect of tryptophan on reducing meal size in rats, Stevens et al.
(62,63) have described a specific tryptophan receptor for inhibition

of gastric emptying in the dog. As in the dog, tryptophan inhibited gastric emptying, which was independent of stimulation of acid secretion, in the cat (64). Thus this effect by tryptophan on the stomach may influence the quantity of diet consumed. In humans tryptophan has been utilized to decrease appetite (65).

Many experimental studies have been concerned with the effect of tryptophan on aggressive behavior and reactivity to enteroceptive and novel stimuli. It has been demonstrated that muricidal activity, by which some strains of rats stereotypically kill mice, can be influenced in laboratory rats by manipulations of tryptophan intake. Decreased or increased intake of tryptophan has been demonstrated to respectively facilitate or reduce the spontaneous and strain-dependent mouse-killing by rats (66-69). In addition, rats fed a tryptophan-free diet demonstrate an increased reactivity of the acoustic startle reflex (70), of the startle responses to air-puffs (68), and to the novelty of the open field (68). These behavior effects have been attributed to alterations in the levels of cerebral serotonin as influenced by tryptophan intake.

A number of other studies have been concerned with the effect of feeding a tryptophan-free diet on the behavioral activities in rats. After reporting that the administration of an amino acid mixture lacking tryptophan or a tryptophan-free diet to rats resulted in a rapid, selective depletion of brain serotonin and 5-hydroxyindoleacetic acid (71,72), Fratta et al. (73) found that feeding such a diet to rats and rabbits caused a marked increase in male-to-male mounting behavior. Administration of serotonin or 5-hydroxyindoleacetic acid (73) but not of L-tryptophan (74) suppressed the copulatory behavior in all the animals. These findings support the hypothesis that brain serotonin plays an inhibitory role in controlling male sexual behavior.

Increased sensitivity to painful stimuli has been reported in animals subjected to dietary deprivation of tryptophan (75-77). Also some studies have revealed that in humans tryptophan causes decreased pain perception (52).

Mental and neuromuscular symptoms have been described in a number of diseases in which a tryptophan deficiency occurs (78). Pellagra and carcinoidosis are two diseases that have been known to involve tryptophan deficiency. In pellagra, the deficiency is due to a low content of tryptophan in the diet, to an insufficient intake of essential nutritional substances in relation to requirements, or to diseases of the alimentary tract. In carcinoidosis, the deficiency is due to an increased metabolism of tryptophan in the tumor cells, often aggravated by losses of tryptophan in the stools because of frequent diarrhea. A disturbance of tryptophan metabolism occurs also in certain diseases due to inborn errors of metabolism, as in phenylketonuria, Hartnup and Down's diseases, the blue diaper syndrome, and maple syrup disease (78). In these diseases the tryptophan disturbances are secondary to other metabolic insufficiencies. Mental symptoms are also associated with these disorders.

For many years the question of the role of tryptophan and serotonin in the biochemistry of mental disease, especially depressive illness, has been raised in the literature. There appears to be much conflict, not only on the nature of the facts, but also on their interpretation. Because of the complexities of relationships between cause and effect and the lack of reliable experimental models, it is most difficult to obtain reliable data. A few review articles are cited for coverage of this area (79-81).

VI. EFFECT ON THE LIVER

A. Induction of Enzymes in the Liver

Many studies have demonstrated that the administration of tryptophan causes an enhancement in the activity of a variety of liver enzymes. Table 1 summarizes a number of hepatic enzymes that have reportedly become enhanced in activity because of tryptophan. Tryptophan has specific effects on the activities of many hormonally and nutritionally sensitive enzymes, not necessarily related to tryptophan metabolism itself (82-86). The mechanisms by which tryptophan acts to affect these enzyme levels are not clear. In some instances, there is

Table 1 Hepatic Enzymes Induced by Tryptophan

Enzyme	Reference
Alanine aminotransferase	85
Aniline hydroxylase	243
Aspartate aminotransferase	85
Cytochrome b_5	242
Cytochrome P450	241-243
DNA-dependent RNA polymerase (I and II)	85,131-134
Histidase	82,89
Nitrosodimethylamine demethylase	240
Ornithine σ-transaminase	82,85,86
Ornithine decarboxylase	256
Nucleoside triphosphatase	164
Phosphoenolpyruvate carboxykinase	82,84,85,89
Phosphoprotein phosphohydrolase	155
Protein phosphokinase	155
Serine (threonine) dehydratase	82,85,89
Tryptophan dioxygenase	83,85,89
Tyrosine α-ketoglutarate transaminase	85,89,91

evidence that the specific regulation by tryptophan involves enzyme degradation rather than synthesis (84). Deguchi and Barchas (87) suggested that a metabolite, rather than tryptophan itself, may be the actual active component. Such is indeed the case for tryptophan-induced hypoglycemia in vivo (88). In searching for a unifying mechanism, Smith et al. (89) conducted a systematic study involving the activities of a number of tryptophan-sensitive enzymes, tyrosine and tryptophan transaminases, tryptophan dioxygenase, histidase, serine dehydratase, and phosphoenolpyruvate carboxykinase. Their findings did not support the concept of a single simple mechanism for explaining the action of tryptophan on enzyme activities. They concluded that a variety of mechanisms may be involved.

In considering the mechanisms by which tryptophan causes an increase in certain hepatic enzymes, Kenney (90) has suggested that tryptophan may increase the rates of synthesis of tyrosine aminotransferase, phosphopyruvate carboxylase, and others, indirectly via cyclic AMP. Indeed, cAMP will induce these enzymes. Also, Kato (91) indicated that the mechanism of tryptophan-mediated induction of tyrosine aminotransferase includes a cAMP accumulating process that is sensitive to phentolamine, an adrenergic-blocking agent. Schumm and Webb (92) recently have reported that the addition of physiological concentrations of cAMP stimulated the release of RNA from isolated prelabeled rat liver nuclei to a fortified cytosol in a cell-free system.

B. Effect on Carbohydrate Metabolism in the Liver

Among the many metabolic effects of tryptophan, the induction of hypoglycemia has been an important finding. Gullino et al. (93), in 1955, first described this effect, as it has since been confirmed by others (88,94). The mechanism by which tryptophan induces hypoglycemia has been of interest to a number of investigators. Mirsky et al. (95) claimed that several indolic and other metabolites could also elicit decreases in blood sugar concentrations and ascribed these actions to changes in the availability of circulating insulin. On the other hand, Lardy and coworkers (96,97) demonstrated that glucose synthesis in the isolated perfused liver was inhibited by tryptophan and metabolites derived through the kynurenine pathway. This finding suggested that impaired gluconeogenesis, arising from increases in hepatic quinolinate concentrations, caused hypoglycemia in the intact animal. More recently, Smith and Pogson (88) proposed that the hypoglycemic action of tryptophan is mediated through formation of intracellular serotonin and is unrelated to the inhibition of gluconeogenesis. They consider that this increased synthesis of serotonin does not involve either the central nervous system or the adrenal glands. Thus our understanding of the mechanism by which tryptophan induces hypoglycemia is still unresolved.

C. Effect on Lipid Metabolism in the Liver

It was first reported by Hirata et al. (98), and later confirmed by Ramakrishna Rao et al. (99), that the administration of high levels (50-200 mg/100-g body weight) of L-tryptophan to fasted (20 hr) rats induced a fatty liver within 2.5 to 4 hr. The liver revealed a significant increase in total lipids, chiefly in triglycerides, with no significant alterations in hepatic cholesterol and phospholipids. The fatty liver, in which the accumulated lipids were localized in the peripheral zones of the lobules, was resistant to added choline, L-methionine, inositol, or vitamin B_6 (98). However Sakurai et al. (100,101) and Fears and Murrell (102) using lower doses of L-tryptophan (5-50 mg/100-g body weight) reported no changes in hepatic total lipids. The fatty liver induced by high doses of L-tryptophan has been attributed to an increased mobilization of free fatty acids as reflected in the elevated levels of plasma free fatty acids and to increased hepatic fatty acid synthesis (99). A number of investigators (99,101-103) have reported that hepatic fatty acid synthesis is increased following low (5 mg/100-g body weight) or high (75 mg/100-g body weight) doses of L-tryptophan in both fasted and fed rats. In attempting to explain the enhanced hepatic fatty acid synthesis, the following mechanisms have been considered: Miyazawa et al. (103) proposed the activation of the lipogenic enzyme acetyl-CoA carboxylase (EC 6.4.1.2) by citrate, and Fears and Murrell (102) proposed that the stimulation of insulin secretion by tryptophan increases the activities of hepatic fatty acid synthetase and adipose tissue lipoprotein lipase.

Concerning the fatty livers due to tryptophan described by some investigators, Hirata et al. (98) reported a decrease in the level of ATP in the liver caused by tryptophan administration that they considered to play a role in the induction of the fatty liver. However Sakurai et al. (101) failed to observe a change in the hepatic ATP level caused by tryptophan. Dianzani (29) speculated that the fatty liver provoked by treatment with tryptophan was due to impaired synthesis of apolipoprotein, possibly related to a diversion of synthesis directed to other proteins. Thus, currently, the mechanism whereby tryptophan affects lipid metabolism in the liver is not clear. The need for very high doses of L-tryptophan to induce fatty liver in the rat suggests that it may be related to toxicity. Indeed, Gullino et al. (104) reported that the LD_{50} level for L-tryptophan in the rat is 162 mg/100-g body weight.

D. Effect of Tryptophan on Protein Synthesis

1. *Introduction*

The concept that metabolism within the liver is stimulated following a meal has prevailed for many years. Yet details regarding the mechanisms by which this occurs are becoming unraveled only in recent years. A number of studies have been concerned with the important physiological phenomenon, the stimulation of hepatic protein synthesis in response to food intake. These studies have focused upon learning how specific nutrients may act in regulating hepatic protein synthesis. It is of special interest that tryptophan has been found to have a unique role.

Many reports have described how the livers of fasted animals have responded rapidly to a single feeding of good-quality protein, or of complete amino acids, with enhanced protein synthesis and a shift in polyribosomes toward heavier aggregation (105-112). In general, a good correlation has been demonstrated between the protein synthetic ability of cells and tissues and the state of polyribosomal aggregation (113). Subsequently, some investigators have determined how animals respond to a single feeding of an incomplete amino acid mixture (a complete amino acid mixture free of single essential amino acids). Such studies (107,114) revealed that when complete amino acid mixtures devoid of single essential amino acids (arginine, histidine, isoleucine, leucine, lysine, methionine, phenylalanine, threonine, or valine) were tube-fed once to fasted animals the response was the same as that found after feeding the complete amino acid mixture. One notable exception was observed, that of the deletion of tryptophan, where the hepatic polyribosomes did not respond but remained as those of the fasted animals (106,107,114,115). Next, studies were undertaken relating to the role that tryptophan alone may have on hepatic polyribosomes and protein synthesis.

In two early reports (106,107), our laboratory demonstrated that tryptophan alone elicited a response of hepatic polyribosomes toward heavier aggregation as well as an enhancement of protein synthesis as measured in vivo or in vitro. Tryptophan administered alone, but not a single administration of one of three other essential amino acids (threonine, methionine, or isoleucine), elicited a stimulatory hepatic response that was similar to the one obtained with a complete amino acid mixture (107). In an earlier study, Feigelson et al. (116) reported that the parenteral administration of tryptophan in the rat produced a transient elevation in hepatic protein synthesis as determined by measuring $2[^{14}C]$glycine incorporation into hepatic protein. Subsequent studies from other laboratories (114,117-122) have confirmed the stimulatory effect of tryptophan on hepatic protein synthesis.

Several studies have been conducted to determine which proteins were involved in the enhanced hepatic protein synthesis induced by tryptophan. Increased synthesis of albumin has been reported by Rothschild et al. (118) and by Jorgensen and Majumdar (121,123). The latter investigators have also found that tryptophan increased the synthesis of transferrin, fibrinogen, and ferritin. Thus a number of extracellular and intracellular proteins that are synthesized by the liver are affected by tryptophan. Consistent with this finding are data from our laboratory (124) which demonstrated that after tryptophan administration both the free and the membrane-bound polyribosomes of the liver showed a shift toward heavier aggregation, more marked for free polyribosomes than for membrane-bound polyribosomes. Also, in vitro protein synthesis revealed a greater increase with free polyribosomes than with membrane-bound polyribosomes of the livers of tryptophan-treated animals in comparison with similar fractions of livers of control animals.

2. Mechanism of Action

Many studies have been concerned with elucidating the mechanism by which tryptophan stimulates hepatic protein synthesis. Although the answer is still incomplete, sufficient information has accumulated to allow one to develop an overall picture of the dynamics that occur in the liver following the administration of tryptophan. A review of the findings of experiments concerned with the effects of tryptophan on regulatory control mechanisms follows.

Tryptophan as the Limiting Amino Acid

Initially, the special role of tryptophan in hepatic protein metabolism was attributed to the very low levels of free tryptophan in the liver and blood, especially after fasting (125). Studies by Hori et al. (126) and by Hunt et al. (127), using reticulocytes, suggested that tryptophan tRNA may become limiting. Similarly, it was considered that tryptophan tRNA may become limiting in livers of fasted animals because of no dietary tryptophan intake and therefore would limit the rate of hepatic protein synthesis. However data from other studies (128,129) suggested that the effect of tryptophan on hepatic polyribosomes and protein synthesis involved a more complex process than one consisting merely of raising the low levels of tryptophan in blood or liver. When tryptophan was administered to fed mice in which the blood and hepatic levels of tryptophan were much higher than in fasted mice, the hepatic polyribosomes shifted toward heavier aggregation, and hepatic protein synthesis (in vitro) was enhanced, similar to that which occurred when using fasted animals (128). Also, Allen et al. (130) reported no differences in the levels of tryptophanyl-tRNA in fasted and fed rats, suggesting that this was not rate limiting in fasted animals. In addition, tryptophan has been effective in enhancing hepatic protein synthesis in animals treated with selected hepatotoxic agents where the tryptophan levels in blood and liver were not changed but were similar to those in control animals (129).

Alteration in Translational Control

Evidence exists that tryptophan can act at the translational level of control of hepatic protein synthesis in rats or mice. This conclusion was based upon data from experiments in which actinomycin D was administered first to inhibit RNA synthesis followed by the administration of tryptophan. Tryptophan was still found to have a stimulatory

effect on hepatic polyribosomes and in vitro protein synthesis (107).
Also, studies by Jorgensen and Majumdar (123) revealed that, when
well-fed adrenalectomized rats were pretreated with actinomycin D,
the administration of tryptophan caused an increase in [^{3}H]leucine
incorporation into ferritin, albumin, transferrin, and fibrinogen, compared with that in water-fed controls. Without using the inhibitor,
tryptophan caused a somewhat larger increase of incorporation into
the same components.

Alteration in Transcriptional Control

Other findings suggested that tryptophan administration may act at
the transcriptional level of control of hepatic protein synthesis. Tryptophan administration has been reported to cause an increase of DNA-
dependent RNA polymerase activity (131-133), nuclear RNA synthesis
(131,132,134), polyribosomal RNA synthesis (131,135), and cytoplasmic
mRNA (136,137) in the liver. It has been reported that the nuclei of
rat liver contain at least three forms of DNA-dependent RNA polymerase
that play specific roles in genetic transcription (138,139). Polymerase
I is presumed to transcribe ribosomal genes and is Mg^{2+}-dependent and
amanitin-resistant. Polymerase II and III are considered to synthesize,
respectively, heterogeneous and smaller molecular weight RNAs. Polymerase II is $Mn^{2+}/(NH_4)_2SO_4$-dependent and amanitin-sensitive. Vesley
and Cihak (132) and Majumdar and Jorgensen (140,141) reported that
following tryptophan administration the activities of both polymerase I
and II were rapidly elevated and this was unrelated to adrenal secretion.
Also, Majumdar (133) studied the activities of the engaged (chromatin-
bound) and free states of nuclear RNA polymerases in young and adult
rats tube-fed tryptophan and observed marked increases in the activities of both engaged and free polymerases in 2-week-old rats but only
an increase in the engaged polymerases of the adult rats.

Alteration in Posttranscriptional Control

A number of reports from our laboratory have demonstrated that tryptophan has an effect at the posttranscriptional level of control of hepatic protein synthesis. Since our laboratory has attached much
significance to the effect of this level, our evidence will be reviewed in
some detail. After first observing that there was an elevation in hepatic cytoplasmic mRNA following tryptophan administration (136,137),
we explored whether this effect could be due, in part or totally, to
increased nucleocytoplasmic transport of mRNA. In experiments using
fasted mice, hepatic RNA was prelabeled with [6-^{14}C]orotic acid and
then the animals were treated with actinomycin D to inhibit further RNA
synthesis. Next the animals were tube-fed either tryptophan or water.
The livers showed elevated levels of cytoplasmic mRNA and a shift in

polyribosomes toward heavier aggregates caused by tryptophan (136).
This suggested that in the absence of RNA synthesis, tryptophan may
act to stimulate the transfer of mRNA into the cytoplasm of hepatic
cells. In further studies we observed that the administration of tryptophan to fasted animals resulted in significant increases in the amounts
of hepatic polyadenylic acid [poly(A)] and poly(A)-mRNA in the cytoplasm and this stimulation also occurred even following the administration of cordycepin (an inhibitor of poly(A) synthesis) and/or actinomycin D (an inhibitor of RNA synthesis) (137). Administration of tryptophan to fasted rats pretreated with cordycepin or actinomycin D, or
both, induced a shift in hepatic polyribosomes toward heavier aggregates
and an increase in in vitro protein synthesis. Also, fasted rats that
received [U-^{14}C]adenosine to prelabeled hepatic poly(A) and then were
treated with cordycepin or actinomycin D, or both, before tube-feeding
tryptophan revealed increased hepatic levels of labeled polyribosomal
poly(A) in comparison with controls (137). The administration of tryptophan to fasted rats pretreated with cordycepin and actinomycin D led
to decreased levels of nuclear poly(A)-mRNA and a concomitant increase in the levels of polyribosomal poly(A)-mRNA in the cytoplasm
(142). These findings suggested that tryptophan plays a role in stimulating the rate of translocation of poly(A)-mRNA from the nucleus into the cytoplasm of the livers of normal animals (136,137) and in those
treated with selected inhibitors (129,143,144).

The possible role that informosomal mRNA may play in the stimulation of
hepatic protein synthesis induced by tryptophan was recently investigated (145). While tryptophan induced an increase (20%) in the amount
of polyribosome-associated poly(A)-mRNA in the liver within 1 hr, the
amount of the informosomal poly(A)-mRNA revealed no significant increase or decrease. Since the size of the increase in the polyribosome-
associated poly(A)-mRNA was equal to the entire amount of informosomal
poly(A)-mRNA present in the hepatic cells, it was concluded that the
failure to detect a significant decrease in the size of the informosomal
mRNA pool indicated that the increase in the polyribosome-associated
poly(A)-mRNA must be due to a different mechanism, such as enhanced
nucleocytoplasmic translocation of poly(A)-mRNA, which has been reported earlier and will be discussed in the following paragraphs. Hybridization studies with DNA/RNA were also conducted using hepatic
polyribosome-associated poly(A)-mRNA from tryptophan-treated and
control rats to determine if any qualitative or quantitative changes
occurred in the RNA sequences along with the increase in poly(A)-
mRNA following tryptophan administration (145). Although qualitatively
no new species of mRNA were detected in the mRNA from the tryptophan-
treated rats, kinetic analysis of the hybridization curves indicated that
there was a shift or accumulation of hepatic poly(A)-mRNA belonging
to the intermediate- and possibly the high-frequency classes of polyribosome-associated poly(A)-mRNA in the livers of the tryptophan-
treated rats.

It has been established that posttranscriptional controls are operating within the nucleus, at the nuclear membrane, and/or in the cytoplasm to regulate the rate of nucleocytoplasmic translocation of ribonucleoprotein particles both qualitatively and quantitatively (146,147). Evidence for modification of nucleocytoplasmic transport of RNA owing to chemical carcinogens (148,149), in transformed cells (150,151), and during aging (152) has been demonstrated. In view of this background information, further studies relating to the enhanced hepatic nucleo-cytoplasmic translocation of mRNA occurring following tryptophan administration were undertaken. Experiments were conducted using a cell-free system, modeled after that used by Schumm and Webb (153), in which the release of RNA from isolated nuclei into a defined medium could be studied. These workers observed that the release of RNA from isolated hepatic nuclei was influenced by dialyzed cell sap (cytosol) and suggested that the cell sap contained nondialyzable components that act posttranscriptionally to regulate the nuclear processing and/or transport of mRNA to the cytoplasm. Using this assay system, we obtained evidence that tryptophan affected both the nucleus and the cytoplasm in the process of enhancing intracellular transport of RNA. First, using control liver cell sap, we found that there was greater release of labeled poly(A)-mRNA (prelabeled in vivo, with [^{14}C]orotic acid) into the medium from isolated hepatic nuclei of tryptophan-treated (10 min) rats than from nuclei of control animals (142). This effect was detected as early as 3 or 6 min after the administration of tryptophan (154). In view of this rapid response, we investigated how early some of the other hepatic changes occurred after tryptophan administration. We observed that tryptophan concentrations became elevated in the plasma and in the liver (including hepatic nuclei) within 5 min (154) and that increased hepatic protein synthesis was evident within 10 min (107). To determine if the increased hepatic protein synthesis was involved in the nuclear effect of the tryptophan-treated animals, we investigated whether or not puromycin treatment before (10 min) tryptophan administration would affect the nuclei. Using our in vitro assay system, hepatic nuclei of rats that received tryptophan following puromycin exhibited greater release of labeled RNA (poly(A)-mRNA) into the medium in comparison with control nuclei of puromycin-treated rats (154).

Next, we determined whether or not the liver cell sap of the tryptophan-treated animals was influential. Since in vivo hepatic protein synthesis was already stimulated within 10 min after tryptophan administration (107), it was conceivable that some regulatory protein(s) in the cytosol might play a role in the transport activity. To test for this possibility, cell saps were prepared from livers of rats tube-fed water or tryptophan 10 min before killing, and their effects on in vitro release of labeled poly(A)-mRNA from control hepatic nuclei in the presence of added tryptophan in the medium were investigated. There was more release of total RNA and poly(A)-mRNA from control hepatic nuclei incubated with liver cell sap of tryptophan-treated rats compared with release using liver cell sap of control rats (154). The addition of tryptophan to the incubation medium was essential for the increased release of labeled RNA by the experimental liver cell sap. Since the dialyzed cell saps used in the incubation medium lacked tryptophan, its addition was important.

To determine if the rapidly stimulated hepatic protein synthesis following tryptophan administration was involved in this process, we next examined whether or not pretreatment of the animals with inhibitors of protein synthesis would influence the increased transporting effect of the liver cell sap of tryptophan-treated rats. Animals were pretreated with cycloheximide for 2-1/2 hr or with puromycin for 20 min before the tryptophan or water administration which was given 10 min before killing. The effects of liver saps prepared from the control and experimental groups on in vitro release of labeled RNA from hepatic nuclei were investigated. Liver cell saps from rats treated with cycloheximide or puromycin before tryptophan administration were not able to stimulate the release of labeled RNA, as could liver cell saps of tryptophan-treated rats (154).

A dichotomy exists between the effect of inhibition of hepatic protein synthesis, as induced by puromycin or cycloheximide, and the action of tryptophan on the nuclear cytosol involvement in the enhanced RNA transport activity as measured in vitro. While inhibition of protein synthesis does not affect the stimulatory effect of nuclei, it inhibits the stimulatory effect of liver cytosol of tryptophan-treated rats. In an attempt to explain these differences one may speculate that, in animals treated with inhibitors of protein synthesis, tryptophan may be able to act in one of two ways: (a) to stimulate the synthesis of regulatory proteins in the nucleus or (b) to enhance the transfer of the existing amounts of regulatory factors in the cytosol to the nuclear membrane. The question of whether protein synthesis does or does not occur in the nucleus as yet is not definitively resolved. Some investigators have implicated the nucleus and nuclear envelope as sites of protein synthesis and have reported that nuclear protein synthesis is inhibited to a lesser extent by puromycin and cycloheximide than is cytoplasmic protein synthesis (155,156). Evidence suggesting that preexisting proteins of the cytosol may act on the nuclear envelopes has been obtained in experiments described in a subsequent section (157). It is quite difficult to unravel the complex mechanisms that come into play following the use of tryptophan alone in normal animals and possibly even more so when tryptophan is used in combination with a variety of inhibitors. Indeed, inhibitors that are considered to have one primary action often cause a variety of secondary effects in vivo in animals. Although inhibitors of RNA and protein synthesis have been of great value in experiments attempting to unravel mechanisms,

caution must be exercised in interpreting data based upon their use
in normal animals.

Summarizing the above data, it appears that the enhanced nucleocytoplasmic translocation of mRNA of the livers of tryptophan-treated
animals is manifested in effects on the nucleus as well as on the cytosol.
The nuclear effect appears to be independent of changes in hepatic
protein synthesis, while the cytosol effect is dependent on hepatic
protein synthesis. Thus while both (nuclear and cytosol) effects are
probably operative in normal animals treated with tryptophan, only
one effect, that on the nucleus, is probably operative in animals treated
with hepatotoxic agents that act to inhibit hepatic protein synthesis.
The latter mechanism could be invoked in explaining the stimulatory
effect of tryptophan on hepatic polyribosomal aggregation and protein
synthesis of rats treated with inhibitors of protein synthesis such as
ethionine, puromycin, or hypertonic NaCl (143,158,159). Indeed, experimental studies with ethionine, puromycin, and hypertonic NaCl
support this conclusion (129).

In probing further into the mechanism whereby tryptophan stimulates
hepatic nucleocytoplasmic translocation of mRNA, it became necessary
to investigate special components of the nucleus—the nuclear envelope
and the nuclear pore complex—which are considered to play a key role
in the regulation of nucleocytoplasmic RNA translocation (160). A
nucleoside triphosphatase (NTPase) has been identified in the mammalian liver nuclear envelope, and this enzyme appears to be involved
in the nucleocytoplasmic translocation of RNA (161-163). Furthermore,
following treatment of rats with thioacetamide or CCl_4, a parallelism
between alterations in nuclear RNA transport and nuclear envelope
NTPase activity in the liver has been demonstrated (161). Therefore
we investigated whether or not the enhanced nucleocytoplasmic translocation of mRNA caused by tryptophan was related to an alteration in
the activity of nuclear envelope NTPase. The results demonstrated
that the levels of nuclear envelope NTPase activity were significantly
elevated in the livers of rats tube-fed tryptophan at 10, 30, or 60 min
before killing (164). As described earlier, concomitant with this rapid
(10 min) increase in the NTPase activity, there was a greater release
of labeled RNA from isolated hepatic nuclei of livers of tryptophan-
treated rats than from those of control rats. The parallel increases in
both NTPase activity of nuclear envelopes and the translocation of RNA
suggested that these two processes were associated with, or related to,
one another. In further experiments we found that the administration
of tryptophan was able to stimulate the levels of hepatic nuclear envelope NTPase activity in rats pretreated with puromycin, similar to
the increases in the control rats that received tryptophan alone (164).
This finding added further support to the view that the increased
activity of nuclear envelope NTPase was related to, or probably responsible for, the enhanced RNA transport activity hepatic nuclei of

puromycin plus tryptophan-treated rats compared with that of the hepatic nuclei of control rats.

Recently, activities of two other enzymes, protein phosphokinase and phosphoprotein phosphohydrolase, have been identified on the mammalian nuclear envelope (165-167). These investigators suggested that the levels of phosphorylation and dephosphorylation of nuclear envelope protein protein by these two enzymes may regulate nucleo-cytoplasmic RNA translocation (168). Since these nuclear envelope-associated enzymes may play a key role in the regulation of nuclear RNA transport, a study was conducted to investigate if the administration of tryptophan would influence the phosphorylation and dephosphorylation process in the hepatic nuclear envelopes, which may then modulate nucleocytoplasmic transport of RNA. In this study (155), the activities of protein phosphokinase (PK) and phosphoprotein phosphohydrolase (PH) were investigated in the livers of rats that received a single tube-feeding of tryptophan 10 min before killing. The hepatic nuclear envelope activities of both enzymes were found to be increased. Furthermore, tryptophan administration increased the in vivo incorporation of [^{3}H]leucine into proteins of the nuclear envelopes (+ 83%) and also into proteins of the other subcellular fractions (+ 34 to + 43%) of the liver compared with that into proteins of the corresponding fractions of the control rats. Rats that received [^{3}H]-leucine to prelabel hepatic proteins and then were treated with puromycin to inhibit further protein synthesis followed by tube-feeding of tryptophan, revealed greater radioactivity associated with nuclear envelope proteins than did controls. The latter findings suggest that tryptophan may act to stimulate the movement or availability of proteins to the vicinity of the nuclear envelope, possibly specific regulatory proteins, such as NTPase, PK, and PH, which show increases in activities and may be responsible for the increase in the rate of nucleocytoplasmic translocation of mRNA.

In view of the probable importance of the nuclear envelope in controlling active nucleocytoplasmic transfer of RNP particles, several workers have examined the ultrastructure of the nuclear envelope. The intact nuclear envelope is composed of inner and outer nuclear membranes and nuclear pore complexes (169). Treatment of nuclei with nonionic detergents, such as Triton X-100, completely removes the outer nuclear membrane, leaving intact nuclei with preservation of nuclear pore complexes. Evidence suggests that the nuclear pores are the major sites of nucleocytoplasmic transfer of macromolecules (170, 171). Therefore we examined the effect of removing the outer nuclear membranes (by Triton X-100 treatment of isolated hepatic nuclei) of control and tryptophan-treated rats on the capacity of the nuclei to transport RNA in vitro and on the activity of nuclear envelope NTPase. Following treatment with Triton X-100, there was greater release of

labeled RNA from the isolated hepatic nuclei of tryptophan-treated
rats compared with that from control hepatic nuclei (164). These
findings were similar to those observed with nuclei not treated with
Triton X-100. Similar increases in the activity of nuclear envelope
NTPase were found with the experimental compared with the control
samples (164). Our findings that the increased translocation of mRNA
occurred along with the increased activity of nuclear envelope NTPase
in the livers of our experimental rats are in general agreement with
data of others which demonstrate that: (a) detergent treatment was
not deleterious to the ability of nuclei to transport RNA, (b) the transported RNA was of intranuclear origin, and (c) nuclear pore complexes
and NTPase activity were probably responsible for the nucleocytoplasmic translocation of mRNA (162,171,172).

Since tryptophan administration can rapidly enhance nucleocytoplasmic translocation of RNA in the liver, it became of interest to learn
if tryptophan deprivation would be inhibitory to this process. A few
experimental studies suggest that indeed this may be the case. Bocker
et al. (35), using rats fed ad libitum a tryptophan-free or complete
diet (pair-fed controls) for 15 days, studied the incorporation of
[^{3}H]orotic acid into RNA fractions separated from the hepatic nucleus
and cytoplasm. The incorporation of orotic acid into the high-weight
components of the RNA of the nucleus was increased, but no increase
was found in the cytoplasmic RNA. These findings were interpreted
as indicating that RNA in the nuclei in the livers of the experimental
rats was synthesized but not delivered to the cytoplasm. A similar observation was made based upon experiments in which rats were fed an
amino acid-deficient diet (6% casein) by Wannemacker et al. (173), who
suggested that the nuclear RNA was synthesized but not properly
processed and therefore not transported to the cytoplasm. Further
studies are needed to elucidate if indeed long-term feeding of tryptophan-devoid diets impair hepatic nucleocytoplasmic translocation of
mRNA.

Mechanisms other than those considered thus far also may be involved
in the enhanced nuclear RNA-transport activity due to tryptophan.
Schumm and Webb (92) have suggested that cyclic nucleotides can exert
an influence on the posttranscriptional events of RNA processing and
transport since they found that the addition of cyclic AMP or GMP stimulated the release of RNA from isolated hepatic nuclei. Subsequently,
we have found in preliminary experiments that the addition of cAMP
to the cell-free system composed of cell saps of livers of control rats,
but not composed of cell saps of livers of tryptophan-treated rats,
caused increases in the release of labeled RNA from the liver nuclei
of control rats. This response in the cell-free system to added cAMP
probably reflects the in vivo concentrations of cAMP in the tissues from
which the cytosol was prepared. Thus these preliminary findings suggest that tryptophan may elevate in vivo the cAMP levels in liver

cytosol, and this may be of importance in the enhanced nucleocytoplasmic translocation of mRNA in liver owing to tryptophan. However
preliminary assays of cAMP activities in the livers of control and tryptophan-treated rats have failed to reveal significant differences. Further studies are being conducted to determine whether or not changes
in cAMP are involved.

3. *Speculation as to How Tryptophan Acts*

Currently, in an attempt to explain mechanistically how tryptophan acts,
we attach much significance to its ability to rapidly enhance nucleocytoplasmic translocation of mRNA in the livers of normal rats or of
rats pretreated with puromycin and/or actinomycin D. We speculate
that L-tryptophan itself, possibly without incorporation into peptides
of protein(s), can act at the level of the hepatic nuclei (nuclear membranes). Indeed, it is possible that tryptophan itself, or tryptophan
bound to a cytoplasmic carrier protein, may act at receptor sites of
the nuclei and/or nuclear membranes. Experiments are in progress
that should enable us to determine whether or not our speculation is
valid.

Consideration That L-Tryptophan May Act Similar to a Hormone

Some of our current considerations are that L-tryptophan itself or in
relation with a protein(s) (presently undefined) may act on hepatic
cells in a way similar to that of a hormone, such as insulin or steroid
hormones. Analogous to insulin, its first step may be the binding of
L-tryptophan to a specific receptor protein on the surface of the target cells. As described for insulin (174,175), L-tryptophan may bind
at specific receptors on the plasma membranes of target cells such as
liver. After binding, the tryptophan-receptor complex may lead to
many of the subsequent actions of L-tryptophan. On the other hand,
based upon recent findings with insulin, there is evidence that insulin
has a direct effect upon nuclei. Goldfine and Smith (176,177) have reported that purified nuclei from both rat liver and cultured lymphocytes contain specific binding sites for insulin. Also, Vigneri et al.
(178) have reported that the nuclear envelope is the major site of
insulin binding to the cell nucleus. Goldfine et al. (179,180) have reviewed the evidence that nuclear envelopes contain specific high-
affinity binding sites for insulin, that insulin stimulates nuclear envelope NTPase activity, and the insulin directly stimulates the release of
mRNA from isolated nuclei. Roth and Cassell (181) have recently reported, in studies using highly purified preparations of insulin receptor,
that the insulin receptor itself may be a protein kinase. Since we now
hypothesize that L-tryptophan may be acting directly on the nuclei
(nuclear membranes) of liver cells, we have begun to explore whether
or not L-tryptophan may have some effects on hepatic nuclei similar to
those of insulin.

It is now widely accepted that steroid hormones act on target cells essentially through a sequence of early events involving (a) the penetration of the hormone into the target cell, (b) its binding to specific receptor proteins, and (c) the temperature-dependent activation of the steroid-receptor complex, activation necessary for the migration of the complexes to the nucleus. Finally, the attachment of the activated complexes to chromatin is probably responsible for the alteration of transcription of specific genes (182,183). The finding of specific receptors for glucocorticoids in the cytosol of most of the target organs including liver is considered one of the strongest pieces of evidence in support of this model (184,185).

Markovic and Petrovic (186) have reported that the radioactive profiles on sucrose density-gradient analysis of the macromolecular liver cytosol fraction incubated with tritiated hydrocortisone and tryptophan were closely similar. This resemblance suggested to them that both compounds bind to rat liver proteins of similar size and that a competition may exist between hydrocortisone and tryptophan for receptor proteins of rat liver cytosol. Baker et al. (187) have reported that tryptophan methyl ester, a competitive inhibitor of chymotrypsin, is also a competitive inhibitor of dexamethasone binding to the glucocorticoid receptor in HTC cells, which suggests that the binding sites of tryptophan methyl ester and dexamethasone are partially contiguous. The findings by Markovic and Petrovic (186) pointed to the possibility that the maximal biological activity of one compound might be impaired by the presence of the other. Majumdar and Jorgensen (188) observed, using well-fed adrenalectomized rats, that while the stimulatory effect on hepatic protein synthesis of tryptophan was not fully expressed in the presence of cortisol, the reverse is true for cortisol in that the amino acid blocks the cortisol-mediated stimulation of hepatic protein synthesis to a great extent, as measured by [^{3}H]leucine incorporation into plasma albumin, fibrinogen, and liver ferritin in vivo. However we observed, using normal rats, that tryptophan administration to rats, that have had hepatic protein synthesis enhanced by previous treatment with cortisone acetate, further stimulated hepatic polyribosomes toward heavier aggregation and hepatic protein synthesis in vivo (189).

Consideration That L-Tryptophan May Bind to Cytosol and Nuclear Proteins

Possible similarities to the binding between L-tryptophan and serum albumin: It has long been known that L-tryptophan binds to serum albumin which then transports tryptophan in the blood. Thus tryptophan is present within blood in a bound (approximately 85%) and in a free (approximately 15%) form. McMenamy (190) has reviewed the relationship of the binding between tryptophan and albumin. Indeed, the association constant for the binding of tryptophan to albumin plays a role in the distribution of tryptophan throughout the body and thus in influencing its biological activity, especially on the brain. Also, a number of small molecules and drugs bind to serum albumin, some of which have been found to compete with and/or influence the level of tryptophan binding to albumin. Much information is now available regarding the ligand-binding sites on albumin (191).

Since L-tryptophan in the serum binds with albumin, we investigated whether or not a similar binding may be occurring with proteins in the cytosol of liver cells, and whether or not the administration of tryptophan orally would influence such binding. Following the tube-feeding of tryptophan (30 mg/100-g body weight) to fasted rats, total tryptophan concentrations after 15 min became markedly elevated in serum (+ 727%) and in liver (+ 797%), and after 2 hr the levels were still elevated in serum (+ 458%) and in liver (+ 95%). Similar changes have been reported earlier (154). In several experiments we determined the free- and bound-tryptophan levels of serum and of liver homogenates or postmitochondrial supernatants of rats treated with water or tryptophan 15 min or 2 hr before killing. The methodology employed was that of Badawy and Smith (192) using supernatants after trichloroacetic acid (TCA) precipitation to determine total-tryptophan levels and using ultrafiltrates prepared by centrifuging liver homogenates through Amicon Centriflo membrane cones and measuring free-tryptophan levels on pass throughs. Protein-bound tryptophan was determined by differences. While in the control rats the serum contained 10%-free and 90%-bound tryptophan and the liver contained 76%-free and 24%-bound tryptophan, after tryptophan tube-feeding (15 min or 2 hr) the serum changed to 36%-free and 64%-bound and the liver changed to 38%-free and 62%-bound. Combining the data of the increases in total tryptophan levels with the free and bound percentages, it appears that the serum shows greater increases in the amounts of free than of bound tryptophan, while in the liver there are greater increases in bound than in free tryptophan. We are currently evaluating the possible significance of these changes in the liver.

Next, we determined, in the absence of protein synthesis, the in vitro binding of [³H]tryptophan into proteins of dialyzed liver cell saps of control and tryptophan-treated (10 min) rats and found a similar percentage of binding in the free and bound fractions of control and experimental groups. Only in the TCA-precipitable fractions of dialyzed cell saps (where only 5% of the total counts resided) was there a significant increase in the experimental over the control groups.

When [³H]leucine binding in vitro was measured instead of [³H]tryptophan under the same experimental conditions, there were no differences in the binding to TCA-precipitable proteins of the dialyzed liver cell saps of control and experimental (tryptophan-treated, 10 min) groups.

Thus the [³H]tryptophan binding to TCA-precipitable proteins under

our experimental conditions does not appear related to the normal binding of tryptophan to serum albumin or to general hepatic proteins.

Possible other types of bindings: In consideration that L-tryptophan may rapidly bind to, or become absorbed by, certain proteins of
the cytosol or nuclei of livers, we have conducted a number of experiments designed to determine whether or not this reaction occurs. Initially, these experiments were designed to determine if such an effect
could be demonstrated. Secondly, if an effect was found, it was necessary to determine which proteins were involved.

Tryptophan rapidly becomes incorporated into proteins and it also
binds to proteins of liver cells, particularly to proteins of the nuclear
envelopes (193). Increased in vitro binding of [^{3}H]tryptophan to proteins (TCA-precipitable) of nuclei and cytosols of livers of tryptophan-
treated (10 min) rats was observed, and this reaction appears to correlate with enhanced in vitro release of hepatic nuclear RNA and increased nuclear NTPase and protein phosphokinase activities (193).

In vitro [^{3}H]tryptophan binding to proteins of cytosols or nuclei of rat
livers was decreased by the addition of cold insulin, and in vitro [^{125}I]-
insulin binding to proteins of hepatic nuclei was increased when incubated with cytosols of livers of rats treated in vivo with tryptophan in comparison with cytosols of livers of control rats treated with water. Pretreatment of rats with puromycin before tryptophan administration prevented the increased in vitro binding of [^{3}H]tryptophan to nuclear and
cytosol proteins caused by cytosols of experimental rats, but it did not
prevent the increased in vitro binding of [^{3}H]tryptophan to nuclear
proteins resulting from nuclei of experimental rats. Preincubation of
hepatic nuclei with concanavalin A prevented the increased in vitro
binding of [^{3}H]tryptophan to nuclear proteins, prelabeled nuclear RNA
release, and nuclear NTPase activity of livers of tryptophan-treated
rats. The results suggest that tryptophan rapidly binds with hepatic
proteins (possibly glycoproteins) associated with the nuclear membrane,
where there is an increase in the activities of enzymes involved in
phosphorylation and dephosphorylation along with release of nuclear
mRNA into the cytoplasm.

Consideration of the possible importance of glycoproteins: For a
number of years our laboratory has been interested in the acute effects
of administering hypertonic NaCl solutions on hepatic polyribosomes
and on protein synthesis (194-196). In general, the acute administration (orally or IP) of hypertonic NaCl leads to a rapid disaggregation
of hepatic polyribosomes and inhibition of hepatic protein synthesis.
This response is in marked contrast to the stimulatory effect of tryptophan. Nonetheless, we have recently observed that the dialyzed liver
cell sap of hypertonic NaCl-treated (10.7%) (10 min) rats leads to enhanced (+ 70%) release in vitro of labeled RNA from control nuclei,

similar to that reported with dialyzed liver cell sap of tryptophan-treated rats. However the isolated hepatic nuclei of hypertonic NaCl-treated rats incubated with dialyzed liver cell saps of control rats revealed a marked decrease (− 63%) in release of labeled RNA in vitro from these nuclei compared with control nuclei which is unlike that observed with hepatic nuclei of tryptophan-treated rats. Most recently we have investigated the in vitro binding or adsorption of [^{3}H]tryptophan to dialyzed liver cell sap of hypertonic NaCl-treated rats with isolated control nuclei. To our surprise, we found in three experiments that in the TCA-precipitable proteins there was a 44% increase in counts in the dialyzed cell sap fraction and a 64% increase in the nuclear fraction when compared with using dialyzed cell saps of control rats.

A recent paper has cast some light on our results and may offer an explanation for the findings described. LeJohn and Stevenson (197) reported using *Achlya* (a freshwater mold), that tryptophan binds to a cell wall membrane proteoglycan. The tryptophan uptake was considered to be a binding process, because while uptake of methionine and phenylalanine was inhibited by metabolic poisons (NaN$_3$, dinitrophenol, and Hg^{2+}), that of tryptophan was not. What the tryptophan could be binding to was diciphered by LeJohn in the following way. When germlings were osmotically shocked (198), they lost their ability to take up tryptophan by a purine analogue-enhanced process. The proteoglycan that was isolated from the osmotic shock fluid has been shown to bind tryptophan (199). Thus, considering these studies, we speculate that the hepatic cytosol protein that may play a role in enhancing nucleocytoplasmic translocation of mRNA may be a glycoprotein. In the livers of rats treated with hypertonic NaCl this proteoglycan is liberated into the liver cytosol from membranes (possibly nuclear membranes) and then has the ability to stimulate nuclear transport of RNA in normal control nuclei. In the livers of rats treated with tryptophan, this proteoglycan is stimulated by increased synthesis or activation in the liver cytosol and then leads to the stimulation of nuclear RNA transport.

Given the above hypothesis, we are currently investigating whether or not tryptophan may enhance the presence of glycoprotein(s) in the liver (nuclear membranes and cytosol) of rats. This glycoprotein(s) may have a binding affinity for [^{3}H]tryptophan and may be involved in nucleocytoplasmic translocation of mRNA.

A recent report by Baglia and Maul (200) indicated that a glycoprotein, identified as lamin B, is a major component of the nuclear envelope and nuclear matrix. This glycoprotein not only may be a structural nuclear protein but also may have NTPase activity, which is essential in nucleocytoplasmic transport. We are currently investigating for changes in the lamin B of liver nuclear envelopes of rats treated with tryptophan.

Summary of Current Views

Figure 2 presents a schematic diagram relating to our hypothesis
about how tryptophan acts to stimulate hepatic protein synthesis.
Our current thoughts are that tryptophan stimulates protein synthesis in either one or both of two ways: (a) enhancement of mRNA
synthesis and (b) increased nucleocytoplasmic translocation of mRNA.
In normal animals, tryptophan probably acts in both mechanisms.   In
abnormal animals as when animals are treated with inhibitors of RNA
metabolism, such as cordycepin and actinomycin D, it acts only by
the second mechanism.   Also, in abnormal animals, as when animals
are treated with other inhibitors that primarily inhibit protein synthesis such as ethionine, puromycin, or CCl_4, the second mechanism

may be mainly involved, but a full explanation is not yet available.

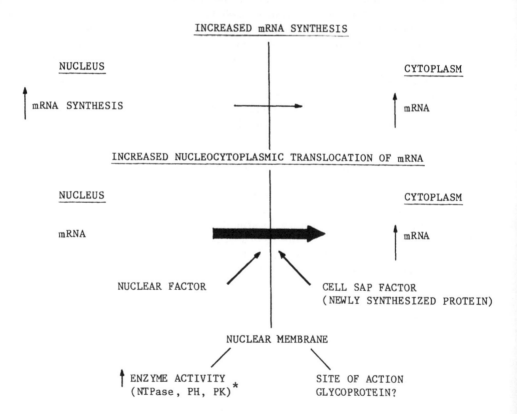

Figure 2   Mechanism of tryptophan effect (*NTPase, nucleoside triphosphatase; PH, phosphoprotein phosphohydrolase; PK, protein
phosphokinase).

Concerning the phenomenon of nucleocytoplasmic translocation of mRNA, it is not yet fully understood how the posttranscriptional controls within the nucleus and cytoplasm regulate the flow of RNA from nucleus to cytoplasm. Our studies suggest that in normal animals, tryptophan affects both the nucleus as well as the cytosol in leading to an increase in the levels of cytoplasmic mRNA.

On the basis of the findings in earlier studies (106,107,124,128,136, 137,142,154,164) and more recent studies (155,193), we speculate that tryptophan acts at the level of posttranscriptional control in the following manner in stimulating hepatic protein synthesis. Following the tube-feeding of tryptophan, tryptophan rapidly enters the liver and individual hepatocytes. Within the hepatocytes tryptophan goes directly, or via binding to a preexisting cytoplasmic protein, to the nuclear membrane (envelope) where it then binds with a glycoprotein. This triggers in the nuclear membrane, possibly at the nuclear pore complex, an increase in the activities of enzymes involved in phosphorylation and dephosphorylation and release of nuclear mRNA into the cytoplasm. Among the mRNAs released is one that codes for a "carrier" protein (?glycoprotein) which enhances additional transport of tryptophan from the cytoplasm to the nuclear membrane. This process hastens the further release of nuclear mRNAs into the cytoplasm, leading to increased hepatic polyribosomal aggregation and synthesis of hepatic proteins. Recent experimental results using concanavalin A (193) suggest that a nuclear glycoprotein might be the site where tryptophan binds and acts to stimulate nuclear mRNA release. This glycoprotein may be one such as lamin B, which was demonstrated by Baglia and Maul (200) to be intimately involved in nucleocytoplasmic translocation of mRNA.

VII. EFFECT ON THE LIVERS OF RATS TREATED WITH HEPATOTOXIC AGENTS

Earlier studies from our and other laboratories have explored whether or not the acute administration of L-tryptophan may have a beneficial effect on hepatic protein synthesis in animals suffering from liver injury induced by a variety of hepatotoxic agents. Some of the agents used have been actinomycin D (201), CCl_4 (144), ethanol (202), ethionine (159), hypertonic NaCl (143) and puromycin (158). In general, tryptophan improved hepatic protein synthesis when administered in conjuction with (mainly following shortly thereafter) the hepatotoxic agent. In a recent report (129) we have expanded the information concerning tryptophan and its effect with other hepatotoxic agents, such as alpha-amanitin, cordycepin, NaF, sparsomycin, aflatoxin B_1, dimethylnitrosamine, and galactosamine. Thus we have now gathered much information about the inhibitory effects on protein synthesis by a variety of hepatotoxic agents and determined whether tryptophan may act to prevent, to cure, or both, the inhibitory effect of each of the toxic agents. Since the selected hepatotoxic agents act to interfere with hepatic protein synthesis by a variety of mechanisms (203), it was felt that the results of such studies could provide clues about how tryptophan may act in a stimulatory manner in certain diseased livers and possibly not in others. The results of our findings are summarized in Table 2. It is apparent that, under the experimental conditions used in our studies, tryptophan has a preventive and curative effect with alpha-amanitin, cordycepin, puromycin, and hypertonic NaCl; a preventive effect only with CCl_4 and NaF; a curative effect only with actinomycin D and ethionine; and no effect with sparsomycin. Also, we have observed a curative effect with aflatoxin B_1, dimethylnitrosamine, and galactosamine (129). Earlier, Rothschild et al. (202,204) reported that the addition of tryptophan to isolated perfused livers of rabbits treated with ethanol or CCl_4 aided in the recovery of albumin synthesis. Kroger et al. (205) have reported that the administration of D-galactosamine-HCl induces alterations in livers of normal or adrenalectomized rats, histologically resembling hepatitis, and that pretreatment with DL-tryptophan can prevent this effect.

In some experiments concerned with the mechanism by which tryptophan may act to improve hepatic protein synthesis after toxic injury, we measured the ability of tryptophan to stimulate hepatic mRNA synthesis, nucleocytoplasmic translocation of RNA in vitro, and nuclear envelope nucleoside triphosphatase activity after hepatotoxic injury. Nucleoside triphosphatase (Mg^{2+}-dependent adenosine triphosphatase, EC 3.6.1.3.1) is present in mammalian liver nuclear envelopes and there is evidence that this enzyme is involved in nucleocytoplasmic

Table 2 Effect of Tryptophan on Hepatic Polyribosomes and Protein Synthesis in Animals Treated with Selected Hepatotoxic Agents

Preventive and curative	Preventive only	Curative only	No effect
α-Amanitin (129)	CCl_4 (144)	Actinomycin D (201)	Sparsomycin (129)
Cordycepin (129,136)	NaF (129)	Ethionine (129,159)	
Puromycin (129,158) Hypertonic NaCl (129,143)			

translocation of RNA. It is of interest that these three parameters were elevated significantly by tryptophan following the administration of agents, such as actinomycin D, cordycepin, ethionine, puromycin, and hypertonic NaCl, which demonstrated a curative effect by tryptophan, but were not elevated after tryptophan administration following CCl_4, NaF, and sparsomycin, which demonstrated no improvement

with tryptophan. These findings emphasize the importance of the role tryptophan plays in stimulating the availability of cytoplasmic mRNA, and this effect occurs even after liver injury by certain toxic agents. On the other hand, it is unclear why after CCl_4, NaF, or

sparsomycin administration, this effect on hepatic mRNA, as well as on hepatic polyribosomes and protein synthesis, is absent.

VIII. EFFECT ON OTHER ORGANS

In addition to reports that tryptophan stimulates hepatic and blood protein synthesis as described in a previous section, it has also been reported that tryptophan stimulates protein synthesis in other organs. Gacad et al. (206) have reported that tryptophan administration to food-deprived rabbits 45 min before killing induces an increase in

protein synthesis ([^{14}C]leucine incorporation into protein) by lung slices. Jorgensen and Majumdra (123) reported that a single tube-feeding of L-tryptophan to well-fed adrenalectomized rats stimulated

in vivo incorporation of [^{3}H]leucine into brain and kidney proteins as well as liver proteins. Also, Majumdar and Nakhla (207) reported that tube-feeding tryptophan to well-fed adrenalectomized rats induced an increase in the activity of cerebral acetylcholinesterase, which could be prevented when the animals were pretreated with actinomycin D. Whether or not the enhanced protein synthesis in the brain following tryptophan administration may be related to a rise in brain serotonin level caused by increased plasma tryptophan concentration is not known.

Since starvation lowers protein synthesis in a number of organs or tissues and refeeding reverses the situation, Majumdar (208) investigated the importance of tryptophan in regulating protein synthesis in one such tissue, the gastric mucosa. He reported that refeeding a nutritionally complete diet to fasted rats stimulated the ability of gastric mucosal polyribosomes to synthesize protein in a cell-free system. In contrast, a tryptophan-free diet (otherwise nutritionally complete) was ineffective. Also, Majumdar (209) demonstrated that tube-feeding tryptophan to well-fed adrenalectomized rats stimulated in vivo amino acid incorporation into gastric total proteins. In consideration that tryptophan may influence enzymes of the gastrointestinal tract, specifically the stomach and small intestine, Majumdar reported that tube-feeding tryptophan to adrenalectomized rats increased the activities of gastric mucosal pepsin (209) and disaccharidases (lactase and maltase) in the jejunum and ileum (210). Also, Majumdar reported that, using fasted (2 days) rats, refeeding of a complete diet, but not a tryptophan-free diet, increased the activities of small-intestinal alkaline phosphatase and disaccharidases (maltase and sucrase) to normal levels (211).

IX. TRYPTOPHAN AND HORMONES

Since the administration of single amino acids elicit certain metabolic reactions indirectly via stimulation through hormones from the adrenal cortex, experiments were performed with adrenalectomized animals to determine whether or not tryptophan could still stimulate hepatic protein synthesis. Using mice which had been adrenalectomized 2 days before and then fasted overnight, we observed that the stimulation of hepatic protein synthesis was independent of adrenal cortical hormones (107). Also, we reported that tryptophan administered to rats that had hepatic protein synthesis stimulated by the administration of cortisone acetate showed further stimulation of hepatic protein synthesis (in vitro) and a shift of hepatic polyribosomes toward heavier aggregation (189). Other investigators have reported that tube-feeding of tryptophan to well-fed adrenalectomized and adrenalectomized-diabetic rats stimulated amino acid incorporation into plasma albumin, transferrin, and fibrinogen, and into liver ferritin in vivo (121,123). On the other hand, Cammarano et al. (117) reported that there was involvement of adrenal steroids in the changes of polyribosome organization during feeding of high levels of L-tryptophan.

A number of studies have been concerned with whether or not tryptophan may alter the secretion of hormones. Ajdukiewicz et al. (212) reported that nonfasted humans showed a significant rise in plasma insulin 30 min after ingesting tryptophan. Floyd et al. (213) showed a small rise in plasma insulin following intravenous tryptophan. Fahmy et al. (214) demonstrated that oral tryptophan induced an elevation of plasma insulin in normal subjects and in adult-onset diabetics. Modlinger et al. (215,216) reported that tryptophan administration to fasted patients induced elevations in the blood levels of aldosterone, renin, cortisol, and ACTH. Yokogoshi and Yoshida (217) have reported that fasted (48 hr) rats tube-fed tryptophan for 2 hr revealed no change in serum insulin levels but increases in serum glucocortocoid. However they also observed in fasted rats increases in serum glycocorticoid levels after tube-feeding other single amino acids (methionine, threonine, or leucine), yet each of these, unlike tryptophan, did not shift the hepatic polyribosomes toward heavier aggregation. Many investigators have reported that an increased plasma tryptophan concentration elevates the brain serotonin level (39-41,218).

The role of serotonin in the regulation of the secretion of anterior pituitary hormones via effects on hypothalamic hypophyseal releasing and release-inhibiting factors has been investigated for many years. Direct or indirect pharmacological evidence exists that the secretion of prolactin, growth hormone, luteinizing-releasing hormone, thyrotropin, corticotropin, and perhaps also aldosterone, beta-endorphin, and renin, may be influenced by serotonergic mechanisms in some species (219).

It has been reported that increased hepatic polyribosomal aggregation and protein synthesis occurs in fasted animals after giving carbohydrates (220) or after giving a protein-free diet (221). However this effect is different from that induced by administering tryptophan to fasted animals. In the case of animals fed the protein-free diet, the rats must be fasted for a long period and the response to the administered carbohydrates took hours to occur (222). On the other hand, the hepatic response to tryptophan does not require lengthy fasting, or fasting at all, and the effect is rapid (minutes) in onset (222). It is speculated that hormonal stimulation, probably via insulin, accounts for the effect observed in fasted animals that receive the carbohydrates.

X. TRYPTOPHAN AND CANCER

In recent years the relationship between nutrition and cancer has attracted much attention (223-225). On the basis of epidemiological studies and also animal experimentation, a variety of nutrients have been considered to play a role in the induction of certain tumors. In this review the focus will be on the possible relationship between L-tryptophan and cancer.

A. Relationship with Tumor Induction

Review of the literature reveals that tryptophan has been implicated in carcinogenesis of the bladder in earlier (226-229), as well as in more recent (230-232), experimental studies. On the other hand, the effect of tryptophan on liver tumorigenesis is conflicting; an enhancing effect has been reported by some workers with diethylnitrosamine and N-2-fluorenylacetamide (226,233), while a decreasing effect has been reported by other workers with 3'-methyl-4-dimethylaminoazobenzene, diethylnitrosamine, and dibutylnitrosamine (234,235). Also, indole has been reported to have a suppressive effect on hepatocarcinogenesis in animals caused by N-2-fluorenylacetamide (236-238). Thus the possible role of tryptophan in carcinogenesis of the bladder or of the liver is still not clear.

Wattenberg (239) has reported that a number of indole-containing compounds present in the diet, such as indole-3-acetonitrile, indole-3-carbinol, and 3,3'-diindolylmethane, have the capacity to increase the activity of the microsomal mixed-function oxidase system and also have inhibitory effects on chemical carcinogenesis in animals. Whether the indole-containing essential amino acid tryptophan may act in a similar manner is unknown. It has been reported from many laboratories that L-tryptophan can stimulate the activities of a number of hepatic enzymes (see earlier section). Evarts and Mostafa (240) reported that rats fed a diet containing 1% L-tryptophan for 6 to 12 days revealed an increase in liver nitrosodimethylamine demethylase activity. but such was not the case with similar additions of other amino acids. Also, these investigators (241) reported that 1% L-tryptophan significantly increased the microsomal protein concentration and the P-450 concentration in rat livers. In our laboratory, we reported that rats force-fed for 3 days an elevated (1%) tryptophan-containing diet revealed increased hepatic cytochrome P-450 and b_5 activities (242).

Also, Jorgensen and Majumdar (243) reported that force-feeding of tryptophan caused within 24 hr an increase in microsomal cytochrome P-450 and in aniline hydroxylase activity in rat livers.

Matsukura et al. (244) have reported that the charred parts of broiled meat and fish contain a series of new heterocyclic amines, in the pyrolysate of amino acids and proteins, that are mutagenic, and some—3-amino-1,4-dimethyl-5H-pyrido-(4,3-b)indole and 3-amino-1-methyl-5H-pyrido(4,3-b)indole—from tryptophan pyrolysates, are carcinogenic in animals. Thus there is evidence in the literature that tryptophan-related compounds may affect the liver in a variety of ways that may influence the course toward carcinogenesis. Much needs to be unraveled before we understand what and how this occurs.

B. Studies with Hepatomas

In view of the dramatic effect of the administration of tryptophan on hepatic polyribosomes and protein synthesis in normal animals (fasted or fed) (107,124,128), it was important to determine if a similar effect may occur in hepatocellular carcinoma. Therefore we utilized intra-hepatically transplanted hepatomas (H5123 and 19) in rats and then determined how the host liver, as well as the hepatoma, responded to an administration of L-tryptophan. The results revealed that the transplantable hepatomas showed little or no changes after tryptophan administration (124,245). At the same time, the host livers of the tumor-bearing rats revealed only mild to moderate (mainly statistically insignificant) changes. This is in marked contrast to the effects of tryptophan after its administration to normal animals where many parameters (dealing with RNA and protein metabolism) of the livers revealed marked (statistically significant) changes. These findings are summarized in Table 3 (246).

In consideration that the hepatoma, because of its rapid cell division and growth, may become resistant to the effects of tryptophan, we investigated whether or not regenerating livers following partial hepatectomy (1 or 2 days) would respond to the administration of tryptophan.

Table 3 Effect of Tryptophan on Livers of Normal Rats and on Host Livers and Intrahepatically Implanted Hepatomas of Rats[a]

Parameter	Normal rat liver	Hepatoma-bearing rats	
		Host liver	Hepatoma
Aggregation of polyribosomes	1+	0	1−
Protein synthesis (in vitro)	4+	1+	0
Poly(A)-mRNA synthesis	3+	3+	1+
Nucleocytoplasmic translocation of RNA			
Nuclear effect	4+	3+	1−
Cell sap effect	4+	0	0
Nuclear enzyme activity changes			
Nucleoside triphosphatase	4+	2+	0
RNA polymerase I	3+	2+	2−
RNA polymerase II	3+	1+	0

[a]Changes as increases (+), decreases (−) or little or no change (0): 4+ or −, >60%; 3+ or −, 40-59%; 2+ or −, 20-39%; and 1+ or −, 10-19%; 0, 0-9%.

The regenerating livers responded to tryptophan with a shift toward heavier aggregation of polyribosomes and an increase in in vivo protein synthesis (245) similar to that observed in livers of normal animals. Thus the inhibitory or resistant effect found in the hepatoma probably rests in the anaplasia of the cells rather than merely in the rapid division or growth of the cells.

In another study (246) we investigated if tryptophan would influence the polyribosomes and protein synthesis of host liver and of intrahepatically transplanted hepatomas of rats that were treated with two hepatotoxic agents, hypertonic NaCl and CCl_4, which had been reported to affect these parameters adversely in the tumor-bearing rats (247). While treatment with hypertonic NaCl or CCl_4 caused disaggregation of polyribosomes and inhibition of protein synthesis in both host liver and in hepatoma, the subsequent administration of tryptophan caused some improvement in both parameters in host liver but not in hepatoma. Thus, even after hepatotoxic injury to the hepatoma, it is not able to respond to tryptophan as are the livers of normal rats (129) or host livers of tumor-bearing rats.

C. Studies with Hepatic Ornithine Decarboxylase Activity

Ornithine decarboxylase (ODC) is the rate-controlling enzyme in the
biosynthesis of polyamines (248) and its increase in activity correlates with cell growth (249,250). The ODC activity is markedly
elevated in proliferating tissues (249,251), in L1210 leukemia cells
(252), and in experimental hepatomas (253). The activity of ODC
has been postulated to be associated with or related to promotion in
skin carcinogenesis (254). Indeed, a single topical application of a
promoter, croton oil or its active component, 12-*O*-tetradecanoly-
phorbol-13-acetate, results in a rapid and substantial increase in
mouse epidermal ODC activity (255). However since a variety of
agents have been demonstrated to stimulate ODC activity in the
liver, it is indeed questionable if this alteration in the liver may in
any way be related to subsequent carcinogenesis. Nonetheless, it
became of interest to determine if the administration of tryptophan
could alter the activity of ODC in the liver. We observed that a
single tube-feeding of varying levels of L-tryptophan (2.5-30 mg/
100-g body weight) to overnight-fasted rats 1 hr before killing
caused increases in the hepatic ODC activities. This elevation began after 1 hr and peaked at 2 hr (6.5-fold increase over controls)
(256). This effect was similar to that observed with a variety of
agents including hormones, drugs, and hepatotoxins, where the
stimulation of hepatic ODC activity occurred maximally at about 4 hr
(257). Although some of these agents have been implicated as being
promoters of liver carcinogenesis, others have not. Thus the significance of the tryptophan-induced stimulation of hepatic ODC activity
is at present not clear and needs further elucidation.

XI. TRYPTOPHAN AND PREGNANCY

Studies with experimental animals have demonstrated that the nutritional status relative to the amount of dietary protein during prepartum and postpartum development has an important role in determining
subsequent brain tryptophan metabolism. Indeed, alterations in brain
tryptophan, serotonin, and 5-hydroxyindoleacetic acid concentrations
occurred in rats as a consequence of dietary protein insufficiencies
during prenatal and/or postnatal development (258,259). Because
the availability of tryptophan to brain and peripheral tissues may be
a limiting factor in protein synthesis during early development (260,
261), alterations in the utilization of this essential amino acid under
conditions of prenatal and postnatal protein malnutrition have been
considered to possibly cause permanent alterations in the organism.
    The influence of prepartum and postpartum nutritional status on
brain indoleamine metabolism indicated that both play important roles
in determining this process (258,259,262). The increased brain levels

of tryptophan and serotonin as a consequence of protein inadequacies can best be related to the increased amounts of free plasma tryptophan available for their brain metabolism even while total plasma tryptophan levels are decreased. These rats seem to be shunting more tryptophan to the brain at the expense of the periphery. From a teleologic viewpoint, the increases in brain serotonin and 5-hydroxyindoleacetic acid may be only the consequence of a need to assure the availability of tryptophan for adequate brain protein synthesis and development in early life.

Zamenhof et al. (263) studied the effect of dietary deprivation of tryptophan on rats during the second half of pregnancy. The omission of tryptophan from a chemically defined amino acid diet caused a drop in maternal body weight and liver weight. In the neonate, there were decreases in body weight and in all cerebral parameters (weight, DNA, and protein). Similar changes were also observed with the dietary omission of lysine or methionine. These findings were essentially similar to those produced by total dietary protein deprivation in the comparable period of pregnancy (264). Thus it appears that the absence of a single dietary essential amino acid, such as tryptophan, even during half of pregnancy, may be as harmful as total absence of dietary proteins during this period, at least as far as prenatal brain development is concerned. However this study did not indicate any effects specific for dietary deficiency of tryptophan.

Matsueda and Niiyama (265) studied the effects of diets (6% casein), each with an excess (5%) of one essential amino acid, on the maintenance of pregnancy and reproductive performance, including fetal growth in rats from day 1 to day 14 or 21 of pregnancy. Judging from the total food consumption and body weight gain during pregnancy, they found that methionine had the most severe effects followed by leucine and tryptophan. In comparison with pair-fed controls, pregnant rats fed the excess tryptophan diet had a 20% loss of fetuses; of maintained nitrogen balance, carcass protein, and fat; and had lowered weights of females and decreased fetal brain weights and amounts of DNA, RNA, and protein. Excess dietary tryptophan induced no appreciable changes in maternal plasma free amino acids but two- to three-fold increases in acidic and aromatic amino acids in the fetal brain. The results of experiments on feeding diets with excess of a single essential amino acid, such as tryptophan, have been considered in connection with studies on inborn errors of amino acid metabolism. The influences of hyperaminoacidemia on fetal development, in pregnant animals, especially or neuronal development, merit further exploration.

Meier and Wilson (266) have recently reported that added dietary tryptophan during pregnancy reduces embryo and neonate survival in the golden hamster. When using 1.8 to 3.7% tryptophan in a high

(23%)-protein diet or using 8.0% tryptophan in a moderate (16%)-protein diet during pregnancy, litter size and neonate weights were reduced and mortality of neonates after 1 week was increased in comparison with controls. Although the mechanism by which increased dietary tryptophan acts on the embryo and neonate is not clear, consideration has been given that it may be related to the elevation of serotonin. Exogenous serotonin has been reported to cause abortions in several vertebrate species, to reduce litter sizes, to increase still births and neonate abnormalities, and otherwise to adversely influence pregnancy (267-269). Since the availability of tryptophan is probably the most important rate-limiting factor in serotonin synthesis (270), and since studies have revealed that increases (271) or decreases (272) of dietary tryptophan lead to concomitant changes in serotonin levels, it is likely that tryptophan may act via serotonin. Currently many humans use tryptophan to decrease appetite (65) and to promote sleep (55) and, therefore it may be important to determine whether or not increased tryptophan intake during pregnancy may be potentially harmful. Further studies are needed to clarify if this danger exists.

XII. TRYPTOPHAN AND AGING

Dietary manipulation has been used by a number of investigators in an attempt to alter the rate of mammalian aging. Some workers have indeed demonstrated the life-extending effect of caloric restriction and have shown that the rate of aging can be retarded under certain conditions (273-276). Segall and Timiras (277) have undertaken experimental studies to examine the effect of long-term dietary tryptophan restriction on the process of aging in the rat. This deficiency was selected since rats fed tryptophan-deficient diets have lowered brain serotonin levels and can be kept in a state of maturational and growth arrest for long periods. Their preliminary findings indicated that growth-retarded rats fed a tryptophan-deficient diet can reach normal body weight when subsequently fed a normal diet. They showed a delay in the age of onset of visible tumors and indicated a slight increase in their average lifespan at late ages. Their pair-fed control rats and the caloric restricted rats in experiments of others (273-276) also showed a prolonged lifespan and a delayed onset of tumors. Thus, at present, it is difficult to attribute any specific effects to the long-term tryptophan deficiency. Further data are needed.

XIII. TRYPTOPHAN AND HEPATIC COMA

Chronic liver disease in man often is associated with hepatic encephalopathy, a neuropsychiatric syndrome usually beginning with changes in mood and signs of intellectual impairment and leading to confusion, slurred speech, drowsiness, hypersomnia, stupor, and coma as the condition worsens. Liver dysfunction is also characterized by a known number of metabolic abnormalities including increased concentrations of aromatic amino acids in plasma and cerebrospinal fluid (278). These changes include an increase in plasma free tryptophan and are associated with raised CSF levels of tryptophan and 5-hydroxyindoleacetic acid, a precursor and the terminal metabolite of serotonin. Indeed, many workers (216,279-283) have considered tryptophan to be implicated in the pathogenesis of hepatic coma. However, currently, an elevation of plasma tryptophan is not generally considered to be pathognomonic of hepatic encephalopathy (278).

The involvement of tryptophan in hepatic coma may be explained by the action of one of its metabolites, serotonin, which is known to exert profound effects on the central nervous system (284). Although it is generally agreed that the turnover of brain serotonin is directly related to the concentration of brain tryptophan, questions remain regarding the regulation of brain tryptophan levels. The apparent increased turnover of brain serotonin in patients with hepatic encephalopathy and coma cannot be explained by corresponding increases in plasma total tryptophan levels because the reported values in these patients range from low, normal, to mildly elevated (279,281,283,285). However all reports seem to agree that plasma free-tryptophan levels become increased. The raised free-tryptophan concentration is important because it is this fraction that is available for transport into the brain and that plays a key role in regulating the entry of tryptophan into the brain (39,40). In attempting to explain the elevation in plasma free-tryptophan levels, there are a few probable mechanisms which operate singly or in combination: (a) a rise in plasma concentration of unesterified fatty acids occurs (286) which releases tryptophan from plasma protein (44) thus increasing its availability to the brain, and (b) the drop in plasma albumin in chronic liver disease may account for some of the increase in the free/bound tryptophan ratio. Also, brain tryptophan concentration is influenced by the ratio of its plasma concentration to the sum of the concentrations of five other amino acids (tyrosine, phenylalanine, leucine, isoleucine, and valine) which compete with tryptophan for uptake into the brain (46,47).

Patients with liver cirrhosis and hepatic encephalopathy appear to have an altered plasma amino acid pattern: low concentrations of the branched-chain amino acids (leucine, valine, and isoleucine) and high levels of the aromatic amino acids (tryptophan, tyrosine, and phenylalanine) and methionine (287-289). These changes, together with an increased blood-brain barrier permeability (290,291), may augment the influx of aromatic amino acids onto the brain, causing an imbalance in the synthesis of neurotransmitters, which may in turn contribute to the disturbed brain function. An increased concentration of the

branched-chain amino acids in the blood has been proposed to normal-
ize these reactions (292). Experimental studies have indeed demonstrated that the administration of branched-chain amino acids to patients with liver cirrhosis is accompanied by reduced arterial blood
levels and diminished brain uptake of aromatic amino acids (289). Although some case reports have suggested that intravenous or oral
administration of branched-chain amino acids may be beneficial in patients with liver cirrhosis and hepatic encephalopathy (287,293,294),
a multicenter study with 50 patients reported that such therapy, while
reducing the concentration of plasma aromatic amino acids, did not
appear to improve cerebral function or decrease mortality in patients
with hepatic encephalopathy (295). Thus the possible pathogenetic
importance of the observed derangements of plasma amino acid levels
is still in doubt. In attempting to propose a unified hypothesis of
hepatic encephalopathy, Fischer and Bower (296), have considered
ammonia, plasma amino acid imbalance, deranged hormonal profile, and
a deranged plasma amino acid pattern as all contributing to the deranged brain amino acid profile, resulting in distortion of the
aminergic neurotransmitter profile within the central and peripheral
nervous system.

Although there is much evidence relating to the association between
changes in plasma and brain tryptophan levels with hepatic coma, the
underlying cause of the coma is still not clear.. In experimental animals
it has been established that tryptophan is a very toxic amino acid in
terms of lethality (104). In fact, the ingestion of tryptophan by humans is able to induce significant central nervous system signs and
symptoms (297). However these manifestations are different from, but
have been confused with, hepatic coma (278). In humans a tryptophan
dose of 100 mg/kg body weight per os is usually well tolerated with the
exception of some minor gastric disturbances (vomiting, etc.) (298).
A simple toxicity of tryptophan itself can hardly be responsible, as
up to 15 g/day of DL-tryptophan has been given by mouth in the treatment of depression (299), and plasma free tryptophan of human subjects has been raised almost 100-fold by tryptophan infusion without
grossly apparent effect (300). In normal rats injection of a lethal dose
of tryptophan (510-775 mg/100-g body weight) produced dyspnea, de-
hydration, prostration, and death without an intervening phase of
coma while the plasma and brain concentrations increased more than
300-fold (278). However it is conceivable that raised serotonin levels
in the brain might enhance the toxicity in the central nervous system
of other substances accumulating in subjects with liver disease. Thus
tryptophan may certainly be implicated in hepatic coma, but the true
extent of its effects or actions on the brain needs further clarification.

XIV. OTHER ESSENTIAL AMINO ACIDS AND THEIR POSSIBLE UNIQUE ACTIONS

This review has concentrated on the unique effects of tryptophan, particularly on hepatic protein synthesis. However it might be appropriate to review briefly some interesting, recent developments where other essential amino acids have been implicated as having specific effects on protein synthesis in other tissues, specifically muscle tissue. In general, amino acids in relation to the regulation of protein turnover in skeletal muscle have been investigated much less extensively than in liver.

A number of experimental studies have suggested that branched-chain amino acids, particularly leucine, are involved in the regulation of muscle protein synthesis (301). Unlike other dietary amino acids, the branched-chain amino acids are metabolized in peripheral tissues, predominantly muscle, rather than in the liver (302,303). After a protein meal, the branched-chain amino acids effectively bypass the liver and provide 60 to 90% of amino acid nitrogen taken up by muscle (304). These characteristics suggest that these amino acids may have unique therapeutic benefits, particularly in the presence of a diseased liver. Using rat diaphragm incubated in vitro, Buse and Reid (305), Fulks et al. (306) and Goldberg and Chang (307) have demonstrated a direct stimulation of protein synthesis in muscle tissue. They provided evidence for a role of leucine in regulating the synthesis and the degradation of protein, which was not shared with other branched-chain amino acids and which was not due only to providing energy through oxidation of leucine. Using a perfused hemicorpus (308) and a perfused heart (309), others have shown that leucine induces similar direct and rapid changes in muscle protein synthesis.

The role that the branched-chain amino acids may have in regulating muscle protein metabolism in intact animals and in humans has also been investigated. Although this evidence is less direct, the ability of branched-chain amino acids to stimulate muscle protein synthesis has been inferred. In intact animals, stimulation of muscle protein synthesis has been assumed based upon experiments revealing an increase in the proportion of aggregated polyribosomes of isolated muscles of starved rats (310) and based upon improved nitrogen balance in response to branched-chain amino acid administration following trauma associated with laparotomy (311). In humans, a beneficial role of these amino acids has been demonstrated in starvation (312,313), after surgery (314), and in patients with hepatic encephalopathy (287,293,294). Although these changes in nitrogen balance could be due to changes either in synthesis or in degradation of muscle protein, Sherwin (313) suggested that leucine acts through changes in synthesis, since

degradation of muscle protein, as evaluated by 3-methylhistidine excretion, remained unchanged. McCollough et al. (315) has recently
reviewed the data by others in consideration that branched-chain
amino acids have a beneficial place in the nutritional therapy of liver
disease.

To determine more about the conditions whereby leucine acts on
muscle in vivo, McNurlan et al. (316) studied the effects of leucine
in vivo on protein synthesis in a number of tissues in rats that were
either fed or losing body nitrogen as a result of starvation or lack of
dietary protein. The effect of 100 μmol of leucine on muscle protein
synthesis was assessed in the intact rat. They found that leucine
had no immediate effect on protein synthesis in gastrocnemius muscle,
heart, as well as liver in rats which were fed, starved for 2 days, or
deprived of dietary protein for 9 days. Thus it is not now clear
whether the ability of leucine to stimulate muscle protein synthesis is
a general phenomenon, or whether it may occur under rather special
conditions. Further studies are needed to establish if the branched-
chain amino acids, particularly leucine, have special effects on muscle
protein synthesis in a general way.

XV. POTENTIAL IMPORTANCE OF TRYPTOPHAN AND ITS UNIQUE ACTIONS

As more and more information is gained about the effects and actions
of tryptophan on the liver, it becomes obvious that this knowledge
may have important implications and uses particularly in relation to
liver injury and disease. It is possible that tryptophan may be utilized
to prevent and/or cure certain forms of toxic liver injury. Likewise,
it might be used to detoxify or deactivate certain hepatocarcinogens.

Much information has also been gathered relating tryptophan and the
central nervous system. Indeed, tryptophan has been demonstrated
to influence, probably through serotonin, a number of parameters,
such as sleep, appetite, and pain sensitivity. More data are needed
to define more clearly how these actions occur. Also, the possible
toxic effects of high doses of tryptophan on the brain, liver, and
embryo or fetus have been surfacing in a variety of reports. Future
research on tryptophan holds much promise. It is indeed a unique
dietary component of great importance, much of which is still to be
determined. The knowledge available today regarding tryptophan and
its actions should encourage investigators to pursue further studies
in a number of directions. The author's objective in this review article
has been to describe the current state of knowledge regarding tryptophan and to encourage others to become involved in this interesting
and exciting field.

ACKNOWLEDGMENTS

Many of the experimental studies reported in this chapter were supported by U.S. Public Health Service Research Grants AM-27339 from the National Institute of Arthritis, Diabetes and Digestive and Kidney Diseases and CA-26557 from the National Cancer Institute. The following investigators collaborated with the author in the experimental studies: M. Bongiorno, R. Hornseth, C. N. Murty, E. Myers, D. S. R. Sarma, and E. Verney.

REFERENCES

1. F. G. Hopkins and S. W. Cole, On the proteid reaction of Adamkiewicz, with contributions to the chemistry of glyoxylic acid. *Proc. R. Soc.* 68:21 (1901).
2. A. Ellinger and C. Flamand, Uber die Konstitution der Indolgruppe im Eiweiss. IV. Vorlaufige Mitteilung. Synthese des racemischen Tryptophans. *Ber.* 40:3029 (1907).
3. E. G. Willcock and F. G. Hopkins, The importance of individual amino acids in metabolism. Observations on the effect of adding tryptophane to a dietary in which zein is the sole nitrogenous constituent. *J. Physiol.* 35:88 (1906).
4. T. B. Osborne and L. B. Mendel, Amino acids in nutrition and growth. *J. Biol. Chem.* 17:325 (1914).
5. W. C. Rose, G. F. Lambert, and M. J. Coon, The amino acid requirement of man. VII. General procedures: The tryptophan requirement. *J. Biol. Chem.* 211:815 (1954).
6. E. S. Nasset and M. T. Ely, Nitrogen balance of adult rats fed amino acids low in L-, DL-, and D-tryptophan. *J. Nutr.* 51:449 (1953).
7. C. D. Bauer and C. P. Berg, The amino acids required for growth in mice and the availability of their optical isomer. *J. Nutr.* 26:51 (1943).
8. E. T. Mertz, W. M. Beeson, and H. D. Jackson, Classification of essential amino acids for the weanling pig. *Arch. Biochem. Biophys.* 38:121 (1952).
9. W. C. Rose and E. E. Rice, The significance of the amino acids in canine nutrition. *Science* 90:186 (1939).
10. J. R. Totter and P. L. Day, Cataract and other ocular changes resulting from tryptophane deficiency. *J. Nutr.* 24:159 (1942).
11. A. Ferraro and L. Roizin, Ocular involvement in rats on diets deficient in amino acids. I. Tryptophan. *Arch. Ophthalmol.* 38:331 (1947).
12. V. P. Sydenstricker, W. K. Hall, L. L. Bowles, and H. L. Schmidt, Jr., The corneal vascularization resulting from deficiencies of amino acids in the rat. *J. Nutr.* 34:481 (1947).

13. L. von Sallman, M. E. Reid, P. A. Grimes, and E. M. Collins, Tryptophan-deficiency cataract in guinea pigs. *Arch. Ophthalmol. 62*:662 (1959).
14. R. van Heyningen, Experimental studies on cataract. *Invest. Ophthalmol. 15*:685 (1976).
15. A. A. Albanese, L. E. Holt, Jr., C. N. Kajdi, and J. E. Frankston, Observations on tryptophane deficiency in rats. Chemical and morphological changes in the blood. *J. Biol. Chem. 148*:299 (1943).
16. A. S. Cole and P. P. Scott, Tissue changes in the adult tryptophan-deficient rat. *Br. J. Nutr. 8*:125 (1954).
17. E. B. Scott, Histopathology of amino acid deficiencies. IV. Tryptophan. *Am. J. Pathol. 31*:1111 (1955).
18. F. B. Adamstone and H. Spector, Tryptophan deficiency in the rat. Histologic changes induced by forced feeding of an acid hydrolyzed casein diet. *Arch. Pathol. 49*:173 (1950).
19. L. T. Samuels, H. C. Goldthorpe, and T. F. Dougherty, Metabolic effects of specific amino acid deficiencies. *Fed. Proc. 10*:393 (1951).
20. J. F. Van Pilsum, J. F. Speyer, and L. T. Samuels, Essential amino acid deficiency and enzyme activity. *Arch. Biochem. 68*:42 (1957).
21. H. Spector and F. B. Adamstone, Tryptophan deficiency in the rat induced by forced feeding of an acid hydrolyzed casein diet. *J. Nutr. 40*:213 (1950).
22. H. Patrick and L. K. Bennington, Biochemical changes in the rat produced by feeding a tryptophan deficient ration. *Proc. West Va. Acad. Sci. 41*:161 (1969).
23. T. M. Kloppel and G. Post, Histological alterations in tryptophan-deficient rainbow trout. *J. Nutr. 105*:861 (1975).
24. H. A. Poston and G. L. Rumsey, Factors affecting dietary requirement and deficiency signs of L-tryptophan in rainbow trout. *J. Nutr. 113*:2568 (1983).
25. H. Sidransky, Chemical and cellular pathology of experimental acute amino acid deficiency. *Meth. Achie. Exp. Pathol. 6*:1 (1972).
26. G. Wilfred and T. N. Sekhara Varma, The mechanism of hepatic fatty infiltration in acute threonine deficiency. *Biochim. Biophys. Acta 187*:442 (1969).
27. H. Sidransky, D. S. Wagle, M. Bongiorno, and E. Verney, Chemical pathology of acute amino acid deficiencies. Studies on hepatic enzymes in rats force-fed a threonine-devoid diet. *J. Nutr. 100*:678 (1970).
28. B. Lombardi, Considerations on the pathogenesis of fatty liver. *Lab. Invest. 15*:1 (1966).
29. M. U. Dianzani, Biochemical aspects of fatty liver. *Biochem. Soc. Trans. 1*:903 (1973).

30. Y. S. Lee, H. Naito, and M. Kametaka, The effect of force feeding a tryptophan-deficient diet on arterio-venous difference of plasma amino acids across skeletal muscle of rats. *J. Nutr.* *109*:119 (1979).

31. H. Sidransky and E. Verney, Enhanced hepatic protein synthesis in rats force-fed a tryptophan-devoid diet. *Proc. Soc. Exp. Biol. Med. 135*:618 (1978).

32. H. Naito and M. Kandatsu, Effect of essential amino acid-deficiency on protein nutrition. Part II. Effect of tryptophan-deficiency on turnover of tissue proteins in rat. *Nippon Nogeikagaku Kaishi 41*:623 (1967).

33. H. Naito and M. Kandatsu, Effect of essential amino acid-deficiency on protein nutrition. Part III. Incorporation of ^{35}S-methionine into the tissues of rats fed on tryptophan-deficiency. *Agric. Biol. Chem. 34*:1078 (1970).

34. M. E. Nimni and L. A. Bavetta, Dietary composition and tissue protein synthesis. I. Effect of tryptophan deficiency. *Proc. Soc. Exp. Biol. Med. 108*:38 (1961).

35. R. Bocker, I. K. Jones, and W. Kersten, Metabolism of protein and RNA in liver of rats deprived of tryptophan. *J. Nutr. 107*:1737 (1977).

36. R. J. Block and K. W. Weiss, The amino acid composition of proteins, in *Amino Acid Handbook*. Charles C. Thomas, Springfield, 1956, p. 288.

37. P. E. Schurr, H. T. Thompson, L. M. Henderson, and C. A. Elvehjem, Determination of free amino acids in rat tissues. *J. Biol. Chem. 182*:39 (1950).

38. R. M. McMenemy, C. C. Lund, and J. L. Oncley, Unbound amino acid concentrations in human blood plasmas. *J. Clin. Invest. 36*:1672 (1957).

39. A. Tagliamonte, G. Biggio, L. Vargiu, and G. L. Gessa, Free tryptophan in serum controls brain tryptophan level and serotonin synthesis. *Life Sci. 12*:277 (1973).

40. P. J. Knott and G. Curzon, Free tryptophan in plasma and brain tryptophan metabolism. *Nature 239*:452 (1972).

41. G. Curson, The control of brain tryptophan concentration. *Acta Vitaminol. Enzymol. 29*:69 (1975).

42. J. D. Fernstrom and R. J. Wurtman, Elevation of plasma tryptophan by insulin. *Metabolism 21*:337 (1972).

43. M. Ikeda, H. Tsuji, S. Nakamura, A. Ichiyama, Y. Nishizuka, and O. Hayaishi, Studies on the biosynthesis of nicotinamide adenine dinucleotide. II. A role of picolinic carboxylase in the biosynthesis of nicotinamide adenine dinucleotide from tryptophan in mammals. *J. Biol. Chem. 240*:1395 (1965).

44. G. Curzon, J. Friedel, and P. J. Knott, The effect of fatty acids on the binding of tryptophan to plasma protein. *Nature 242*: 198 (1973).

45. J. D. Fernstrom, Acute effects of tryptophan and single meals on serotonin synthesis in the rat brain. *Adv. Biochem. Psychopharmacol. 34*:85 (1982).

46. J. D. Fernstrom and R. J. Wurtman, Brain serotonin content: Physiological regulation of plasma neutral amino acids. *Science 178*:414 (1972).

47. J. H. James, J. M. Hodgman, J. M. Funovics, N. Yoshimura, and J. E. Fisher, Brain tryptophan, plasma free tryptophan and distribution of plasma neutral amino acids. *Metabolism 25*: 471 (1976).

48. G. L. Gessa, G. Biggio, F. Fadda, G. U. Corsini, and A. Tagliamonte, Tryptophan-free diet: A new means for rapidly decreasing brain tryptophan content and serotonin synthesis. *Acta Vitaminol. Enzymol. 29*:72 (1975).

49. J. D. Fernstrom and R. J. Wurtmann, Nutrition and the brain. *Sci. Am. 230*:84 (1974).

50. R. J. Wurtman and J. D. Fernstrom, Control of brain neurotransmitter synthesis by precursor availability and nutritional state. *Biochm. Pharmacol. 25*:1691 (1976).

51. R. J. Wurtman, Nutrients that modify brain function. *Sci. Am. 246*:50 (1982).

52. G. Kolata, Food affects human behavior. *Science 218*:1209 (1982).

53. R. J. Wyatt, K. Engelman, D. J. Kupfer, D. H. Fram, A. Sjoerdsma, and F. Snyder, Effects of L-tryptophan (a natural sedative) on human sleep. *Lancet 2*:842 (1970).

54. W. J. Griffiths, B. K. Lester, J. D. Coulter, and H. L. Williams, Tryptophan and sleep in young adults. *Psychophysiology 9*:345 (1972).

55. E. Hartmann and C. L. Spinweber, Sleep induced by L-tryptophan. Effect of dosage within the normal dietary intake. *J. Nerv. Ment. Dis. 167*:497 (1979).

56. E. Hartmann and R. Chung, Sleep-inducing effects of L-tryptophan. *J. Pharm. Pharmacol. 24*:252 (1972).

57. M. Jouvet, Biogenic amines and the states of sleep. *Science 163*:32 (1969).

58. W. J. Wojcik, C. Fornal, and M. Radulovacki, Effect of tryptophan on sleep in the rat. *Neuropharmacology 19*:163 (1980).

59. C. J. Latham and J. E. Blundell, Evidence for the effect of tryptophan on the pattern of food consumption in free feeding and food deprived rats. *Life Sci. 24*:1971 (1979).

60. A. M. Barrett and L. McSharry, Inhibition of drug-induced anorexia in rats by methysergide. *J. Pharm. Pharmacol. 27*: 889 (1975).

61. S. B. Weinberger, S. Knapp, and A. J. Mandell, Failure of tryptophan load-induced increases in brain serotonin to alter food intake in the rat. *Life Sci. 22*:1594 (1978).

62. J. R. Stephens, R. F. Woolson, and A. R. Cooke, Effect of essential and nonessential amino acids on gastric emptying in the dog. *Gastroenterology 69*:920 (1975).

63. J. R. Stephens, R. F. Woolson, and A. R. Cooke, Osmolyte and tryptophan receptors controlling gastric emptying in the dog. *Am. J. Physiol. 231*:848 (1976).

64. A. R. Cooke, Gastric emptying in the cat in response to hypertonic solutions and tryptophan. *Am. J. Dig. Dis. 23*:312 (1978).

65. J. E. Blundell, Serotonin and skin diseases, in *Serotonin in Health and Disease*, Vol. 5, (W. B. Essman, ed.), S. P. Medical and Scientific Books, New York, 1979, pp. 403-450.

66. J. L. Gibbons, G. A. Barr, W. H. Bridger, and S. F. Leibowitz, Manipulations of dietary tryptophan: Effect on mouse killing and brain serotonin in the rat. *Brain Res. 169*: 139 (1979).

67. K. M. Kantak, L. R. Hegstrand, J. Whitman, and B. Eichelman, Effects of dietary supplements and a tryptophan-free diet on agressive behavior in rats. *Pharmacol. Biochem. Behav. 12*:173 (1980).

68. M. Vergnes and E. Kempf, Tryptophan deprivation: Effects on mouse-killing and reactivity in the rat. *Pharmacol. Biochem. Behav. 14* (Suppl. 1):19 (1981).

69. L. Valzelli, S. Bernasconi, and M. Dalessandro, Effect of tryptophan administration on spontaneous and P-CPA-induced muricidal aggression in laboratory rats. *Pharmacol Res. Commun. 13*:891 (1981).

70. J. K. Walters, M. Davis, and M. H. Sheard, Tryptophan-free diet: Effect on the acoustic startle reflex in rats. *Psychopharmacology 62*:103 (1979).

71. G. L. Gessa, G. Biggio, F. Fadda, G. U. Corsini, and A. Tagliamonte, Effect of the oral administration of tryptophan-free amino acid mixtures on serum tryptophan, brain tryptophan and serotonin metabolism. *J. Neurochem. 22*:869 (1974).

72. G. Biggio, F. Fadda, P. Fanni, A. Tagliamonte, and G. L. Gessa, Rapid depletion of serum tryptophan, brain tryptophan, serotonin and 5-hydroxyindoleacetic acid by a tryptophan-free diet. *Life Sci. 14*:1321 (1974).

73. W. Fratta, G. Biggio, and G. L. Gessa, Homosexual mounting behavior induced in male rats and rabbits by a tryptophan-free diet. *Life Sci. 21*:379 (1977).

74. G. L. Gessa and A. Tagliamonte, The role of serotonin and dopamine in male sexual behavior, in *Sexual Behavior* (M. Sandler and G. L. Gessa, eds.). Raven Press, New York, 1975, p. 120.

75. L. D. Lytle, R. B. Messing, L. Fisher, and L. Phebus, Effects of long-term corn consumption on brain serotonin and the response to electric shock. *Science 190*:692 (1975).

76. R. B. Messing, L. A. Fisher, L. Phebus, and L. D. Lytle, Interaction of diet and drugs in the regulation of brain 5-hydroxyindoles and the response to painful electric shock. *Life Sci.* *18:*707 (1976).
77. J. D. Fernstrom and L. D. Lytle, Corn malnutrition, brain serotonin and behavior. *Nutr. Rev.* *34:*257 (1976).
78. J. Lehmann, Mental and neuromuscular symptoms in tryptophan: Pellagra, carcinoidosis, phenylketonuria, Hartnup disease and disturbances of tryptophan metabolism induced by p-chlorophenylalamine, levodopa and α-methyldopa. *Acta Psychiatr. Scand.* *237:*4 (1972).
79. A. Coppen and K. Wood, Tryptophan and depressive illness. *Psychol. Med.* *8:*49 (1978).
80. G. Curzon, Study of disturbed tryptophan metabolism in depressive illness. *Ann. Biol. Clin.* *37:*27 (1979).
81. D. M. Achaw, R. Blazek, S. E. Tidmarsh, and G. J. Riley, Amino acids, amines and affective disorder, in *Origin, Prevention and Treatment of Affective Disorders* (M. Schore and E. Stromgren, eds.). Academic Press, New York, 1979, pp. 139-153.
82. J. H. Kaplan and H. C. Pitot, The regulation of intermediary amino acid metabolism in animal tissues, in *Mammalian Protein Metabolism*, Vol. 4 (H. N. Munro, ed). Academic Press, New York, 1970, pp. 387-443.
83. R. T. Schimke, E. W. Sweeney, and C. M. Berlin, The roles of synthesis and degradation in the control of rat liver tryptophan pyrrolase. *J. Biol. Chem.* *240:*322 (1965).
84. F. J. Ballard and M. F. Hopgood, Phosphopyruvate carboxylase induction by L-tryptophan: Effects on synthesis and degradation of the enzyme. *Biochem. J.* *136:*259 (1973).
85. A. Cihak, L-tryptophan action on hepatic RNA synthesis and enzyme induction. *Mol Cell. Biochem.* *24:*131 (1979).
86. P. Y. Chee and R. W. Swick, Effect of dietary protein and tryptophan on the turnover of rat liver ornithine aminotransferase. *J. Biol. Chem.* *251:*1029 (1976).
87. T. Deguchi and J. Barchas, Induction of hepatic tyrosine aminotransferase by indole amines. *J. Biol. Chem.* *246:*7217 (1971).
88. S. A. Smith and C. I. Pogson, Tryptophan and the control of plasma glucose concentrations in the rat. *Biochem. J.* *168:*495 (1977).
89. S. A. Smith, F. A. O. Marston, A. J. Dickson, and C. I. Pogson, Control of enzyme activities in rat liver by tryptophan and its metabolites. *Biochem. Pharmacol.* *28:*1645 (1979).
90. F. T. Kenney, Hormonal regulation of synthesis of liver enzymes, in *Mammalian Protein Metabolism* Vol. 4 (H. N. Munro, ed.) Academic Press, New York, 1970, pp. 131-176.

91. K. Kato, Selective repression of benzoate or tryptophan mediated induction of liver tyrosine aminotransferase by Phentolamine in adrenalectomized rats. *FEBS Lett.* 8:316 (1970).

92. D. E. Schumm and T. E. Webb, Effect of adenosine 3':5'-monophosphate and guanosine 3':5'-monophosphate on RNA release from isolated nuclei. *J. Biol. Chem.* 253:8513 (1978).

93. P. Gullino, M. Winitz, S. M. Birnbaum, J. Cornfield, M. C. Otey, and J. P. Greenstein, The toxicity of individual essential amino acids and their diastereoisomers in rats and the effect on blood sugar levels. *Arch. Biochem.* 58:253 (1955).

94. H. G. McDaniel, B. R. Boshell, and W. J. Reddy, Hypoglycemic action of tryptophan. *Diabetes* 22:713 (1973).

95. I. A. Mirsky, G. Perisutti, and R. Jinks, The hypoglycemic action of metabolic derivatives of L-tryptophan by mouth. *Endocrinology* 60:318 (1957).

96. P. D. Ray, D. O. Foster, and H. A. Lardy, Paths of carbon in gluconeogenesis and lipogenesis. IV. Inhibition of L-tryptophan of hepatic gluconeogenesis at the level of phosphoenolpyruvate formation. *J. Biol. Chem.* 241:3904 (1966).

97. C. M. Veneziale, P. Walter, N. Kneer, and H. A. Lardy, Influence of L-tryptophan and its metabolites on gluconeogenesis in the isolated, perfused liver. *Biochemistry* 6:2129 (1967).

98. Y. Hirata, T. Kawachi, and T. Sugimura, Fatty liver induced by injection of L-tryptophan. *Biochim. Biophys. Acta* 144:233 (1967).

99. P. Ramakrishna Rao, A. Bhaskar Rao, and S. Ramakrishnan, Biochemical mechanism of induction of fatty liver by tryptophan. *Indian J. Exp. Biol.* 18:1335 (1980).

100. T. Sakurai, S. Miyazawa, and T. Hashimoto, Effect of tryptophan on fatty acid synthesis in rat liver. *FEBS Lett.* 36:96 (1973).

101. T. Sakurai, S. Miyazawa, Y. Shindo, and T. Hashimoto, The effect of tryptophan administration on fatty acid synthesis in the liver of the fasted normal rat. *Biochim. Biophys. Acta* 360:275 (1974).

102. R. Fears and E. A. Murrell, Tryptophan and the control of triglyceride and carbohydrate metabolism in the rat. *Br. J. Nutr.* 43:349 (1980).

103. S. Miyazawa, T. Sakurai, Y. Shindo, M. Imura, and T. Hashimoto, The effect of tryptophan administration on fatty acid synthesis in the livers of rats under various nutritional conditions. *J. Biochem.* 78:139 (1975).

104. P. Gullino, M. Winitz, S. M. Birnbaum, J. Cornfield, M. C. Otey, and J. P. Greenstein, Studies on the metabolism of amino acids and related compounds in vivo. I. Toxicity of essential amino acids individually and in mixtures, and the protective effect of L-arginine. *Arch. Biochem. Biophys.* 64:319 (1956).

105. A. Fleck, J. Shepherd, and H. N. Munro, Protein synthesis in rat liver. Influence of amino acids in diet on microsomes and polysomes. *Science 150*:628 (1965).
106. H. Sidransky, M. Bongiorno, D. S. R. Sarma, and E. Verney, The influence of tryptophan on hepatic polyribosomes and protein synthesis in fasted mice. *Biochem. Biophys. Res. Comm. 27*:242 (1967).
107. H. Sidransky, D. S. R. Sarma, M. Bongiorno, and E. Verney, Effect of dietary tryptophan on hepatic polyribosomes and protein synthesis in fasted mice. *J. Biol. Chem. 243*:1123 (1968).
108. H. C. Sox, Jr. and M. B. Hoagland, Functional alterations in rat liver polysomes associated with starvation and refeeding, *J. Mol. Biol. 20*:113 (1966).
109. T. Staehelin, E. Verney, and H. Sidransky, The influence of nutritional change on polyribosomes of the liver. *Biochim. Biophys. Acta 145*:105 (1967).
110. T. E. Webb, G. Blobel, and V. R. Potter, Polyribosomes in rat tissues. III. The response of the polyribosomes pattern of rat liver to physiologic stress. *Cancer Res. 26*:253 (1966).
111. S. H. Wilson and M. B. Hoagland, Physiology of rat-liver polysomes. The stability of messenger ribonucleic acid and ribosomes. *Biochem. J. 102*:556 (1967).
112. W. H. Wunner, J. Bell, and H. N. Munro, The effect of feeding with a tryptophan-free amino acid mixture on rat liver polysomes and ribosomal ribonucleic acid. *Biochem. J. 101*:417 (1966).
113. H. Noll, T. Staehelin, and F. O. Wettstein, Ribosomal aggregate engaged in protein synthesis. Ergosome breakdown and messenger ribonucleic acid transport. *Nature 198*:632 (1963).
114. A. W. Pronezuk, B. S. Baliga, J. W. Triant, and H. N. Munro, Comparison of the effect of amino acid supply on hepatic polysome profiles in vivo and in vitro. *Biochim. Biophys. Acta 157*:204 (1968).
115. B. Pamart, A. Girard-Globa, and G. Bourdel, Induction of tyrosine aminotransferase and depression of protein synthesis in rat liver by a tryptophan-free mixture of amino acids. *J. Nutr. 104*:1149 (1974).
116. P. Feigelson, M. Feigelson, and C. Fancher, Kinetics of liver nucleic acid turnovers during enzyme induction in the rat. *Biochim. Biophys. Acta 32*:133 (1959).
117. P. Cammarano, G. Chinali, S. Gaeteni, and M. A. Spandoni, Involvement of adrenal steroids in the changes of polysome organization during feeding of imbalanced amino acid diets. *Biochim. Biophys. Acta 155*:302 (1968).
118. M. A. Rothschild, M. Oratz, J. Mongelli, L. Fishman, and S. S. Schreiber, Amino acid regulation of albumin synthesis. *J. Nutr. 98*:395 (1969).

119. M. Oravec and T. L. Sourkes, Inhibition of hepatic protein synthesis by α-methyl-DL-tryptophan in vivo. Further studies on the glyconeogenic action of α-methyltryptophan. *Biochemistry* 9:4458 (1970).

120. O. J. Park, L. M. Henderson, and P. B. Swan, Effects of the administration of single amino acids on ribosome aggregation in rat liver. *Proc. Soc. Exp. Biol. Med. 142*:1023 (1973).

121. A. J. F. Jorgensen and A. P. N. Majumdar, Bilateral adrenalectomy: Effect of a single tube-feeding of tryptophan on amino acid incorporation into plasma albumin and fibrinogen in vivo. *Biochem. Med. 13*:231 (1975).

122. A. P. N. Majumdar, Tryptophan requirement for protein synthesis. A review. *Nutr. Rep. Int. 26*:509 (1982).

123. A. J. F. Jorgensen and A. P. N. Majumdar, Bilateral adrenalectomy: Effect of tryptophan force-feeding on amino acid incorporation into ferritin, transferrin, and mixed proteins of liver, brain and kidneys in vivo. *Biochem. Med. 16*:37 (1976).

124. H. Sidransky, E. Verney, and D. S. R. Sarma, Effect of tryptophan on polyribosomes and protein synthesis in liver. *Am. J. Clin. Nutr. 24*:779 (1971).

125. H. N. Munro, Role of amino acid supply in regulating ribosome function. *Fed. Proc. 27*:1231 (1968).

126. M. Hori, J. M. Fisher, and M. Rabinowitz, Tryptophan deficiency in rabbit reticulocytes: Polyribosomes during interrupted growth of hemoglobin chains. *Science 155*:83 (1967).

127. R. T. Hunt, A. R. Hunter, and A. J. Munro, The control of hemoglobin synthesis. Factors controlling the output of alpha and beta chains. *Proc. Nutr. Soc. Engl. Scot. 28*:248 (1969).

128. D. S. R. Sarma, E. Verney, M. Bongiorno, and H. Sidransky, Influence of tryptophan on hepatic polyribosomes and protein synthesis in non-fasted and fasted mice. *Nutr. Rep. Int. 4*:1 (1971).

129. H. Sidransky, C. N. Murty, and E. Verney, Effect of tryptophan on the inhibitory action of selected hepatotoxic agents on hepatic protein synthesis. *Exp. Mol. Pathol. 37*:305 (1982).

130. R. E. Allen, R. L. Raines, and D. M. Regen, Regulatory significance of transfer RNA charging levels. I. Measurements of changing levels in livers of chow-fed rats, fasting rats, and rats fed balanced and imbalanced mixtures of amino acids. *Biochim. Biophys. Acta 190*:323 (1969).

131. A. R. Henderson, The effect of feeding with a tryptophan-free amino acid mixture on rat liver magnesium ion-activated deoxyribonucleic acid-dependent ribonucleic acid-dependent polymerase. *Biochem. J. 120*:205 (1970).

132. J. Vesley and A. Cihak, Enhanced DNA-dependent RNA polymerase and RNA synthesis in rat liver nuclei after administration of L-tryptophan. *Biochim. Biophys. Acta 204*:614 (1970).

133. A. P. N. Majumdar, Effect of tryptophan on hepatic nuclear free and engaged RNA polymerases in young and adult rats. *Experimentia 34*:1258 (1978).

134. M. Oravec and A. Korner, Stimulation of ribosomal and DNA-like RNA synthesis by tryptophan. *Biochim. Biophys. Acta 247*:404 (1971).

135. W. H. Wunner, The time sequence of RNA and protein synthesis in cellular compartments following acute dietary challenge with amino acid mixtures. *Proc. Nutr. Soc. 27*:153 (1967).

136. C. N. Murty and H. Sidransky, The effect of tryptophan on messenger RNA of the livers of fasted mice. *Biochim. Biophys. Acta 262*:328 (1972).

137. C. N. Murty, E. Verney, and H. Sidransky, Effect of tryptophan on polyriboadenylic acid and polyadenylic acid-messenger ribonucleic acid in rat liver. *Lab. Invest. 34*:77 (1976).

138. S. P. Blatti, C. J. Ingels, T. J. Lindell, P. M. Morris, R. F. Weaver, F. Weinberg, and W. J. Rutter, Structure and regulatory properties of eukaryotic RNA polymerase. *Cold Spring Harbor Symp. Quant. Biol. 35*:649 (1970).

139. R. G. Roeder and W. Rutter, Specific nucleolar and nucleoplasmic RNA polymerase. *Proc. Nat. Acad. Sci. 65*:675 (1970).

140. A. P. N. Majumdar and A. J. F. Jorgensen, Responses of well-fed adrenalectomized rats to tryptophan force-feeding on hepatic protein and RNA synthesis. *Biochem. Med. 16*:266 (1976).

141. A. P. N. Majumdar, Effects of fasting and tryptophan force-feeding on the activity of hepatic nuclear RNA polymerases in rats. *Scand. J. Clin. Lab. Invest. 39*:61 (1979).

142. C. N. Murty, E. Verney, and H. Sidransky, The effect of tryptophan on nucleocytoplasmic translocation of RNA in rat liver. *Biochim. Biophys. Acta 474*:117 (1977).

143. H. Sidransky, E. Verney, and C. N. Murty, Effect of tryptophan on hepatic polyribosomal disaggregation due to hypertonic sodium chloride. *Lab. Invest. 34*:291 (1976).

144. H. Sidransky, E. Verney, and C. N. Murty, Effect of tryptophan on hepatic polyribosomes and protein synthesis in rats treated with carbon tetrachloride. *Toxicol. Appl. Pharmacol. 39*:295 (1977).

145. C. T. Garrett, V. Cairns, C. N. Murty, E. Verney, and H. Sidransky, Effect of tryptophan on informosomal and polyribosome-associated messenger RNA in rat liver. *J. Nutr. 114*:50 (1984)

146. G. Brawerman, Eukaryotic messenger RNA. *Annu. Rev. Biochem. 43*:621 (1974).

147. R. P. Perry, Processing of RNA. *Annu. Rev. Biochem. 45*:605 (1976).

148. E. A. Smuckler and R. M. Koplitz, Polyadenylic acid content and electrophoretic behavior of in vitro released RNA's in chemical carcinogenesis. *Cancer Res. 36*:881 (1976).
149. G. A. Clawson, C. H. Woo, and E. A. Smuckler, Independent responses of nucleoside triphosphatase and protein kinase activities in nuclear envelope following thioacetamide treatment. *Biochem. Biophys. Res. Comm. 95*:1200 (1980).
150. N. T. Patel, D. S. Folse, and V. Holoubek, Release of repetitive nuclear RNA into the cytoplasm in liver of rats fed 3'-methyl-4-dimethylaminoazobenzene. *Cancer Res. 39*:4460 (1979).
151. R. W. Shearer, Altered RNA transport without derepression in rat kidney tumor induced by dimethylnitrosamine. *Chem. Biol. Int. 27*:91 (1979).
152. A. Yannarell, D. E. Schumm, and T. E. Webb, Age-dependence of nuclear RNA processing. *Mech. Aging Dev. 6*:259 (1977).
153. D. E. Schumm and T. E. Webb, Modified messenger ribonucleic acid release from isolated hepatic nuclei after inhibition of polyadenylate formation. *Biochem. J. 139*:191 (1974).
154. C. N. Murty, E. Verney, and H. Sidransky, In vivo and in vitro studies on the effect of tryptophan on translocation of RNA from nuclei of rat liver. *Biochem. Med. 22*:98 (1979).
155. H. Ono and H. Terayama, Amino acid incorporation into proteins in isolated rat liver nuclei. *Biochim. Biophys. Acta 166*: 175 (1968).
156. H. Ono, T. Ono, and O. Wada, Amino acid incorporation by nuclear membrane fraction of rat liver. *Life Sci. 18*:215 (1976).
157. C. N. Murty, R. Hornseth, E. Verney, and H. Sidransky, Effect of tryptophan on enzymes and proteins of hepatic nuclear envelopes of rats. *Lab. Invest. 48*:256 (1983).
158. D. S. R. Sarma, M. Bongiorno, E. Verney, and H. Sidransky, Effect of oral administration of tryptophan or water on hepatic polyribosomal disaggregation due to puromycin. *Exp. Mol. Pathol. 19*:23 (1973).
159. H. Sidransky, E. Verney, and D. S. R. Sarma, Effect of tryptophan on hepatic polyribosomal disaggregation due to ethionine. *Proc. Soc. Exp. Biol. Med. 149*:633 (1972).
160. J. R. Harris, The biochemistry and ultrastructure of the nuclear envelope. *Biochim. Biophys. Acta 515*:55 (1978).
161. G. A. Clawson, J. James, C. H. Woo, D. S. Friend, D. Moody, and E. A. Smuckler, Pertinence of nuclear envelope nucleoside triphosphatase activity to ribonucleic acid transport. *Biochemistry 19*:2748 (1980).
162. P. Agutter, B. McCaldin, and H. J. McArdle, Importance of mammalian nuclear-envelope nucleoside triphosphatase in nucleoplasmic transport of ribonucleoproteins. *Biochem. J. 182*:811 (1979).

163. A. Vorbrodt and G. G. Maul, Cytochemical studies on the relation of nucleoside triphosphatase activity to ribonucleoproteins in isolated rat liver nuclei. *J. Histochem. Cytochem. 28:*28 (1980).

164. C. N. Murty, E. Verney, and H. Sidransky, Effect of tryptophan on nuclear envelope nucleoside triphosphatase activity in rat liver. *Proc. Soc. Exp. Biol. Med. 163:*155 (1980).

165. K. S. Lam and C. B. Kasper, Selective phosphorylation of a nuclear envelope polypeptide by an endogenous protein kinase. *Biochemistry 18:*307 (1979).

166. R. C. Steer, M. J. Wilson, and K. Ahmed, Protein phosphokinase activity of rat liver nuclear membrane. *Exp. Cell. Res. 119:* 403 (1979).

167. R. C. Steer, M. J. Wilson, and K. Ahmed, Phosphoprotein phosphatase activity of rat liver nuclear membrane. *Biochem. Biophys. Res. Commun. 89:*1082 (1979).

168. J. R. McDonald and P. S. Agutter, The relationship between polyribonucleotide binding and the phosphorylation and diphosphorylation of nuclear envelope protein. *FEBS Lett. 116:*145 (1980).

169. R. P. Aaronson and G. Blobel, On the attachment of the nuclear pore complex. *J. Cell Biol. 62:*746 (1974).

170. W. W. Franke, Structure, biochemistry and functions of the nuclear envelope. *Int. Rev. Cytol. 4:*71 (1974).

171. S. E. Stuart, G. A. Clawson, F. M. Rottman, and R. J. Patterson, RNA transport in isolated myeloma nuclei. Transport from membrane-denuded nuclei. *J. Cell Biol. 72:*57 (1977).

172. T. Palayoor, D. E. Schumm, and T. E. Webb, Transport of functional messenger RNA from liver nuclei in a reconstituted cell-free system. *Biochim. Biophys. Acta 654:*201 (1981).

173. R. W. Wannemacker, Jr., W. K. Cooper, and M. B. Yatvin, The regulation of protein synthesis in the liver of rats. Mechanisms of dietary amino acid control in the immature animal. *Biochem. J. 107:*615 (1968).

174. J. Roth, Peptide hormone binding to receptors: A review of direct studies in vitro. *Metabolism 22:*1059 (1973).

175. C. R. Kahn, Membrane receptors for polypeptide hormones. *Meth. Membr. Biol. 3:*81 (1975).

176. I. D. Goldfine and G. J. Smith, Binding of insulin to isolated nuclei. *Proc. Natl. Acad. Sci. 73:*1427 (1976).

177. I. D. Goldfine, G. J. Smith, K. Y. Wong, and A. L. Jones, Cellular uptake and nuclear binding of insulin in human cultured lymphocytes: Evidence for potential intracellular sites for insulin action. *Proc. Natl. Acad. Sci. 74:*1368 (1977).

178. R. Vigneri, I. D. Goldfine, K. Y. Wong, G. J. Smith, and V. Pezzino, The nuclear envelope. The major site of insulin binding in rat liver nuclei. *J. Biol. Chem. 253:*2098 (1978).

179. I. D. Goldfine, F. Purrello, G. A. Clawson, and R. Vigneri, Insulin binding sites on the nuclear envelope: Potential relationship to mRNA metabolism. *J. Cell. Biochem. 20*:29 (1982).

180. I. D. Goldfine, G. A. Clawson, E. A. Smuckler, F. Purrello, and R. Vigneri, Action of insulin at the nuclear envelope. *Mol. Cell. Biochem. 48*:3 (1982).

181. R. A. Roth and D. J. Cassell, Insulin receptor: Evidence that it is a protein kinase. *Science 219*:299 (1983).

182. J. Gorski and F. Gannon, Current models of steroid hormone action: A critique. *Annu. Rev. Physiol. 38*:425 (1976).

183. K. R. Yamamoto and B. M. Alberts, Steroid receptors: Elements for modulation of eukaryotic transcription. *Annu. Rev. Biochem. 45*:721 (1976).

184. P. L. Ballard, J. D. Baxter, S. J. Higgins, G. G. Rousseau, and G. M. Tomkins, General presence of glucocorticoid receptors in mammalian tissues. *Endocrinology 94*:998 (1974).

185. G. Giannopoulos, Z. Hassan, and S. Solomon, Glucocorticoid receptors in fetal and adult rabbit tissues. *J. Biol. Chem. 249*: 2424 (1974).

186. R. Markovic and J. Petrovic, Competition of tryptophan and hydrocortisone for receptor proteins of rat liver cytosol. *Int. J. Biochem. 6*:47 (1975).

187. M. E. Baker, D. A. Vaughn, and D. D. Fanestil, Competitive inhibition of dexamethasone binding to the glucocorticoid receptor in HTC cells by tryptophan methyl ester. *J. Steroid Biochem. 13*:993 (1980).

188. A. P. N. Majumdar and A. J. F. Jorgensen, Influence of cortisol on tryptophan-mediated stimulation of hepatic protein synthesis in well-fed adrenalectomized rats. *Biochem. Med. 17*:116 (1977).

189. E. Verney and H. Sidransky, Further enhancement by tryptophan of hepatic protein synthesis stimulated by phenobarbital or cortisone acetate. *Proc. Soc. Exp. Biol. Med. 158*:245 (1978).

190. R. H. McMenamy, Albumin binding, in *Albumin Structure, Function and Uses* (V. M. Rosenoer, M. Oratz, and M. A. Rothschild, eds.). Pergamon Press, New York, 1977, pp. 143-158.

191. U. Kragh-Hansen, Relations between high-affinity binding sites for L-tryptophan, diazepam, salicylate and phenol red on human serum albumin. *Biochem. J. 299*:135 (1983).

192. A. A. B. Badawy and M. J. H. Smith, Changes in liver tryptophan and tryptophan pyrrolase activity after administration of salicylate and tryptophan to the rat. *Biochem. Pharmacol. 21*: 97 (1972).

193. H. Sidransky, C. N. Murty, and E. Verney, Nutritional control of protein synthesis: Studies relating to tryptophan-induced stimulation of nucleocytoplasmic translocation of mRNA in rat liver. *Am. J. Pathol.* In Press.

194.   J. K. Lynn, D. S. R. Sarma, and H. Sidransky, Response of hepatic polyribosomes of the mouse to the administration of anisotonic solutions. *Life Sci. 10*:385 (1971).

195.   J. K. Lynn, C. N. Murty, and H. Sidransky, Osmoregulation of ribosomal function in mouse liver. *Biochim. Biophys. Acta 299*:444 (1973).

196.   J. K. Lynn and H. Sidransky, Effect of changes of osmotic pressure of portal blood on hepatic protein synthesis. *Lab. Invest. 31*:332 (1974).

197.   H. B. LeJohn and R. M. Stevenson, Inhibition of amino acid transport and enhancement of tryptophan binding in a water mold by a range of natural and synthetic cytokinins. *Can. J. Microbiol. 28*:1165 (1982).

198.   L. E. Cameron and H. B. LeJohn, On the involvement of calcium in amino acid transport and growth of the fungus *Achlya. J. Biol. Chem. 247*:4729 (1972).

199.   H. B. LeJohn, A rapid and sensitive auxin-binding system for detecting N^6-substituted adenines, and some urea and thiourea derivatives that show cytokinin activity in cell division tests. *Can. J. Biochem. 53*:768 (1975).

200.   F. A. Baglia and G. G. Maul, Nuclear ribonucleoprotein release and nucleoside triphosphatase activity are inhibited by antibodies directed against one nuclear matrix glycoprotein. *Proc. Nat. Acad. Sci. 80*:2285 (1983).

201.   H. Sidransky and E. Verney, Effect of diet and tryptophan on hepatic polyribosomal disaggregation due to actinomycin. *Exp. Mol. Pathol. 17*:233 (1972).

202.   M. A. Rothschild, M. Oratz, J. Mongelli, and S. S. Schreiber, Alcohol-induced depression of albumin synthesis: Reversal by tryptophan. *J. Clin. Invest. 50*:1812 (1971).

203.   M. U. Dianzani, Toxic liver injury by protein synthesis inhibitors. *Prog. Liver Dis. 5*:232 (1976).

204.   M. A. Rothschild, M. Oratz, and S. S. Schreiber, Effect of tryptophan on the hepatotoxic effects of alcohol and carbon tetrachloride. *Trans. Assoc. Am. Physiol. 84*:313 (1971).

205.   H. Kroger, R. Gratz, C. Museteanu, and J. Haase, The influence of nicotinamide, tryptophan, and methionine upon galactosamine-induced effects in the liver. *Arzneim. Forsch. 31*:987 (1981).

206.   G. Gacad, K. Dickie, and D. Massaro, Protein synthesis in lung: Influence of starvation on amino acid incorporation into protein. *J. Appl. Physiol. 33*:381 (1972).

207.   A. P. N. Majumdar and A. M. Nakhla, Influence of tryptophan on the activity of acetylcholinesterase in the brain of well-fed normal and adrenalectomized rats. *Biochem. Biophys. Res. Commun. 76*:71 (1977).

208. A. P. N. Majumdar, Effect of fasting and subsequent feeding of a complete or tryptophan-free diet on the activity of DNA-synthesizing enzymes and protein synthesis in gastric mucosa of rats. *Ann. Nutr. Metab. 26*:264 (1982).

209. A. P. N. Majumdar, Bilateral adrenalectomy: Effect of tryptophan on protein synthesis and pepsin activity in the stomach of rats. *J. Gastroenterol. 14*:949 (1979).

210. A. P. N. Majumdar, Effect of adrenalectomy and tryptophan force-feeding on the activity of intestinal disaccharidases in adult rats. *Scand. J. Gastroenterol. 15*:225 (1980).

211. A. P. N. Majumdar, Influence of dietary tryptophan on the activity of intestinal digestive enzymes. *Nutr. Rep. Int. 24*: 1067 (1981).

212. A. B. Ajdukiewicz, P. Keane, J. Pearson, A. E. Read, and P. R. Salmon, Insulin releasing activity of oral L-tryptophan in fasting and non-fasting subjects. *Scand. J. Gastroenterol. 3*:622 (1968).

213. J. C. Floyd, Jr., S. S. Fajans, J. W. Conn, R. F. Knopf, and J. Rull, Stimulation of insulin secretion by amino acids. *J. Clin. Invest. 45*:1487 (1966).

214. K. A. Fahmy, S. M. Moustapha, M. K. Salama, M. Khattab, and H. G. Basta, Tryptophan—an insulinotropic nutrient. *Nutr. Rep. Int. 28*:653 (1983).

215. R. S. Modlinger, J. M. Schonmuller, and S. P. Arora, Stimulation of aldosterone, renin, and cortisol by tryptophan. *J. Clin. Endocrinol. Metab. 48*:599 (1979).

216. R. S. Modlinger, J. M. Schonmuller, and S. P. Arora, Adrenocorticotropin release by tryptophan in man. *Clin. Endocrinol. Metab. 50*:360 (1980).

217. H. Yokogoshi and A. Yoshida, Effect of feeding or intubation of tryptophan or methionine and threonine in hepatic polysome profiles in rats under meal-feeding or fasting conditions. *Agric. Biol. Chem. 47*:373 (1983).

218. H. N. Munro, J. D. Fernstrom, and R. J. Wurtman, Insulin, plasma amino-acid imbalance, and hepatic coma. *Lancet 1*:722 (1975).

219. H. Y. Meltzer, B. Wiita, B. J. Tricou, H. M. Simonovic, V. Fang, and G. Manov, Effect of serotonin precursors and serotonin agonists on plasma hormone levels. *Adv. Biochem. Psychopharmacol. 34*:117 (1982).

220. J. S. Wittman III, K. L. Lee, and O. N. Miller, Dietary and hormonal influences on rat liver polysome profile: Fat, glucose and insulin. *Biochim. Biophys. Acta 174*:536 (1969).

221. H. Sidransky and E. Verney, Studies on hepatic polyribosomes and protein synthesis in rats force-fed a protein-free diet. *J. Nutr. 101*:1153 (1971).

222. H. Sidransky, Regulatory effect of amino acids on polyribosomes and protein synthesis of liver. *Prog. Liver. Dis.* *4*:31 (1972).

223. M. Winick, ed., *Nutrition and Cancer, Current Concepts in Nutrition*, Vol. 6, John Wiley & Sons, New York, 1977.

224. B. S. Reddy, L. A. Cohen, G. D. McCoy, P. Hill, J. H. Weisburger, and E. L. Wynder, Nutrition and its relationship to cancer. *Adv. Cancer Res.* *32*:237 (1980).

225. G. R. Newell and N. M. Ellison, eds., *Cancer and Nutrition: Etiology and Treatment, Progress in Cancer Research and Therapy*, Vol. 17, Raven Press, New York, 1981.

226. W. F. Dunning, M. R. Curtis, and M. E. Maun, The effect of added dietary tryptophan on the occurrence of 2-acetylamino-fluorene induced liver and bladder cancer in rats. *Cancer Res.* *10*:454 (1950).

227. G. T. Bryan, The role of urinary tryptophan metabolites in the etiology of bladder cancer. *Am. J. Clin. Nutr.* *24*:841 (1971).

228. J. L. Radomski, E. M. Glass, and W. B. Deichmann, Transitional cell hyperplasia in the bladder of dogs fed DL-tryptophan. *Cancer Res.* *31*:1690 (1971).

229. M. Miyakawa and O. Yoshida, DNA synthesis of the urinary bladder epithelium in rats with long-term feeding of DL-tryptophan added and pyridoxine-deficient diet. *Gann.* *64*:411 (1973).

230. M. Matsushima, The role of the promoter L-tryptophan on tumorigenesis in the urinary bladder. 2. Urinary bladder carcinogenicity of FANFT (initiating factor) and L-tryptophan (promoting factor) in mice. *Jpn. J. Urol.* *68*:731 (1977).

231. S. M. Cohen, M. Arai, J. B. Jacobs, and G. H. Friedell, Promoting effect of saccharin and DL-tryptophan in urinary bladder carcinogenesis. *Cancer Res.* *39*:1207 (1979).

232. S. Fukushima, G. H. Friedell, J. B. Jacobs, and S. M. Cohen, Effect of L-tryptophan and sodium saccharin on urinary tract carcinogenesis initiated by *N*-(40(5-nitro-2-furyl)-2-thiazoly)-formamide. *Cancer Res.* *41*:3100 (1981).

233. T. Kawochi, Y. Hirata, and T. Sugimura, Enhancement of *N*-nitro-diethylamine hepatocarcinogenesis by L-tryptophan in rats. *Gann* *59*:523 (1968).

234. R. P. Evarts and C. A. Brown, Effect of L-tryptophan on diethylnitrosamine and 3'-methyl-4-*N*-dimethyl-aminoazobenzene hepatocarcinogenesis. *Food Cosmet. Toxicol.* *15*:431 (1977).

235. E. Okajima, T. Hiramatsu, Y. Motomiya, K. Iriya, M. Ijuim, and N. Ito, Effect of DL-tryptophan on tumorigenesis in the urinary bladder and liver of rats treated with nitrodibutylamine *Gann.* *62*:163 (1971).

236. J. A. Miller and E. C. Miller, The metabolic activation of carcinogenic aromatic amines and amides. *Prog. Exp. Tumor Res.* *11*:273 (1969).

237. R. Oyasu, H. Sumie, and H. E. Burg, Neoplasms of urinary bladders of hamsters treated with 2-acetylaminofluorene and indole. *J. Natl. Cancer. Inst.* 45:853 (1970).

238. R. Oyasu, D. A. Miller, J. H. McDonald, and G. M. Hass, Neoplasms of rat urinary bladder and liver. *Arch. Pathol.* 75: 184 (1963).

239. L. W. Wattenberg, Inhibitors of chemical carcinogenesis. *Adv. Cancer Res.* 26:197 (1978).

240. R. P. Evarts and M. H. Mostafa, The effect of L-tryptophan and certain other amino acids on liver nitrodimethylamine demethylase activity. *Food Cosmet. Toxicol.* 16:585 (1978).

241. R. P. Evarts and M. H. Mostafa, Effect of indole and tryptophan on cytochrome P-450, dimethylnitrosamine demethylase and arylhydrocarbon hydroxylase activities. *Biochem. Pharmacol.* 30:517 (1981).

242. H. Sidransky, E. Verney, and C. N. Murty, Effect of elevated dietary tryptophan on protein synthesis in rat liver. *J. Nutr.* 111:1942 (1981).

243. A. J. F. Jorgensen and A. P. N. Majumdar, Influence of tryptophan on the level of hepatic microsomal cytochrome P-450 in well-fed normal, adrenalectomized and phenobarbital-treated rats. *Biochim. Biophys. Acta* 444:453 (1976).

244. N. Matsukura, T. Kawachi, K. Morino, H. Ohgaki, and T. Sugimura, Carcinogenicity in mice of mutagenic compounds from a tryptophan pyrolyzate. *Science* 213:346 (1981).

245. H. Sidransky and E. Verney, Effect of nutritional alterations on protein synthesis in transplantable hepatomas and host livers of rats. *Cancer Res.* 39:1995 (1979).

246. H. Sidransky, E. Verney, and C. N. Murty, Effect of tryptophan on hepatoma and host liver of rats. Influence after treatment with hypertonic sodium chloride and carbon tetrachloride. *Exp. Mol. Pathol.* 35:124 (1981).

247. H. Sidransky and E. Verney, Effect of inhibitory and stimulatory agents on protein synthesis in hepatomas and host livers of rats. *J. Natl. Cancer Inst.* 63:81 (1979).

248. D. V. Maudsley, Regulation of polyamine biosynthesis. *Biochem. Pharmacol.* 28:153 (1979).

249. D. H. Russell and S. H. Snyder, Amine synthesis in regenerating rat liver: Extremely rapid turnover of ornithine decarboxylase. *Mol. Pharmacol.* 5:253 (1969).

250. E. Holtta and J. Janne, Ornithine decarboxylase activity and the accumulation of putrescine at early stages of liver regeneration. *FEBS Lett.* 23:117 (1972).

251. J. Janne, E. Holtta, and S. K. Guha, Polyamines in mammalian liver during growth and development, in *Progress in Liver Diseases* (H. Popper and F. Schaffner, eds.) Grune & Stratton, New York, 1976, p. 100.

252. D. H. Russell and C. C. Levy, Polyamine accumulation and biosynthesis in a mouse L1210 leukemia. *Cancer Res. 31*:248 (1971).
253. H. G. Williams-Ashman, G. C. Coppoc, and G. Weber, Imbalance in ornithine metabolism in hepatomas of different growth rates as expressed in formation of putrescine, spermidine and spermine. *Cancer Res. 32*:1924 (1972).
254. J. L. Marx, Tumor promoters: Carcinogenesis gets more complicated. *Science 201*:515 (1978).
255. T. G. O'Brien, The induction of ornithine decarboxylase as an early possible obligatory event in mouse skin carcinogenesis. *Cancer Res. 36*:2644 (1976).
256. H. Sidransky, C. N. Murty, E. Myers, and E. Verney, Tryptophan-induced stimulation of hepatic ornithine decarboxylase activity in the rat. *Exp. Mol. Pathol. 38*:346 (1983).
257. J. Janne, H. Poso, and A. Raina, Polyamines in rapid growth and cancer. *Biochim. Biophys. Acta 473*:241 (1978).
258. M. Miller, J. P. Leaky, W. C. Stern, P. J. Morgane, and O. Resnick, Tryptophan availability: Relation to elevated brain serotonin in developmentally protein-malnourished rats. *Exp. Neurol. 57*:142 (1977).
259. M. Miller and O. Resnick, Tryptophan availability: The importance of prepartum and postpartum dietary protein on brain indoleamine metabolism in rats. *Exp. Neurol. 67*:298 (1980).
260. K. Aoki and F. L. Siegel, Hyperphenylalaninemia: Disaggregation of brain polyribosomes in young rats. *Science 168*:129 (1970).
261. R. Blazek and D. M. Shaw, Tryptophan availability and brain protein synthesis. *Neuropharmacology 24*:1065 (1975).
262. T. J. Sobotka, M. P. Cook, and R. F. Brodie, Neonatal malnutrition: Neurochemical, hormonal and behavioral manifestations. *Brain Res. 65*:443 (1974).
263. S. Zamenhof, S. M. Hall, L. Grauel, E. Van Marthens, and M. J. Donahue, Deprivation of amino acids and prenatal brain development in rats. *J. Nutr. 104*:1002 (1974).
264. S. Zamenhof, E. Van Marthens, and L. Grauel, DNA (cell number) and proteins in neonatal rat brains: Alteration by timing of maternal dietary protein restriction. *J. Nutr. 101*:1265 (1971).
265. S. Matsueda and Y. Niiyama, The effects of excess amino acids on maintainance of pregnancy and fetal growth in rats. *J. Nutr. Sci. Vitaminol. 28*:557 (1982).
266. A. H. Meier and J. M. Wilson, Tryptophan feeding adversely influences pregnancy. *Life Sci. 32*:1193 (1983).
267. D. Waugh and M. J. Pearl, Serotonin-induced acute nephrosis and renal cortical necrosis in rats. A morphologic study with pregnancy correlations. *Am. J. Pathol. 36*:431 (1960).

268. E. Paulson, J. M. Robson, and F. M. Sullivan, Teratogenic effect of 5-hydroxytryptamine in mice. *Science 141*:717 (1963).

269. R. Hammer, Embryotoxic effects of serotonin during early pregnancy in the rat. *Anat. Rec. 196*:71A (1980).

270. J. D. Fernstrom and R. J. Wurtman, Brain serotonin content: Physiological dependence on plasma tryptophan levels. *Science 173*:149 (1971).

271. H. Green, S. M. Greenberg, R. W. Erickson, J. L. Sawyer, and T. Ellison, Effect of dietary phenylalanine and tryptophan upon rat brain amine levels. *J. Pharmacol. Exp. Ther. 136*: 174 (1962).

272. G. Biggio, F. Fadda, P. Fanni, A. Togliamonte, and G. L. Gessa, Rapid depletion of serum tryptophan, brain tryptophan serotonin and 5-hydroxyindolacetic acid by a tryptophan-free diet. *Life Sci. 14*:1321 (1974).

273. C. M. McCay, Chemical aspects of aging and the effect of diet upon aging, in *Cowdry's Problems of Ageing*, 3rd ed. (A. I. Lansing, ed.), Williams & Wilkins, Baltimore, 1952, p. 139.

274. B. N. Berg, Nutrition and longevity in the rat. I. Food intake in relation to size, health and fertility. *J. Nutr. 71*:242 (1960).

275. B. N. Berg and H. S. Simms, Nutrition and longevity in the rat. II. Longevity and onset of disease with different levels of food intake. *J. Nutr. 71*:255 (1960).

276. M. H. Ross, Length of life and nutrition in the rat. *J. Nutr. 75*:197 (1961).

277. P. E. Segall and P. S. Timiras, Patho-physiologic findings after chronic tryptophan deficiency in rats. A model for delayed growth and aging. *Mech. Ageing Dev. 5*:109 (1976).

278. L. Zieve, Hepatic encephalopathy: Summary of present knowledge with an elaboration on recent developments. *Prog. Liver Dis. 6*:327 (1979).

279. C. Hirayama, Tryptophan metabolism in liver disease. *Clin. Chim. Acta 32*:191 (1971).

280. K. Ogihara, T. Mozai, and S. N. Hirai, Tryptophan as cause of hepatic coma. *N. Engl. J. Med. 275*:1255 (1966).

281. A. J. Knell, A. R. Davidson, R. Williams, B. D. Kantamaneni, and G. Curzon, Dopamine and serotonin metabolism in hepatic encephalopathy. *Br. Med. J. 1*:549 (1974).

282. T. L. Sourkes, Tryptophan in hepatic coma. *J. Neural. Trans. Suppl. 14*:79 (1978).

283. J. Ono, D. G. Hutson, R. S. Dombro, J. U. Levi, A. Livingstone, and R. Zeppa, Tryptophan and hepatic coma. *Gastroenterology 72*:196 (1978).

284. D. W. Woolley, *The Biochemical Basis of Psychoses or the Serotonin Hypothesis About Mental Disease*. John Wiley & Sons, New York, 1962.

285. J. E. Fischer, Hepatic coma in cirrhosis, portal hypertension, and following portacaval shunt. *Arch. Surg. 108*:325 (1974).
286. A. Mortiaux and A. M. Dawson, Plasma free fatty acids in liver disease. *Gut 2*:304 (1961).
287. J. E. Fischer, H. M. Rosen, A. M. Ebeid, J. H. James, J. M. Keane, and P. B. Soeters, The effect of normalization of plasma amino acids on hepatic encephalopathy in man. *Surgery 80*:77 (1976).
288. V. Iob, W. W. Coon, and M. Sloan, Free amino acids in liver, plasma, and muscle of patients with cirrhosis of the liver. *J. Surg. Res. 7*:41 (1967).
289. Y. Sato, S. Eriksson, L. Hagenfeldt, and J. Wahren, Influence of branched-chain amino acid infusion on arterial concentrations and brain exchange of amino acids in patients with hepatic cirrhosis, *Clin. Physiol. 1*:151 (1981).
290. J. H. James, J. Escourrou, and J. E. Fischer, Blood-brain neutral amino acid transport activity is increased after portacaval anastomosis. *Science 200*:1395 (1978).
291. G. Zanchin, P. Rigotti, N. Dussini, P. Vassanelli, and L. Battistin, Cerebral amino acid levels and uptake in rats after portacaval anastomosis. II. Regional studies in vivo. *J. Neurosci. Res. 4*:301 (1979).
292. J. E. Fischer and R. J. Baldessarini, Pathogenesis and therapy of hepatic coma, in *Progress in Liver Diseases*, Vol. 5 (H. Popper and F. Schaffner eds.), Grune and Stratton, New York, 1976, p. 363.
293. H. Freund, N. Yoshimura, and J. E. Fischer, Chronic hepatic encephalopathy: Long term therapy with a BCAA-enriched diet. *J. Am. Med. Assoc. 242*:347 (1979).
294. H. Freund, J. Dienstag, J. Lehrich, N. Yoshimura, R. R. Bradford, H. Rosen, S. Atamian, E. Slemmer, J. Holroyde, and J. E. Fischer, Infusion of branched-chain enriched amino acid solutions in patients with hepatic encephthalopathy. *Ann. Surg. 196*:209 (1982).
295. J. Wahren, J. Denis, P. Desurmont, L. S. Eriksson, J. M. Escoffier, A. P. Gauthier, L. Hagenfeldt, H. Michel, P. Opolon, J. C. Paris, and M. Veyrac, Is intravenous administration of branched chain amino acids effective in the treatment of hepatic encephalopathy? A multicenter study. *Hepathology 3*:475 (1983).
296. J. F. Fischer and R. H. Bower, Amino acids in liver disease, in *The Kidney in Liver Disease*, 2nd ed. (M. Epstein, ed.), Elsevier Biomedical, New York, 1983, pp. 515-534.
297. B. Smith and D. J. Prockop, Central nervous system effects of ingestion of L-tryptophan by normal subjects. *N. Engl. J. Med. 267*:1338 (1962).

298. A. S. Montenero, Sulla tossicita e tollerabilita del triptofaro e di suoi metaboliti. *Acta Vitaminol. Enzymol.* 32:188 (1978).

299. A. Coppen, D. M. Shaw, and J. P. Farrell, Potentiation of the antidepressive effect of monoamine-oxidase inhibitor by tryptophan. *Lancet 1*:79 (1963).

300. G. Curzon, B. D. Kantamaneni, J. Winch, A. Rojas-Bueno, I. M. Murray-Lyon, and R. Williams, Plasma and brain tryptophan changes in experimental acute hepatic failure. *J. Neurochem. 21*:137 (1973).

301. A. L. Goldberg and M. E. Tischler, Regulatory effects of leucine on carbohydrate and protein metabolism, in *Metabolism and Clinical Implications of Branched Chain Amino And Ketoacids* (M. Walser and J. R. Williamson, eds.). Elsevier/North Holland, Amsterdam, 1981, pp. 205-216.

302. R. H. McMenamy, W. C. Shoemaker, J. E. Richmond, and D. Elwyn, Uptake and metabolism of amino acids by the dog liver perfused in situ. *Am. J. Physiol. 202*:407 (1962).

303. P. Felig, Amino acid metabolism in man. *Ann. Rev. Biochem. 44*:933 (1975).

304. J. Wahren, P. Felig, and L. Hagenfeldt, Effect of protein ingestion on splanchnic and leg metabolism in normal man and patients with diabetes mellitus. *J. Clin. Invest. 57*:987 (1976).

305. M. G. Buse and S. S. Reid, Leucine: A possible regulator of protein turnover in muscle. *J. Clin. Invest. 56*:1250 (1975).

306. R. M. Fulks, J. B. Li, and A. L. Goldberg, Effects of insulin, glucose, and amino acids on protein turnover in rat diaphragm, *J. Biol. Chem. 250*:290 (1975).

307. A. L. Goldberg and T. W. Chang, Regulation and significance of amino acid metabolism in skeletal muscle. *Fed. Proc. 37*: 2301 (1978).

308. J. B. Li and L. S. Jefferson, Influence of amino acid availability on protein turnover in perfused skeletal muscle. *Biochim. Biophys. Acta 544*:351 (1978).

309. B. Chua, D. L. Siehl, and H. E. Morgan, Effect of leucine and metabolites of branched-chain amino acids on protein turnover in heart. *J. Biol. Chem. 254*:8358 (1979).

310. M. G. Buse, R. Atwell, and V. Mancusi, In vitro effect of branched chain amino acids on the ribosomal cycle in muscles of fasted rats. *Horm. Metab. Res. 11*:289 (1979).

311. H. Freund, N. Yohimura, and J. E. Fischer, The role of alanine in the nitrogen-conserving quality of the branched-chain amino acids in the postinjury state. *J. Surg. Res. 29*:23 (1980).

312. D. G. Sapir and M. Walser, Nitrogen sparing induced early in starvation by infusion of branched-chain ketoacids. *Metab. Clin. Exp. 26*:301 (1977).

313. R. S. Sherwin, Effect of starvation on the turnover and metabolic response to leucine. *J. Clin. Invest. 61*:1471 (1978).

314.   H. Freund, H. C. Hoover, S. Atamian, and J. E. Fischer,
       Infusion of the branched-chain amino acids in postoperative
       patients. Anticatabolic properties. *Ann. Surg. 190*:18 (1979).
315.   A. J. McCollough, K. D. Mullen, and A. S. Tavill, Branched-
       chain amino acids as nutritional therapy in liver disease:
       Death or surfeit? *Hepathology 3*:269 (1983).
316.   M. A. McNurlan, E. B. Fern, and P. J. Garlick, Failure of
       leucine to stimulate protein synthesis in vivo. *Biochem. J.
       204*:831 (1982).

2

Effects of Nutrition on Transcriptional and Translational Controls of Protein Synthesis in Liver

Challakonda N. Murty[†]

George Washington University Medical Center
Washington, D.C.

I. INTRODUCTION

The liver, being the first organ to receive nutrients from the gastro-intestinal tract, plays a vital role in the metabolism of ingested dietary components. Furthermore, liver cells are actively involved in synthesizing many proteins including an extensive array of enzymes involved in drug metabolism. Also, the liver manufactures most of the plasma proteins. In relation to the important role that the liver plays, it is readily vulnerable to the ingestion of deficient or imbalanced diets and to toxic agents, which may alter its protein metabolism. Disturbances in protein metabolism may rapidly induce pathological changes in the liver and this subject has been the topic of many reviews and symposia (1-8).

The purpose of this chapter is to review current information relating to the effects of dietary disturbances on protein metabolism in the liver. The present review deals mainly with biochemical changes, especially those relating to the transcriptional and/or translational levels of control mechanisms by which liver protein synthesis may become affected by altered nutrition. This survey restricts itself to studies conducted with laboratory animals. Other studies dealing with nutritional intake and injury in humans have recently been reviewed (9).

[†]Deceased.

II. MECHANISMS THAT CONTROL HEPATIC PROTEIN SYNTHESIS

The understanding of the mechanisms involved in protein synthesis
has progressively expanded over the last decade. Because of the subject's complexity, only a brief review of some of the relevant points
will be considered here. Detailed reviews dealing with the controls
that regulate mammalian protein synthesis are available (10-12).

The central dogma of protein synthesis originally proposed by Crick
(13) can be summarized as follows:

DNA—Transcription→RNA—Translation→PROTEIN

The synthesis of a specific protein is ultimately controlled at the level
of the gene. A gene may be defined as a sequence of DNA that is responsible for the synthesis of a specific polypeptide chain. In simplistic terms, gene expression can be modified at either of two major levels:
the transcriptional or the translational.

A. Transcriptional Control

Within the cell the nucleus is the major site for transcriptional events,
and it transfers its transcriptional products, such as mRNP particles,
ribosomal subunits, tRNA, etc., to the cytoplasm. The nucleus synthesizes three types of RNA, each of which has a different role in protein synthesis. This review limits itself only to ribosomal RNA and
messenger RNA.

1. *Ribosomal RNA Synthesis*

The nucleolus is responsible for the synthesis of ribosomal RNA (rRNA),
which constitutes about 85% of total cellular RNA and plays an essential
role in the production of new ribosomes. In general, transcription of
the genome is mediated by specific enzymes, DNA-dependent RNA polymerases. The nucleus contains three different RNA polymerases and
each enzyme is responsible for the synthesis of a specific type of cellular RNA. Ribonucleic acid polymerase I, which is found in association
with the nucleolus, catalyzes the synthesis of rRNA. Kinetic studies
suggest that the biosynthesis of rRNA in mammalian cells is a stepwise
process involving the synthesis initially of a large-molecular-weight
precursor RNA (45S) and the subsequent conversion of this RNA into
two major types of rRNAs: 28S and 18S. The 28S and 18S rRNA molecules, following association with proteins, appear in the cytoplasm as
60S and 40S ribosomal subunits, respectively. These subunits, linked
together with an mRNA strand, are the backbone of the polyribosomal
structure in the cytoplasm. In addition to the two major rRNA species,
two minor rRNA molecules, 5.8S and 5S, have been identified in the
nucleus. The 5.8S rRNA molecule is an integral part of a smaller 40S

ribosomal subunit, whereas the 5S rRNA molecule is closely associated with the larger 60S ribosomal subunit. Although definite functions for the two minor rRNA molecules have not yet been defined, current evidence indicates that they are essential for the function of the ribosome.

2. *Messenger RNA (mRNA) Synthesis*

Messenger RNAs are synthesized in the nucleus (extranucleolar) (14) and function in the cytoplasm as templates for protein synthesis. The specific enzyme, RNA polymerase II, is involved in transcribing DNA into mRNA. In the nucleus, mRNA is synthesized initially as a long precursor molecule, heterogeneous nuclear RNA (HnRNA). The HnRNA then undergoes a series of cleavage reactions to form a specific mRNA product. Most, but not all, of the mammalian mRNA molecules contain a capped structure at the 5'-phosphate terminus and a poly(A) tail at the 3' terminus. These two structures are essential for the functional activity of mRNA in protein synthesis. The 5' cap is added in the nucleus before the transport of the mRNA to the cytoplasm, and its function is presumably to assist in the recognition of the proper site on the template mRNA at which protein synthesis commences. The poly(A) tail, containing approximately 200 adenylic acid residues, is added to the 3' end by a poly(A)polymerase either in the nucleus or in the cytoplasm. Addition of poly(A) sequences appears to be required for normal maturation and/or transport of the bulk of the mRNA to the cytoplasm and for its incorporation into functional polyribosomes. The function of polyadenylation of mRNA in posttranscriptional events has been the concern of an increasing number of studies. Polyadenylation has been shown to have significance in the selection of mRNA transcripts and its subsequent transport to cytoplasm (15). In addition to the poly(A) tract itself, specific proteins associated with poly(A) also have been implicated in the control and regulation of nucleocytoplasmic transport of mRNA (16).

B. Translational Control

In addition to control of protein synthesis at the transcriptional level, other control mechanisms operate at the level of translation. Factors, such as the number of ribosomes in the polyribosome that translate each mRNA, the turnover of the mRNA molecule, the type and the amount of mRNA transported to the cytoplasm, and the cytosol factors that regulate the rate of polypeptide synthesis, may all be important in controlling protein synthesis.

1. *Polyribosomes*

Polyribosomes are aggregates of ribosomes connected by a strand of mRNA and provide the essential framework for the protein-synthesizing machinery. Under the steady state, the size of the ribosomal aggregations correlates well with the protein synthesizing capacity of the cells or tissues (17). Indeed, the size distribution of these polyribosomes appears to be related to the overall degree of protein synthesis (1). The liver, which synthesizes both intracellular and extracellular proteins (plasma proteins), has polyribosomes in the cytoplasm in two forms: (a) membrane-bound polyribosomes, which are attached to the membranes of the endoplasmic reticulum, and (b) free polyribosomes, which are free in the cytoplasm (18,19).

The two polyribosome populations turn over at different rates (20,21). They respond differently to a variety of stimuli, including dietary manipulations (discussed in later sections), and are engaged in the synthesis of different proteins. For example, in rat liver secretory proteins such as serum albumin (22-25) and some integral membrane proteins (e.g., cytochrome P-450) (26-28) have been reported to be synthesized on membrane-bound polyribosomes, whereas intracellular proteins (e.g, ferritin) (29) seem to be preferentially synthesized on free polyribosomes. These findings suggest that each class of hepatic polyribosomes has a distinctive species of mRNA, differing not only in its half-life but also in the message for the primary structures of proteins to be synthesized. In view of the different structural and functional aspects of hepatic free and membrane-bound polyribosomes, each population has an important role in hepatic protein metabolism and may differ from the other under a variety of experimental conditions.

2. *Polypeptide Synthesis*

Each protein consists of a specific arrangement of amino acids, which have been added sequentially to the growing peptide. During the synthesis of a protein on the polyribosomal structure, three phases occur: (a) initiation of synthesis starting from the NH_2 terminus, (b) elongation of the polypeptide chain, and (c) termination of the polypeptide chain.

Initiation

The mechanism of protein chain initiation is a complex process that has been the subject of intensive investigation (30). Formation of a specific initiation complex is essential for protein synthesis to begin. To date, at least eight protein factors have been isolated that are involved in the initiation of the translation of mRNA into protein. Of the amino acids, methionine has been found to be the most abundant at the NH_2 terminus of completed polypeptide chains. Furthermore, two distinct species of tRNA that are capable of accepting methionine are found within the cell. One of these complexes, met-tRNA$_F^{met}$, which can be formylated by an enzyme and recognizes the AUG codon at the start of a mRNA chain is concerned with the initiation of new polypeptide chains. The final initiation complex, 80S-met-tRNA$_f$-mRNA, is assembled through a series of reactions requiring initiation factors, ATP, GTP, 60S and 40S ribosomal subunits, messenger RNA, and initiator met-tRNA.

Elongation

Chain elongation involves a shuttle movement of tRNA between two sites, P (peptidyl) and A (aminoacyl), on the ribosome. Two protein factors are involved in the elongation of the polypeptide chain. The enzyme, elongation factor 1, catalyzes the transfer of an incoming amino acid to the carboxyl end of the nascent peptide chain. The second enzyme, elongation factor 2, catalyzes the movement, known as translocation, of the new tRNA within its nascent peptide from site A to site P.

Termination

Evidence from genetic studies suggest that the triplet codons, UAA, UAG, and UGA, in mRNA act as terminator signals in protein synthesis. In addition, a specific supernatant-protein release factor is necessary for polypeptide chain termination. This factor (RF) has been identified and apparently consists of three components, RF_1, RF_2, and RF_3. RF_1 and RF_2 recognize the codons UAA, UAG, and UGA, whereas RF_3 plays a stimulatory role in the release of the completed protein chains without participating in codon recognition.

3. Nucleocytoplasmic Exchange of Macromolecules

Mechanisms involved in the translocation of macromolecules from nucleus to cytoplasm, and vice versa, in mammalian systems have recently gained much attention. A number of studies suggest that protein biosynthesis might be regulated by posttranscriptional events, in particular at the stage of translocation of mRNA (31).

The phenomenon of nucleocytoplasmic translocation of mRNA and the controls involved in this process, with particular reference to mammalian liver, have been investigated using a cell-free system (32). Current evidence indicates that posttranscriptional controls operate within the nucleus at the nuclear membrane and in the cytoplasm to regulate the outflow of messenger RNP particles from the nucleus into the cytoplasm both qualitatively and quantitatively. Some of these posttranscriptional controls involve: (a) Polyadenylation (14)—the addition of approximately 200 adenylic acid residues to the 3' end of the mRNA molecule appears to be required for mRNA transport. (b) Phosphorylation and dephosphorylation (33,34)—the nuclear envelopes of mammalian liver contain several enzymes, nucleoside triphosphatase (NTPase), protein phosphokinase (PK), and phosphoprotein phosphohydrolase (PH), that are involved in these reactions. Current evidence suggests that these enzymes may influence the rate of nucleocytoplasmic translocation of RNA. According to one proposed scheme, the polypeptides of the nuclear envelopes become phosphorylated during the hydrolysis of ATP (which also supplies energy for RNA transport) by NTPase and/or PK, and RNA binds to phosphorylated polypeptides. Next, ATP (independent of hydrolysis) binds to the nuclear envelope-pore lamina and stimulates the release of bound RNA into the cytoplasm with concomitant dephosphorylation of the phosphorylated nuclear envelope protein by the enzyme PH. (c) Transport factors—specific protein factors, which modulate nuclear RNA transport in the cell-free system (31), have been demonstrated in the cytoplasm. Most of these factors are found to be associated with polyribosomes (35). Currently, the mechanism of action of these proteins is unclear. Studies with cyclic nucleotides suggest that a protein kinase may be involved (36).

Although much information is now available concerning the mechanism of nucleocytoplasmic translocation of mRNA, little is known concerning the mechanism by which proteins are translocated from the cytoplasm to the nucleus. The general consensus is that most, or all, of the nuclear proteins, such as histones and nonhistone chromosomal proteins, and proteins associated with mRNA and rRNA, are synthesized in the cytoplasm, and these proteins then migrate to the nucleus. In support of this view, it has been shown that certain specific proteins of cytoplasmic origin are rapidly translocated to the nucleus, and some of these proteins may exert a profound influence on the regulation of nuclear function (37,38).

III. INFLUENCE OF NUTRITIONAL COMPONENTS ON HEPATIC PROTEIN SYNTHESIS

A. Amino Acids

There is much evidence to indicate that protein synthesis in mammalian cells may be regulated by the supply of certain essential nutrients, particularly amino acids. Studies from our and other laboratories have demonstrated that protein synthesis by the liver can be influenced by the supply of amino acids, which affects the controls at the transcriptional and/or translational levels.

To gain an understanding about how dietary amino acids may influence hepatic protein metabolism, a number of experimental studies have been undertaken. These investigations include approaches utilizing in vivo studies as well as cell-free techniques, perfused livers, isolated hepatocytes, and liver slices. The rates of protein synthesis, as affected by amino acid supply, have been investigated in vivo using labeled amino acids and observing their uptake into the liver proteins of control and experimental animals. In vitro hepatic protein synthesis has been examined employing cell-free systems in which isolated subcellar components, such as microsomes or ribosomes, were tested for their capacity to incorporate labeled amino acids into protein. The cell-free system allows one to investigate which of the subcellular components involved in protein synthesis, such as ribosomes, cell sap, etc., are affected by dietary manipulations. The status of polyribosomal aggregation, as determined by sucrose density-gradient analysis, correlates well with the protein synthetic ability of cells or tissues and is another common index used to evaluate the impact of amino acid supply on liver protein synthesis.

Studies relating to changes in hepatic RNA metabolism following dietary manipulations have also been undertaken. More than 20 years ago, Munro and Clark (39) demonstrated that in protein-depleted rats, coincident with a decrease in liver protein, there was a parallel loss of RNA content from the liver. Since then many investigators have examined the relationship between dietary amino acid supply and RNA metabolism in the liver to gain an understanding of the overall hepatic RNA and protein metabolism in relation to adequate or faulty nutritional intake. In this section, findings will be reviewed relating to the effects of amino acid supply (single amino acids, complete amino acid mixtures, purified diets deficient in amino acids, low-protein or protein-free (amino acid-free) diets) on liver RNA and protein metabolism in intact animals, perfused livers, and isolated hepatocytes.

1. *Studies Using Intact Animals*

Amino Acids

Earlier studies have reported that a single tube-feeding of a complete amino acid mixture or of a complete diet to fasted animals induced a shift in hepatic polyribosomes from lighter to heavier aggregation and to enhanced protein synthesis, as measured in vivo and in vitro (40-43). The stimulatory effect on hepatic protein synthesis observed under these conditions has been attributed to the presence of tryptophan in the complete amino acid mixture. In the absence of tryptophan, but not of other single amino acids, from a complete amino acid mixture, the hepatic polyribosomes did not reveal a stimulatory response to the feeding; it remained similar to that found in fasted control animals (43,44). Indeed, a number of studies from our laboratory (43,45) demonstrated that the administration of tryptophan alone elicited a stimulatory response on hepatic protein synthesis similar to that obtained with a complete amino acid mixture. Subsequent studies from other laboratories have confirmed our findings relating to the action of tryptophan on hepatic polyribosomes and protein synthesis (46-49). These experimental studies established two important points: First, the state of polyribosomal arrangement and protein synthesis in the liver of an experimental animal is sensitive to a feeding of a complete amino acid mixture. Second, among the amino acids, only tryptophan appears to exert a unique effect in regulating hepatic protein synthesis. The unique action of dietary tryptophan, not only on protein metabolism in the liver but also on several other metabolic processes in the liver, is now well recognized. Chapter 1 deals in detail with the unique biological and metabolic properties of tryptophan. Therefore this review presents only a few highlights about how tryptophan acts in stimulating protein synthesis.

Earlier studies with mice treated with actinomycin D to inhibit RNA synthesis followed by tryptophan revealed that tryptophan still had a stimulatory effect on hepatic protein synthesis (43). This finding suggested that tryptophan probably acts on the regulation of protein synthesis at the translational level of control. Further studies revealed that tryptophan administration to fasted animals caused increases in cytoplasmic mRNA (50) and poly(A)-containing mRNA (51), and these increases occurred even if the animals were pretreated with actinomycin D and/or cordycepin, suggesting that tryptophan stimulates both mRNA synthesis and the translocation of nuclear mRNA into the hepatic cytoplasm. Later studies from our laboratory confirmed the stimulatory action of tryptophan on the translocation of mRNA and poly(A)-mRNA from the nucleus to cytoplasm (52,53). Yap et al. (54) presented evidence for another effect of tryptophan on the levels of cytoplasmic mRNA in the liver—stimulated translocation of albumin mRNA from the informosomal pool to the free polyribosomal pool in the cytoplasm. Although a limited shift between specific mRNA of the informosomal and polyribosome-associated mRNA could occur following tryptophan administration, a recent study from our laboratory (55) suggested that the increase in the polyribosome-associated mRNA in the livers of tryptophan-treated rats was not due to this shift but was mainly due to enhanced nuclear-to-cytoplasmic translocation of nuclear mRNA, a finding that has been reported earlier (53). Recently, we demonstrated that an acute administration of tryptophan to fasted animals resulted in enhanced activities of enzymes (nucleoside triphosphatase, protein phosphokinase, and phosphoprotein phosphohydrolase) associated with the hepatic nuclear envelope (56,57). These enzymes have been implicated as being involved in RNA transport (33,34). Thus tryptophan appears to act on the posttranscriptional level of control of protein synthesis.

Other data relating to the effects of tryptophan on hepatic RNA metabolism suggest that tryptophan may act on the transcriptional level of control of protein synthesis. Wunner (58) reported enhanced incorporation of [³H]orotic acid into the hepatic RNA of the polyribosomal fraction (messenger, ribosomal, and some transfer RNA) after feeding a complete amino acid mixture in comparison with one lacking tryptophan. Vesley and Cihak (59), Henderson (60), Oravec and Korner (61) and Majumdar (62) reported increases in the activities of DNA-dependent RNA polymerases and in the synthesis of ribosomal and DNA-like RNA in hepatic nuclei, attributable to tryptophan stimulation. Thus, in conclusion, current experimental evidence indicates that, in normal animals, tryptophan stimulates hepatic polyribosomal aggregation and protein synthesis by acting at three levels of control, i.e., transcriptional, posttranscriptional, and translational.

The role of amino acids in regulating liver protein synthesis in mice and rats has been investigated in experimental studies by feeding purified or synthetic diets. The major advantages of using purified or synthetic diets in investigations dealing with experimental nutritional deficiencies or imbalances are as follows: (a) the composition of all dietary components are defined and therefore one can study the influence of the presence (in variable amounts) or of the absence of a specific nutrient in the diet, and (b) the diets can be suspended or dissolved in water and controlled volumes of complete or deficient diets can be administered via stomach tube. An advantage of tube-feeding the diets is that one can carefully control the amount, as well as the time of each feeding. The advantages of using synthetic or purified diets under force-feeding conditions in comparison with ad libitum-feeding in experimental nutritional deficiency studies have been reviewed in detail (3).

Amino Acid-Deficient Diets

Using the force-feeding technique, Adamstone and Spector (63) described pathological changes, including periportal fatty liver, within a short time (1 to 3 days) after the onset of tube-feeding a purified diet devoid of tryptophan, phenylalanine, or isoleucine. Subsequently, Sidransky and colleagues described the morphological and biochemical changes in young rats force-fed purified diets devoid of threonine, histidine, valine, methionine, leucine, isoleucine, phenylalanine, or lysine (64-69). In this section the effects of selected amino acid-deficient diets on hepatic RNA and protein metabolism will be reviewed, focusing on the findings relating to the mechanism(s) whereby the amino acids may play an essential role in regulating hepatic protein synthesis.

For many years Sidransky and co-workers have been concerned with developing an experimental animal model of kwashiorkor to probe into the pathogenesis of the lesions that occur in this important nutritional deficiency disease in humans. To study the effects of amino acid deficiencies, young rats were force-fed, for 1 to 7 days, purified diets devoid of single essential amino acids, threonine, histidine, valine, methionine, leucine, isoleucine, phenylalanine, or lysine (3,64-69). From these studies in which young rats were force-fed these experimental diets, evidence was obtained indicating that, even though pathological changes in the liver were observed morphologically, which resembled many of those observed in kwashiorkor, there was enhanced overall hepatic protein synthesis in the kwashiorkor-like model (3). Most of the effects on hepatic protein metabolism were similar for each of the purified diets devoid of the single essential amino acids tested, with only minor variations (3,68). Therefore a diet free of threonine was

selected for extensive studies dealing with the biochemical alterations
in the liver in an attempt to gain further insight into the pathogenesis
of the lesions produced by amino acid-deficient diets.

A summary of some of the biochemical changes caused by threonine
deficiency is presented in Table 1. Although total protein per liver
was unchanged in the rats force-fed diets devoid of threonine, in vivo
studies using [^{14}C]amino acid (leucine, isoleucine, or valine) and
measuring incorporation into total hepatic proteins revealed an increase
in protein synthesis in the livers of the experimental animals (65).
These results were valid even after corrections for pool sizes of the
labeled amino acid were taken into consideration. Also, in vitro studies
measuring [^{14}C]amino acid incorporation into hepatic proteins synthesized by cell-free preparations of livers of control and experimental
animals revealed increases in the experimental groups (70,71). Chemical and electron microscopic studies demonstrated that there was an increase in hepatic ribosomes in rats fed the experimental diet, and that
this increase was relatively greater in the free ribosomal fraction than
in the total or membrane-bound ribosomal fraction (72). Chemical studies
on hepatic polyribosomes indicated that hepatic polyribosomes from experimental animals showed a shift toward heavier aggregates than those
from animals force-fed a complete diet (70,71). These findings correlated
well with the increased hepatic protein synthesis observed both in the
in vivo and in vitro incorporation studies.

Since initiation factors associated with ribosomes have been found, in
part, to regulate protein synthesis (30), initiation factors prepared
from hepatic ribosomes of control and experimental (tube-fed a
threonine-devoid diet) rats were investigated. In vitro activities of
initiation factors in the stimulation of polyphenylalanine synthesis and
on the formation of initiation complex were increased in the experimental
rats compared with those of control rats (73). In consideration of the
findings that indicated increased hepatic protein synthesis related to
changes in the hepatic polyribosomes in the experimental animals, we
conducted experimental studies dealing with hepatic RNA metabolism.
While the amounts of DNA in the livers were unchanged, increased
amounts of nuclear, ribosomal (total, free, and membrane-bound), and
soluble (nonsedimentable) RNA have been reported in the livers of experimental animals (70,72). Also, increased activities of DNA-dependent
RNA polymerases I and II have been reported in the experimental animals
(74). In in vivo studies, incorporation of ^{32}P or [^{14}C]orotate into
nuclear and nucleolar RNA (75), ribosomal RNA (74), and mRNA
(76) was increased in the livers of experimental animals in comparison
with controls. More recent studies revealed that [^{14}C]orotic acid incorporation into hepatic poly(A)-mRNA of nuclei and cytoplasmic polyribosomes was increased in the experimental compared with the control

animals (74). Furthermore, using rats force-fed a threonine-devoid diet or complete diet for 3 days, pretreatment with actinomycin D to inhibit RNA synthesis (10 hr before killing) abolished the shift in hepatic polyribosomes toward heavier aggregates, the enhanced protein synthesis, and the increased activities of DNA-dependent RNA polymerases I and II observed in the experimental animals (74,76). The results with actinomycin D indicate that the threonine-devoid diet under these experimental conditions influences the regulation of polyribosomes and protein synthesis at the transcriptional level of control.

In an attempt to determine how the omission of an essential amino acid from the diet caused enhancement in hepatic protein synthesis, several studies were undertaken. It has been well established that protein synthesis is dependent upon the availability simultaneously of all essential amino acids. Thus to enable hepatic protein synthesis to continue when one essential amino acid was missing in the diet, it was necessary for an endogenous source to supply the missing essential amino acids. In our experimental kwashiorkor-like model, there is evidence that the endogenous source was most probably skeletal muscle, a major reservoir of protein and amino acids. Experimental studies with rats force-fed a purified diet free of one essential amino acid indicated that skeletal muscle protein metabolism was rapidly affected; protein synthesis was decreased and protein catabolism was increased (69,77). Analyses of free amino acids of plasma, liver, and gastrocnemius muscle of rats force-fed a complete or threonine-devoid diet for 3 days indicated that while many of the nonessential amino acids were decreased in the experimental groups, the levels of essential amino acids, except for threonine, were markedly elevated (78). Thus skeletal muscle via protein catabolism probably contributes amino acids into the circulation, and these include the essential amino acids. The amino acids from the diet and from skeletal muscle protein catabolism reach the liver where they probably play a role in influencing hepatic protein synthesis.

The biochemical changes described in our experimental kwashiorkorlike model, induced by force-feeding a threonine-devoid diet, were markedly different from those found in animals fed the same diet ad libitum (3,64,65). With the ad libitum-feeding regimen, the rats consumed only small amounts of the experimental diet and developed few or no lesions, particularly in the liver. Thus the quantity of dietary intake appears to be influential in inducing pathological changes. The influence of the quantity of complete or threonine-devoid diet on hepatic protein synthesis was investigated under conditions that were strictly controlled as possible in force-feeding experiments. Young rats (weighing approximately 60 g) force-fed an adequate amount (6.5 g/day) of threonine-devoid diet demonstrated an increase in hepatic protein synthesis and a decrease in skeletal muscle protein synthesis in comparison with rats force-fed the same quantity of complete diet, results similar to those described earlier. In contrast, rats force-fed one-half

Table 1 Effects of Protein or Amino Acids on Liver Protein Synthesis in Intact Animals

Protein or amino acids administered	Responses observed in the	
	Nucleus	Cytoplasm
Amino acid mixture (1 hr after tube-fed)		↑ Heavier polyribosomal aggregates (40, 43) ↑ Protein synthesis in vivo and in vitro (40–43)
Amino acid mixture devoid of single essential amino acids other than tryptophan (1 hr after tube-fed)		↓ Heavier polyribosomal aggregates (43, 44) ↓ Protein synthesis in vivo and in vitro (43, 44)
Amino acid mixture devoid of tryptophan (1 hr after tube-fed)		No response (40–43)
Tryptophan (1 hr after tube-fed)	↑ DNA-dependent RNA polymerase activity (66, 67, 69) ↑ Nuclear RNA synthesis (68)	↑ Heavier total polyribosomal aggregates (43) ↑ Free polyribosomal aggregates > membrane-bound polyribosomes (45) ↑ Protein synthesis in vivo and in vitro (43, 45) ↑ Levels of mRNA, poly(A) and poly(A)-mRNA (50, 51) ↑ Nucleocytoplasmic translocation of mRNA and poly(A)-mRNA (52, 53) ↑ Activities of nuclear envelope-associated enzymes (56, 57)
Threonine-devoid diet (1–3 days after tube-fed)	↑ Activities of RNA polymerases I and II (77)	↑ Protein synthesis in vivo and in vitro (3, 70, 71)

Dietary treatment		Effects
Tryptophan-devoid diet (1 day after tube-fed)	↑ Nuclear rRNA and poly(A)-mRNA synthesis (77,78)	↑ Heavier total polyribosomal aggregates (70,71) ↑ Free polyribosomal aggregates > membrane-bound polyribosomes (72) ↑ Activities of initiation factors (73) ↑ Levels of mRNA and poly(A)-mRNA (74,76)
Tryptophan-devoid diet (15 days after feeding)		↑ Heavier total polyribosomal aggregates (80) ↑ Protein synthesis (80)
Threonine-imbalanced diet[a] (2 or 12 hr after feeding)	↑ rRNA synthesis (88)	↑ Protein synthesis in vivo (84) No change in RNA synthesis (84)
Threonine-imbalanced diet[b] (2 hr after feeding)		No response (88)
Isoleucine- or threonine-imbalanced diet (1 or 4 hr after feeding)		↑ Heavier polyribosomal aggregates (88) ↑ Protein synthesis in vivo (84) No response (87)
Low-protein (6%-casein) diet (28-days feeding)	↑ Activity of RNA polymerase I (89) ↑ RNA synthesis	↑ Protein synthesis in vitro and in vivo (89) ↑ Polyribosomal aggregates ↑ RNA synthesis
Low-protein (6%-casein) diet (28-days feeding)	↑ rRNA synthesis (92) ↑ activities of RNA polymerases I and II	↓ Bound ribosomes but no change in free ribosomes (90) ↓ Protein synthesis in vivo and in vitro ↓ rRNA content
Low-protein (6%-casein) diet (14-days feeding)	↑ Activity of RNA polymerase I (91)	

Table 1 (Continued)

Protein or amino acids administered	Responses observed in the	
	Nucleus	Cytoplasm
Low protein (3%-casein) diet (6-days feeding)	↑ Activities of DNA-dependent RNA of polymerases I and II (93, 94)	↑ Protein synthesis in vitro (93)
Low-protein (6%-casein) diet (7-days feeding)	↓ rRNA synthesis (95)	↓ rRNA content (95) ↓ rRNA synthesis
Protein-free diet (3 to 10-days-feeding)		↓ RNA content (97) ↑ RNA degradation ↓ Ribosomal subunits
Protein-free diet (1-day feeding)		↑ Heavier polyribosomal aggregates (98) ↓ Protein synthesis in vivo and in vitro
Protein-free diet (5- or 10-days feeding)		↓ Heavier polyribosomes (99, 100) ↑ Protein catabolism ↓ Rates of polypeptide initiation and elongation
Protein-free diet (2- or 9-days feeding)		↓ Albumin synthesis (101) ↓ Albumin mRNA concentration ↓ Amounts of RNA, DNA, and protein

Diet	Increase (↑)	Decrease (↓)
Protein-free diet (6-days feeding)		↓ Total protein synthesis (102)
Protein-deprived diet (8-days feeding)		↓ Protein synthesis (103) ↓ RNA content ↓ RNA synthesis
Protein-free diet (5-days feeding)	↑ Activity of RNA polymerase I (107) ↑ Nucleolar size (107)	
Protein-free diet (14-days feeding)	↑ Synthesis of nucleolar RNA (106) ↑ Activity of RNase	↓ Half-lives of rRNA (106,108) ↓ RNA content

↑ = increase, ↓ = decrease.

[a]Rats were previously fed a diet containing a complete amino acid mixture devoid of threonine for one week.

[b]Rats were previously fed a diet containing low protein (6% casein) supplemented with methionine and threonine.

the quantity (3.25 g/day) of the control or experimental diet revealed no, or small, differences in protein synthesis of the liver and gastro-cnemius muscle between the rats force-fed the complete and the threonine-devoid diets (79). Furthermore, rats force-fed one-half the quantity of complete or experimental diet revealed decreased tissue weights and protein contents, disaggregation of polyribosomes, and decreased protein synthesis in liver and skeletal muscle in comparison with rats force-fed the adequate quantity of complete diet (79). This study demonstrated two major important findings. First, the amount of intake of either a complete or a deficient diet has a profound effect upon the regulation of protein metabolism in the liver and in other organs. Second, increased skeletal muscle breakdown in rats force-fed one-half quantity of the complete diet did not in itself stimulate liver protein synthesis. This is in contrast to the earlier findings with young rats that were force-fed an adequate amount of threonine-devoid diet or a complete diet for 1 to 7 days. As mentioned earlier, the increased skeletal muscle breakdown was considered to be a contributory factor in enhancing protein synthesis in the livers of rats force-fed an adequate amount of threonine-devoid diet. However increased skeletal muscle breakdown does not in itself appear to be a specific factor since it occurs in a variety of conditions, such as after fasting or with decreased dietary intake as described previously, yet increased hepatic protein synthesis does not occur. Therefore the increased hepatic protein synthesis in rats force-fed an adequate amount of threonine-devoid diet has been considered to be due to a combination of factors: (a) the release of amino acids by skeletal muscle protein catabolism and (b) the ingested amino acids with one essential amino acid missing in the diet. Both appear to be important in altering the control of hepatic protein synthesis.

In view of the results obtained with a threonine-devoid diet, the influence of force-feeding purified diets devoid of other single essential amino acids such as tryptophan, phenylalanine, or valine and also diets devoid of two (threonine and phenylalanine) or three essential amino acids (threonine, phenylalanine, and valine) was investigated under similar conditions. The results of these studies revealed that young rats, tube-fed these diets for 1 to 3 days, demonstrated heavier poly-ribosomal aggregates and enhanced in vivo protein synthesis in the livers compared with rats force-fed a complete diet (80,81). These results are similar to those obtained after force-feeding a threonine-devoid diet for 1 to 3 days (70,71). Using a tryptophan-devoid diet, other investigators have also observed enhanced hepatic protein synthesis in rats force-fed this diet for 5 days (82) and in rats fed a tryptophan-deficient diet ad libitum for 13 days (83). On the other hand, Bocker et al. (84) observed a diminution in liver protein synthesis in rats fed a tryptophan-deprived diet ad libitum for 5 to 10 days in comparison with rats that received the control diet. These authors also

studied the effects of feeding a tryptophan-deprived diet on hepatic
RNA metabolism and observed that synthesis of nuclear and cytoplasmic
RNA (in vivo [^{3}H]orotic acid incorporation into RNA) remained unaffected up to the 10th day. After 15 days of tryptophan deprivation,
there was an increase in the incorporation of orotic acid into nuclear
RNA but not into cytoplasmic RNA. In these studies where rats received the test diet ad libitum, the body and liver weights and the concentrations of most of the liver free amino acids including tryptophan
were significantly lower than those of control rats. In contrast, in
studies where rats were tube-fed purified diets devoid of tryptophan
or other single essential amino acids, there was an increase in the
levels of essential amino acids in the liver and plasma along with enhanced hepatic protein synthesis (71,80). These studies indicate that
the method of feeding (force-feeding versus ad libitum) the diets has
a significant influence on hepatic protein synthesis. The differences
in the effects of feeding ad libitum versus force-feeding test diets on
hepatic protein metabolism have been further discussed in a number of
studies in the following sections.

It is important to stress that the results obtained after multiple (three)
feedings of a purified diet devoid of tryptophan for 1 day are different
from those obtained in fasted animals (rats or mice) force-fed a single
feeding of a tryptophan-devoid amino mixture. In the latter studies
(43) incorporation of labeled amino acids into hepatic proteins was decreased, but in animals force-fed the tryptophan-devoid diet for 1 day
(three feedings) in comparison with animals force-fed the complete diet,
in vivo hepatic protein synthesis was enhanced (80). Thus the acute
administration of tryptophan and the imbalance induced by force-feeding
this amino acid-deficient diet may influence hepatic protein synthesis
and polyribosomes quite differently. The possible explanation offered
for the enhanced hepatic protein synthesis in rats force-fed a tryptophan-devoid diet for 1 day is similar to the one that is speculated to
be involved in rats force-fed purified diets devoid of other single essential amino acids for 1 or more days. Resembling the changes observed in the skeletal muscle of rats force-fed diets devoid of threonine,
phenylalanine, or valine, decreased skeletal muscle protein synthesis
was observed in rats force-fed a tryptophan-devoid diet for 1 day (80).
Thus the enhanced hepatic protein synthesis induced by feeding a
tryptophan-devoid diet for 1 day is probably related to the supply of
tryptophan and other amino acids released by skeletal muscle catabolism
to the liver.

Munro and Harper and their associates have also been investigating
the effects of ingesting amino acid-imbalanced diets, in particular iso-
leucine- or threonine-imbalanced diets, on hepatic protein synthesis
(85-88). Both groups of investigators have employed complex-feeding
regimens to study the changes in hepatic polyribosomes and protein
synthesis in response to feeding of amino acid-imbalanced diets. In

these studies, rats were trained for about 3 weeks to consume all their
food (about 10 g of 15%-casein diet) within 2 hr daily. On the day of
killing, meal-trained rats were given control or imbalanced diets. Only
those rats that consumed all of the diet offered during a 1-hr period
were used in the experiments. Using this type of feeding regimen,
Yoshida et al. (85) and Benevenga et al. (86) presented evidence for
increased incorporation of the limiting amino acid into liver proteins of
rats fed for 8 or 12 hr a threonine- or histidine-imbalanced diet as
compared with the corresponding control diet. These authors have
suggested that the imbalanced diet causes a stimulation of hepatic protein synthesis, and this is probably related to the more efficient utilization of the most limiting amino acid. On the other hand, Pronczuk
et al. (87) found no statistically significant differences in polyribosome
profiles in livers of rats meal-fed either control or amino acid-imbalanced
diets (isoleucine or threonine) 1 or 4 hr before killing. These authors
have also studied the effect of feeding rats ad libitum control or
threonine-imbalanced diet for 10 hr to 14 days and observed similar
polyribosome profiles in the livers of control and experimental animals
(87). However with the ad libitum-feeding regimen, rats consumed
significantly less experimental diet and therefore lost body weight and
liver weight. Consequently, the data obtained from such ad libitum-
feeding studies are difficult to evaluate in relation to the possible role
of the specific amino acid deficiency in influencing hepatic protein
metabolism.

In further experimental studies using a different approach and design, Ip and Harper (88) reported that when the free threonine pool
of tissues had previously been depleted by feeding rats a diet deficient
in threonine, polyribosome profiles and rates of [^{14}C]leucine incorporation into liver protein were not influenced in response to further feeding of a threonine imbalanced diet. However when the tissue threonine
pool had not been depleted before feeding of the experimental diet, ingestion of the threonine-imbalanced diet 2 hr before sacrifice, stimulated
liver polyribosome aggregation and in vivo [^{14}C]leucine incorporation
into protein more than did the ingestion of the basal diet (88). This
study indicates that polyribosome aggregation and protein synthesis
are influenced by the concentration of threonine when it is the most
limiting amino acid in the diet and that changes in the responses of
these variables to ingestion of a meal of an imbalanced diet can be influenced by the prior nutritional state of the animal. Ip and Harper
(88) have further investigated the effects of a threonine-imbalanced
diet on hepatic polyribosomal profiles, by force-feeding rats a basal or
threonine-imbalanced diet for 4 days. The liver polyribosomal profiles
were almost identical in rats force-fed the basal or the experimental diet.
In this study, the concentrations of most of the plasma and liver free
amino acids were elevated in the experimental animals. This observation

was consistent with our studies with rats force-fed a diet devoid of threonine for 3 days. However their failure to observe more aggregated polyribosome pattern in the livers of rats force-fed the threonine-imbalanced diet may be due to the composition of the diets they employed. Their basal and threonine-imbalance diets were low in protein (6% casein). Furthermore, a mixture of all essential amino acids (except threonine) was added to the experimental diet, whereas in the basal diet this amino acid mixture was omitted. In comparison with the standard protein content (equivalent to 18% casein) of usual animal diets employed by most investigators, the low-protein diets, such as used in Ip and Harper's study, may produce adverse morphological and biochemical effects in the liver. Indeed, some of the studies discussed later indicate that protein restriction causes abnormal changes in hepatic protein and RNA metabolism when compared with animals fed diets containing an adequate supply of protein. Therefore it is difficult to evaluate the significance of the findings in this study because of the complexity of both the control and the experimental diets employed.

Protein-Deficient Diets

The effects of low-protein diets on hepatic RNA and protein metabolism have been investigated for a possible relationship between the amino acid supply via the diet and the protein synthetic ability of the cells. A number of studies have demonstrated that the protein-synthesizing system is adaptable to wide fluctuations in dietary protein. When the quantity of dietary protein is inadequate a series of responses and adaptions, which may act to delay the development of pathological changes, take place in the liver. For example, in rats fed a low-protein (6%-casein) diet for 28 days, the activity of DNA-dependent RNA polymerase I, the synthesis of nuclear and cytoplasmic RNA, and the in vivo and in vitro protein synthesis in the liver were enhanced in comparison with those of rats fed a diet containing adequate protein (20% casein) (89). However, Wannemacher et al. (90) have reported a good correlation between the rates of protein synthesis and the concentration of the free amino acids that are essential for protein synthesis in the livers of rats fed a low-protein (6%-casein) diet. The hepatic cells from the rats fed the low-protein diet revealed a decrease in essential and nonessential free amino acids along with decreased protein synthesis compared with cells from the control rats. In this study protein deprivation affected only the membrane-bound polyribosomes (a decrease in the amount of heavy polyribosomes) and not the free polyribosomal population. This effect suggests that the synthesis of serum proteins, which occurs mainly on membrane-bound polyribosomes, is extremely sensitive to amino acid supply. In addition a similar correlation was observed between the concentration of individual free amino acids and the cellular concentration of ribosomal RNA. The incorporation of [^{3}H]uridine (30-min pulse) into total RNA of the hepatocytes from the amino acid-deprived rats was twice that observed from rats given an adequate diet, and this is consistent with the increased activities of both RNA-polymerases I and II (90). However when the distribution of radioactivity in the RNA of the nucleus and cytoplasm was followed sequentially for 12 hr after a dose of [^{3}H]uridine, the ribosomal RNA fraction in the hepatocytes from the rats given the low-protein diet contained about half the amount of radioactivity in hepatic ribosomes as that from rats given an adequate protein intake. From these studies it was concluded that amino acids may influence RNA synthesis by inhibiting the rate of synthesis of new ribosomes. A similar effect of amino acids on the rate of appearance of newly synthesized ribosomal subunits in the cytoplasm of Landschutz cells have been observed by Shields and Korner (91). These authors suggest that amino acid starvation exerts its effects through an inhibition of synthesis of proteins with a rapid rate of turnover, which is required for ribosome synthesis. A similar mechanism may explain how the concentration of amino acids can regulate the rate of synthesis of ribosomes in liver cells.

Clark and Jacob (92) also have observed an inverse relationship between the protein level in the diet and RNA polymerase activity. The activity of DNA-dependent RNA polymerase I was increased in the nuclei of livers of rats fed a low-protein (6%-casein) diet for 2 weeks compared with rats that received a 18%-casein diet. These authors have proposed that the increase in enzyme activity after feeding a protein-deficient diet was not due to a change in the amount of enzyme but to a possible derepression phenomenon that was occurring during protein deprivation. An increase in template efficiency of endogenous DNA in rats given low-protein diets may be due to loss of deoxyribonucleohistones (93). On the other hand, decreased activity of DNA-dependent RNA polymerases I and II were reported in the livers of rats fed, for 6 days, a diet containing 3% casein compared with rats fed a diet containing 20% casein (94,95). Also, decreased in vitro incorporation of labeled amino acid into proteins was observed in the liver after feeding the rats a 3%-casein diet for 6 days (96). The observed discrepancies as to the effects of protein restriction on the activities of nuclear RNA polymerases may be attributed to differences in the dietary protein levels and/or feeding time employed in these studies.

The results presented by Lewis and Winick (96) indicate that the level of protein in the diet is probably of importance in influencing hepatic nuclear RNA metabolism. These authors have observed increased synthesis of 45S pre-rRNA in the hepatic nuclei of rats fed a 6%-casein diet for 7 days compared with control animals (96). Although cellular cytoplasmic rRNA pool was reduced to 70% of controls, their kinetic studies indicated that the appearance of labeled rRNA in the cytoplasm was the same, but cytoplasmic rRNA-specific activity was greater in the experimental than in the control animals. On the basis of other reports suggesting that nutritional insult enhances liver RNA polymerase I activity (89,90,92), these authors have hypothesized that protein restriction results in an increased rate of rRNA synthesis at the level of increased transcription from DNA. Thus there appears to be some regulatory mechanism whereby a reduction in supply and concentration of essential hepatic free amino acids could control the rate of synthesis of rRNA, its release into the cytoplasm, and the rate of synthesis of cytoplasmic proteins.

Protein-Free Diets

In the preceding sections some of the effects of amino acids, amino acid-deficient diets, and reduced protein intake on hepatic RNA and protein metabolism were reviewed. In this section a few selective studies relating to the effects of total withdrawal of protein (protein-free) from the diet on liver RNA and protein will be reviewed. Enwonwu and Munro (97) were some of the first investigators to study the effects of liver RNA metabolism of feeding a diet free of protein to animals. In their studies on liver RNA turnover, the principal variable was the intake of protein, and therefore it was possible to evaluate the role of protein in the observed responses. When rats were fed a protein-free diet (ad libitum in the form of agar gel) for 3 to 10 days, the animals underwent accelerated breakdown of RNA and decreased RNA synthesis compared with rats receiving a diet adequate in protein (18% casein). In addition, the amount of ribosomal subunits was much lower in the livers of rats fed a protein-free diet for several days. It was suggested that RNA turnover is regulated through changes in the population of free ribosomal subunits in the liver cytoplasm, the abundance of subunits being affected by the amino acid supply. This study, did not examine the observed decrease in RNA metabolism (both synthesis and breakdown) as it would affect hepatic protein synthesis in animals fed a protein-free diet. In an earlier study from our laboratory, we have investigated the effects of a protein-free diet on hepatic polyribosomes and protein synthesis under controlled conditions (98). The results indicated that in young rats force-fed a protein-free diet for 1 day (three feedings) and killed the following morning, there was a shift in hepatic polyribosomes toward heavier aggregation and enhanced hepatic protein synthesis (in vivo and in vitro) in comparison with rats fed a complete diet (98). Also, the response appeared to be intimately related to the decreased synthesis and increased catabolism of skeletal muscle protein. Rats force-fed the protein-free diet for 1 day demonstrated a loss of protein, increased protein catabolism, and a decrease in protein synthesis of the gastrocnemius muscle (98). Thus the overall reaction pattern appeared to be similar for animals force-fed a protein-free diet or a single essential amino acid-devoid diet. However when rats were fed a protein-free diet ad libitum for longer periods (5 days or more), the livers responded differently in comparison with those of rats tube-fed a protein-free diet for 1 day. Thus Sato et al. (99) and

Yokogoshi and Yoshida (100) have observed extensive breakdown of polyribosomes in the livers of rats fed a protein-free diet for 5 or 10 days compared with those of rats fed the control diet. The disaggregation of hepatic polyribosomes by feeding a protein-free diet has been attributed to a depression in both initiation and elongation rates of polypeptide chains (99). Interestingly, the amount and the average size of hepatic poly(A)-mRNA were unchanged during a 10-day period of dietary protein deficiency, indicating high stability of hepatic mRNA under the condition where protein synthesis is decreased. In addition, supplementation of methionine and threonine to the protein-free diet prevented the disaggregation of polyribosomes and loss of body weight and liver weight induced by a protein-free diet (101). These effects occurred because both methionine and threonine were the most-limiting amino acids in rats fed a protein-free diet. In addition to an overall decrease in total liver protein synthesis (101-105), a decrease in albumin synthesis was also observed in rats fed a protein-free diet for 2 or 9 days (101). The concentration of translatable albumin mRNA in the cytoplasm of livers of rats fed a protein-deficient diet was decreased markedly. This observation was consistent with the earlier data of Morgan and Peters (104), who have shown a 60 to 70% fall in the rate of both albumin and transferrin synthesis in the livers of rats fed a protein-free diet for 10 days. Refeeding a complete mixture of amino acids or a complete diet restored albumin synthesis. These findings suggest that the dietary protein supply selectively affects either the synthesis or functional stability, or both, of albumin mRNA in rat liver.

Comments Relating to Amino Acids in the Diet

Although liver protein synthesis is inhibited in animals deprived of protein stimulatory effects have been reported in the nucleus during protein deficiency. These effects appear to be restricted to the nucleolus. Thus the rate of synthesis and the quantity of nucleolar RNA (105,106), nucleolar volume (105,106), and activity of RNA polymerase I (89,93, 106), all become increased in the hepatocytes of rats that are fed a diet deprived of amino acids. In addition, Bailey et al. (107) have observed that omission of methionine, tryptophan, or threonine, but not of other amino acids, from the complete diet, elevated hepatic nucleolar size and the activity of RNA polymerase I, similar to changes observed with a protein-free diet (105,107) or a threonine-devoid diet, which we reported earlier (74,75). These findings suggest that deprivation of certain amino acids affects the nucleolus and may be due to an imbalance in protein metabolism. On the other hand, the inhibitory effect, induced by feeding a protein-free diet, on hepatic protein metabolism has been attributed to the increased rate of rRNA degradation in the cytoplasm (106). A shorter half-life of cytoplasmic rRNA (RNAs of both free and membrane-bound polyribosomes) and decreased total liver RNA content per DNA were reported in the livers of rats fed a protein-free diet ad libitum for 5 or 14 days compared with rats fed a normal diet (106,108). Although a higher rate of rRNA synthesis occurs in the nucleoli in response to feeding a protein-free diet, there is a rapid degradation of nascent rRNA in liver nuclei of rats fed a protein-deficient diet. The observed increase in RNase activity in liver nuclei of rats fed a protein-deficient diet may be one of the factors causing stimulated degradation of nascent rRNA in the liver nuclei of rats fed such a diet (108).

From these studies it is apparent that dietary amino acids have a special role in the regulation of RNA and protein metabolism in the liver. However the question often asked in studies conducted with intact animals is whether or not the responses, in particular the stimulatory effects on hepatic synthesis induced by amino acids, are secondary to hormonal stimulation. This is because of two main reasons. First, a number of hormones including corticosteroids (46,109-114), growth hormone (115-118), glucagon (119,120), insulin (121,122), thyroid hormone (123-126), and sex hormones (127) stimulate the synthesis of specific and/or total hepatic proteins as well as the rate of synthesis of messenger RNA. Second, the administration of certain amino acids indirectly elicits specific metabolic reactions via stimulation through hormones from endocrine glands (128-130). However most of the studies in the literature dealing with the effects of amino acid supply on hepatic polyribosomes and protein synthesis indicate that the observed responses do not seem to be mediated through a hormonal mechanism. In our laboratory, we observed that adrenalectomy did not abolish the stimulation of hepatic protein synthesis in response to tube-feeding a complete amino acid mixture or tryptophan alone, findings similar to those observed when using control, intact animals (43). The results of studies by others support a similar conclusion. Jorgensen and Majumdar (49,131) have reported that tube-feeding of tryptophan to well-fed adrenalectomized and adrenalectomized diabetic rats stimulated in vivo amino acid incorporation into plasma albumin, transferrin, and liver ferritin. Vesley and Cihak (132) found enhanced activity of hepatic nuclear DNA-dependent RNA polymerase after tryptophan administration to adrenalectomized animals; however, these results are not in agreement with those of Cammarano et al. (46) who reported that the stimulatory effects of feeding high levels of tryptophan on hepatic polyribosomes were dependent upon the presence of the adrenal glands. In other experiments, hypophysectomized mice that were treated with tryptophan appeared to respond the same as intact mice (133). In other experiments, young rats, either adrenalectomized (134) or hypophysectomized (135), force-fed a threonine-devoid diet for 1 to 3 days responded with enhanced hepatic protein synthesis, as did intact experimental animals. Ip and Harper (136), using their nutritional model, found that adrenalectomy did not prevent the hepatic polyribosomal aggregation response induced by threonine supplementation. Thus, considering the studies described in the preceding sections, it appears that the observed responses resulting from experimental amino acid feedings were probably not secondary to adrenal cortical or hypophyseal hormone stimulation. Further support of the premise that the hepatic responses to amino acid supply are independent of hormonal influences come from studies (discussed next) using perfused livers and isolated hepatocytes, where no changes in hormonal secretion could be involved.

2. *Studies Using Perfused Livers*

The liver is normally exposed to wide fluctuations in the concentrations of amino acids. To control these in vivo variations, a model system involving isolated perfused livers has been used to investigate the role of amino acids in the regulation of protein synthesis. The perfusion technique uses the liver after it has been isolated from influences of the whole animal. Therefore it enables one to evaluate how controlled levels of circulating amino acids influenced protein synthesis. If adequate concentrations of amino acids are maintained during the perfusion, the rates of synthesis of total protein and albumin (the major secretory protein) in the in vitro system (137) are comparable with those observed in vivo (138). Thus the perfused liver system is considered to be a useful and reliable method for studies concerning the mechanisms of amino acid regulation of hepatic protein synthesis.

Table 2 summarizes selected data pertaining to the influence of amino acid supply on hepatic protein synthesis in the perfused liver. In early studies, Jefferson and Korner (139) noted that, when rat livers were perfused with a medium devoid of amino acids, a breakdown of polyribosomes from heavy to light aggregates occurred along with a decrease in total protein synthesis. The addition of complete amino acids (in multiples of their concentration in normal rat plasma) to the perfusion medium reversed this response. Liver protein synthesis was enhanced linearly with increasing concentrations of amino acids in the perfusion medium. To maintain a maximum rate of protein synthesis along with appropriate polyribosomal profiles in the perfused liver system, similar to that observed in the nonperfused livers of normal, fed animals, it was necessary to provide amino acids at 10 times the concentrations found in arterial plasma of normal-fed rats. This observation has been confirmed by other investigators (140,141) who also showed that perfused amino acid levels had to be kept higher than those found in plasma to maintain polyribosome integrity and active protein synthesis. However these findings have not always been confirmed by others. Using the perfusion medium, which contained amino acids 10 times the normal plasma concentration, Woodside and Mortimore (142) were able to show only a limited stimulation of liver protein synthesis induced by amino acids. Tavill et al. (143) and Peavey and Hansen (144) were unable to demonstrate any kind of improvement in liver protein synthesis in response to perfused amino acids, supplied even up to 20 times the concentrations normally found in normal rat plasma.

Recently, Flaim and co-workers (145,146) have reexamined the influence of amino acids on protein synthesis in perfused livers with the intent of resolving some of the controversy in the literature and providing insight into the underlying mechanisms. The significant feature of these studies was that the concentrations of amino acids present in the perfusate, five times the normal plasma levels, were kept constant during the entire course of perfusion by using a noncirculating medium in which specific activity of the amino acid precursor pool was rigidly maintained. Using this methodology, these authors have demonstrated the sensitivity of total protein and albumin synthesis to the concentrations of amino acids in the perfusate. Thus the failure of some investigators to demonstrate enhanced protein synthesis was probably because of the varying levels of the circulating amino acids and probably resides in the mistaken assumption about the specific activity of the precursor amino acid. By allowing the perfusion medium to recirculate in their studies, they permitted the amino acid concentrations in the liver to vary with time (142,143).

The duration of perfusion appears to play a significant role in the regulation of liver protein synthesis. Protein synthesis was inhibited in livers perfused for up to 20 min with an amino acid-devoid medium. However when perfusion with an amino acid-devoid medium was continued for longer periods (longer than 1 hr), the protein synthetic activity recovered from the insult of acute amino acid deprivation in the medium during the earlier 20 min perfusion. This recovery of the rates of protein synthesis after lengthy perfusion with the amino acid-devoid medium has been attributed to an increase in the accumulation of intracellular amino acids via an increase in the rates of proteolysis. This conclusion is supported by the findings that two potent inhibitors of proteolysis, insulin (147) and glutamine (148), were effective in preventing the recovery of protein synthetic rate in the perfused liver.

The question of whether or not certain single amino acids may have a special role in controlling or regulating liver protein synthesis has been investigated using the perfused liver. In early studies, Jefferson and Korner (139) reported that not one, but a mixture of 11 amino acids (arginine, asparagine, isoleucine, leucine, lysine, methionine, phenylalanine, proline, threonine, tryptophan, and valine, at 10 times the normal plasma concentration) was necessary to improve the polyribosomal profile and to increase incorporation of labeled amino acid into total liver protein. Omission of any one of these 11 amino acids from the mixture added to the perfusate eliminated these stimulatory responses. McGowan et al. (149) have demonstrated that omission of either tryptophan or methionine from the perfusate resulted in disaggregation of polyribosomes and decreased protein synthesis in comparison with the

Table 2 Influence of Amino Acids on Protein Synthesis in Perfused Liver

Nutritional state of donor	Composition and concentration of amino acids in the perfusate	Perfusion time	Responses observed
Fasted	None	15-60 min	→ Total protein synthesis in vivo and in vitro (139) → Total polyribosomal aggregation (139) → Bound polyribosomes > free polyribosomes (145,146) → Albumin synthesis (139)
Fasted or fed	Amino acid mixture (1 X-10 X normal plasma concentration)	30-60 min	↑ Total protein synthesis in vivo and in vitro (139) ↑ Aggregation of polyribosomes (145,146) ↑ Albumin synthesis (145,146)
Fed	Omission of 11 amino acids singly from amino acid mixture (10 X plasma concentration)	60 min	→ Protein synthesis (139) → Polyribosomal aggregates (139)
Fasted	Omission of tryptophan or methionine from amino acid mixture (1 X of plasma concentration)	60 min	→ Protein synthesis (149) → Heavier polyribosomes (149)

88

State	Treatment	Time	Effect
Fed	Addition of tryptophan, lysine, threonine, or isoleucine to amino acids (1 X plasma concentration)	150 min	None: albumin synthesis or ribosomal profiles (151)
Fasted	Addition of tryptophan (0.05-10 mM) isoleucine, arginine, lysine, phenyl-alanine or ornithine (10 mM) to amino acid mixture (1 X plasma level)	150 min	↑ Ribosomal aggregation (151,152) ↑ Albumin synthesis (151,152)
Fed	Omission of tryptophan or threonine from amino acid mixture (1 X plasma level)	12 hr	None: synthesis of secretory proteins, except glycoprotein synthesis with threonine omission (150)
Fasted	Omission of threonine from amino acid mixture	12 hr	None: synthesis of secretory proteins (150)
Fasted	Omission of tryptophan from amino acid mixture (1 X plasma level)		↑ Albumin synthesis (150)

↓ = decrease, ↑ = increase

perfusion medium which contained all of the amino acids (1 × normal
plasma levels). Miller and Griffin (150) also have observed that the
removal of tryptophan, but not of threonine, from the perfusate revealed a small decrease in albumin synthesis by livers from fasted
donors. However in this study removal of either amino acid did not
impair the incorporation of [^{14}C]lysine into total hepatic and other
plasma proteins (fibrinogen, glycoprotein, and globulin) from livers of
both fed and fasted donors (150). The failure to observe diminution
in total protein synthesis by the perfused liver in the absence of tryptophan or threonine was unlike the findings of earlier investigators
and was probably because of the longer duration of perfusion. On the
basis of the increase in overall negative nitrogen balance in the perfused livers, Miller and Griffin (150) concluded that the isolated perfused liver is capable of breaking down its own cell proteins to supply
the needed single amino acid for the maintenance of protein synthesis.

Further studies have been concerned with the underlying molecular
mechanisms by which amino acids regulate protein synthesis. The
inhibition of protein synthesis in livers perfused with an amino acid-
deficient medium has been reported to be accompanied by a greater
breakdown of polyribosomes from the membrane-bound population than
from the free population (146). This breakdown of polyribosomes does
not appear to result from mRNA degradation, because this effect could
be reversed by addition of amino acids to the perfusion medium. Upon
further investigation, it was shown that inhibition of protein synthesis
and the disaggregation of polyribosomes that occurred with an amino
acids-devoid medium was due to a defect in the rate of peptide chain
initiation. In this study, amino acid deficiency did not influence the
charging of tRNA, since no change in the quantities of amino acids
bound to tRNA occurred (145). On the other hand, the block in peptide chain initiation induced by amino acid deprivation was attributed
to sequestration of mRNAs from the polyribosomal fraction into an un-
translatable (i.e., nonpolyribosomal fraction) pool, and further, rein-
troduction of amino acids into the perfusion medium promptly restored
the normal state of polyribosomal aggregation and protein synthesis.
Thus despite a few inconsistencies in the literature, most studies indicate that the process of protein synthesis in the perfused liver is indeed regulated by amino acid availability.

Rothschild et al. (151) have reported that the disaggregated ribosomes in the livers of fasted rabbits were reaggregated in response to
the addition of tryptophan or isoleucine to the perfusion medium. Addition of other amino acids did not have this effect. However in later
studies the same investigators observed that three other amino acids
(phenylalanine, lysine, or arginine), when supplemented to the per-
fusates in which the content of amino acids was regulated, were effective
in stimulating the ribosomal aggregation in the livers of fasted rabbits
(152). Besides promoting ribosomal aggregation, these amino acids also

stimulated albumin synthesis, which was initially depressed in the livers of fasted animals. When some of these "stimulating" amino acids were employed in perfusions of livers from fed donors, they were in-effective in stimulating albumin synthesis. In disagreement with the results of Rothschild et al. (151,152) with fasted rabbit livers, Kelman et al. (141) reported that the addition of single amino acids, such as tryptophan or isoleucine, to the perfusate was not able to stimulate albumin synthesis in the livers of fasted rats. They suggested that this disparity was probably because of inherent differences in liver protein turnover between rats and rabbits. However they observed that livers from fasted rats did respond to the addition of a mixture of 11 amino acids (as 10 times their peripheral blood concentration).

Because of conflicting reports in the literature, which describe the use of perfused liver, the nature of the role of any one particular amino acid in influencing hepatic protein synthesis is as yet unresolved. The differences among these reports are probably related to the diversity of the experimental conditions employed. Notable among these variations were: (a) concentrations of amino acids in the perfusate, (b) the dura-tion of perfusion, and (c) the initial nutritional state of the donor ani-mal. In general, these studies suggest that the isolated liver perfusion system may serve as a useful model in examining the role of amino acids in the regulation of protein synthesis. However it has not been possible, up to now, to resolve the precise role(s) whereby amino acids in the perfusate regulate the rates of hepatic protein synthesis. This is prob-ably due to the diversity of the experimental conditions employed by various investigators.

3. *Studies Using Isolated Hepatocytes*

In recent years, experimental approaches involving both freshly iso-lated and cultured liver cells have been used to assess various metabolic processes in vivo. Such an experimental design is rapidly becoming an important tool in biochemical research. One major application has been the study of amino acid regulation of hepatic protein metabolism. An advantage in studying regulatory mechanisms at the cellular level with isolated cells is that the systematic testing of regulatory events may be carried out without interference from homeostatic controls exerted by the organism. The generally accepted methodology used for the success-ful preparation of viable cells from rat liver in large quantities involves perfusion with collagenase (153-156). Isolated hepatocytes have been shown to actively synthesize and secrete proteins in vitro (157-160). Therefore isolated liver cells have been used for investigating the pos-sible role of amino acids in the control of protein metabolism. The effects of amino acids on protein synthesis and degradation in isolated rat hepatocytes have been investigated mainly by Seglen and colleagues (161-163), and some of their studies are summarized in Table 3. Similar to the findings with isolated perfused liver (139), maximal stimulation

Table 3 Effects of Amino Acids on Protein Synthesis in Isolated Rat Hepatocytes

Amino acid(s) added to the medium	Concentration of amino acid(s) (mmol/L)	Changes in the rate of protein synthesis	Reference
Mixture of all amino acids	17.7	↑	161
Glutamate, aspartate, leucine, isoleucine methionine, or histidine	0.005–30	o	161,164
Alanine, serine, glycine, proline, or threonine	0.5–20	↑	161,162,164
Cysteine, arginine, lysine, or tryptophan	5–30	↓	161
Asparagine, glutamine, phenylalanine, or tyrosine	0.005–2.5	↑	164
Asparagine, glutamine, phenylalanine, or tyrosine	5–30	↓	161
Tryptophan, arginine, or lysine	0.05–0.1	↑	164

↑ = increase, ↓ = decrease, o = no change.

of protein synthesis in isolated hepatocytes was achieved by the addition of a balanced amino acid mixture at concentrations up to 12.5 times the normal plasma levels. To determine to what extent the various amino acids contribute to the protein synthesis-stimulating effect of the amino acid mixture, amino acids have been tested individually in concentrations ranging from 0.005-30 mmol/L. Individual amino acids have different effects on protein synthesis, and therefore they have been divided into four groups (inhibitory, inactive, biphasic, and stimulatory) on the basis of their effects. In the case of some amino acids, the varied effects on protein synthesis were due to the different concentrations of the amino acids added to the incubation medium which contained the isolated hepatocytes. The amino acids, cysteine, arginine, lysine, and tryptophan, were classified as inhibitory since they had no significant effects in the low-dose range (below 5 mmol/L), but at higher concentrations (5-30 mmol/L), they caused significant inhibition of protein synthesis. However in a more recent study, which utilized amino acid concentrations detected in portal blood after a protein meal (164), three (lysine, arginine, and tryptophan) of the four amino acids classified as inhibitory, were able to stimulate protein synthesis when given individually at very low concentrations, ranging from 0.05-0.10 mmol/L. When glutamate, aspartate, leucine, histidine, isoleucine, or methionine were tested at a dose range of 0.005-30 mmol/L, no significant responses were obtained in the rates of protein synthesis, and therefore these were termed as inactive amino acids (161). The amino acids, asparagine, glutamine, phenylalanine, and tyrosine, based upon the concentrations of these amino acids present in the incubation medium, elicited two different types of responses in protein synthesis, one stimulatory and the other inhibitory. These amino acids, classified as biphasic, stimulated protein synthesis (10 to 50%) in isolated hepatocytes at concentrations up to 2.5 mmol/L, while at higher levels (5-30 mmol/L) these amino acids had a pronounced inhibitory effect on protein synthesis (161). Alanine, serine, glycine, proline, and threonine have been grouped under stimulatory amino acids. Each of these amino acids markedly stimulated protein synthesis at all doses (0.1-20 mmol/L) tested (156,160). Alanine, in particular, has been shown to be most effective in stimulating protein synthesis, practically to the same extent as did the complete amino acid mixture (161). The finding that certain single amino acids, when used in an appropriate dose range, can act similarly to a complete amino acid mixture in the stimulation of protein synthesis, suggests that the stimulation may involve a regulatory effect rather than simply reflecting their precursor function for protein synthesis. These observations are in accordance with the liver perfusion experiments of Jefferson and Korner (139), in which a number of individual amino acids greatly stimulated polyribosomal activity.

In searching for the mechanism(s) whereby several amino acids singly can stimulate protein synthesis, consideration has been given to indirect mediation through an amino acid metabolite or through its action as an energy substrate. Aminooxyacetate (164-166), an inhibitor of trans-amination, the first step in the metabolism of most amino acids, com-pletely eliminated the stimulatory effect of a complete amino acid mixture or of a single amino acid such as alanine. This suggests that the prod-ucts of intracellular transamination exert a profound regulatory influ-ence on protein synthesis in isolated hepatocytes. To determine if the stimulatory effect could be due to nonspecific provision of energy, a direct comparison was made with energy substrates such as lactate and pyruvate. These energy substrates were observed to stimulate protein synthesis in isolated hepatocytes to the same extent as did several individual amino acids (164). Furthermore, addition of pyruvate to several amino acids (i.e., alanine, phenylalanine, serine, tyrosine, or glutamine), which stimulate protein synthesis individualy, did not result in any further improvement, suggesting that these amino acids may act as energy substrates. However aminooxyacetate also inhibited the stimulation of protein synthesis induced by energy substrates. This indicates that aminooxyacetate does not merely restrict the utiliza-tion of amino acids for energy production, but that it also blocks some other process involved in the maintenance of protein synthesis.

Hepatic protein synthesis both in vivo (167,168) and in isolated hepa-tocytes (169-171) can be stimulated by cyclic AMP-dependent protein kinases and the cyclic AMP-activating hormones, glucagon and epineph-rine. Hence it is possible that the regulation of hepatic protein synthe-sis by amino acids could be mediated via cyclic AMP-dependent factors.

Protein degradation also plays a significant role in the control of hepatic protein synthesis. In isolated hepatocytes, the amino acids re-quired for protein synthesis are largely derived endogenously via pro-tein degradation. This assumption was supported by the findings that propylamine or leupeptin, both strong inhibitors of lysosomal proteolysis, reduced the rate of protein synthesis and increased the requirement for added amino acid mixtures (164). This indicates that the rate of protein synthesis may be influenced by the rate of protein degradation, partic-ularly when the amino acids are in limited supply. Under these condi-tions endogenous amino acids act as precursors for proteins. Thus studies with isolated hepatocytes add further support to the belief that hepatic protein synthesis is subject to regulation by amino acids. Amino acids can serve as precursors of protein, as energy substrates, or as some other type of effectors in controlling hepatic protein synthesis.

B. Carbohydrates and Fats

Carbohydrates constitute the main bulk of the diet and along with fats serve as the major sources of energy. Webb et al. (172) and Sidransky and Verney (98) reported that feeding a protein-free diet to fasted rats induced a reaggregation of the disaggregated hepatic polyribosomes and was accompanied by increased protein synthesis in the liver. These

studies suggested that dietary carbohydrates, the major energy source
of the protein-free diet, could influence the state of polyribosomal aggregation and the rate of protein synthesis in the liver. The possible
role of diets containing mainly carbohydrates (i.e., protein-free diets)
in regulating hepatic protein metabolism has been presented in the
preceding sections. Here, the effects of administering carbohydrates
or fats alone or diets containing low or high amounts of carbohydrates
on hepatic protein synthesis will be reviewed.

In an earlier study, Wittman et al. (173) found that the disaggregated
liver polyribosomes in 60-hr-fasted rats became reaggregated within 10
hr in response to the ad libitum-feeding of the carbohydrate glucose.
However feeding fat (Crisco), in amounts isocaloric to the glucose, to
fasted rats did not restore the aggregation of the hepatic polyribosomes.
This observation indicates that the type of energy, but not all sources
of energy, may be a determining factor that affects protein synthesis
in the liver. Considering the observed stimulatory effects of glucose,
in concert with the hormone insulin, Wittman et al. (173) speculated
that dietary carbohydrate, through its direct action or through one of
its metabolites, may alter the level of an active hormone, which in turn
influences the rate of hepatic protein synthesis.

More recently, the stimulatory effects of carbohydrates on protein
synthesis have also been demonstrated in isolated hepatoctyes from
fasted rats. Addition of 10 mM glucose, pyruvate, lactate, or xylitol
to isolate cells obtained from fasted rats enhanced the synthesis of
intracellular proteins, as well as of secretory proteins such as albumin
(174). However the addition of each of the these compounds to isolated
cells from fed rats had no effect on the rate of protein synthesis. It
was suggested that the stimulatory response to carbohydrates in hepatocytes from starved rats could be due to its supplying precursors for
the synthesis of nonessential amino acids that were lacking in such cells.
It was earlier demonstrated that in isolated liver cells from starved rats,
the concentrations of such metabolites were lower than in the intact
liver (175,175). This observation applies to the amino acids aspartate,
glutamate, and alanine, and probably to the intermediates in the glycolytic pathway (176). Since intermediates in carbohydrate metabolism are
precursors of nonessential amino acids, the effects of carbohydrates
on hepatic protein synthesis could be due to the amino acids derived
from the precursors. The finding that the addition of the fat oleate
to the isolated hepatocytes had no stimulatory effect on protein synthesis supports this conclusion (174). However in another study it was
shown that when the liver was supplied with excess fatty acids, the
net synthesis of very low-density lipoproteins was enhanced, as estimated by measuring the incorporation of radioactive leucine into these
proteins (177,178). This finding has been demonstrated both in isolated
rat livers perfused with excess fatty acids and in the livers of intact
animals force-fed excess fat (cream containing more than 2 mEq fatty

acids) 3 hr before sacrifice. In further studies it was shown that, in intact rats or in isolated perfused rat livers, fatty acid overload stimulated [^{14}C]orotic acid incorporation into hepatic polyribosomal messenger RNA (179), which was primarily due to a higher incorporation into membrane-bound polyribosomal-associated messenger RNA. These and other studies suggest that de novo RNA synthesis may be involved in the regulation of fatty acid stimulation of lipoproteins.

The effects of intake of either inadequate or excessive amounts of carbohydrate in the diet on hepatic RNA and protein metabolism have been investigated in intact animals. Pfeifer and Szepesi (180) reported enhanced RNA synthesis in the livers of rats fed ad libitum a high-carbohydrate diet. On the other hand, Glick et al. (181) observed that rats fed excess carbohydrates via overfeeding (continuous intragastric infusion) a normal diet for 4 days demonstrated a reduction in the fractional rate of protein synthesis. Overfeeding also resulted in a reduction in the ratio of RNA/protein, indicating a relative loss of ribosomes and thus a reduced capacity for protein synthesis.

Verney and Sidransky (182) were concerned with the influence of reducing the carbohydrate content in the diet on hepatic protein synthesis. In this study, the effect of reducing the carbohydrate intake on hepatic protein synthesis was examined using two types of diets. Young rats were force-fed either a complete diet or a diet free of all amino acids and containing adequate or low amounts of carbohydrate for 1 day. The results revealed that in the livers of rats fed both low-carbohydrate-containing diets, there was a decrease in the amount of heavy polyribosomes and in the rate of protein synthesis in comparison with those of rats force-fed comparable diets containing adequate carbohydrates. Thus these studies stress the importance of the carbohydrate content of the diet on the induction of alterations in hepatic protein metabolism.

C. Minerals and Vitamins

1. *Minerals*

Trace minerals have been known to be essential nutrients for the growth and for a wide array of physiological functions of higher animals (183, 184). In general, trace elements are closely associated with a large variety of proteins and enzymes and therefore are known to act as cofactors or coenzymes, being directly involved in the reactions. Furthermore, the important functions of a number of elements, such as magnesium, calcium, zinc, manganese, etc., in RNA and protein biosynthesis have also been well documented by investigators working with cell-free systems (185-187). These include activation of DNA-dependent RNA polymerases, stabilization of polyribosomal structure, and the rate of translation of proteins. It was not surprising therefore to find that deficiencies of trace elements affect physiological processes involving protein synthesis. Among the deficiencies of trace elements, magnesium and zinc deficiencies most commonly appear to cause chemical problems. These deficiencies are perhaps most frequently seen in association with malnutrition and chronic alcoholism. Especially because of this relationship, it may be of interest to review how dietary magnesium or zinc deficiency may affect hepatic protein metabolism.

The effects of magnesium deficiency on RNA and protein metabolism have been studied in the livers of young, growing rats fed ad libitum diets deficient in magnesium compared with rats fed a complete diet (188,189). A significant reduction in the synthesis of total liver and serum albumin proteins was observed in rats fed the magnesium-deficient diet. Synthesis of liver RNA was also reduced in the experimental animals. However magnesium deficiency did not cause any structural or functional changes in either free or membrane-bound polyribosomes. Using a cell-free system, it was found that the changes in the protein synthetic capacity appeared to reside entirely in the supernatant fraction of the livers of magnesium-deficient rats (188). This fraction includes tRNA, amino acid activating enzymes, and peptide chain initiation, elongation, and termination factors. No further studies have been conducted on the changes in the supernatant factors in response to this nutritional treatment.

In studies dealing with the effects of zinc deficiency on liver protein synthesis, Williams and Chesters (190) reported that protein synthesis was not significantly affected in the livers of rats fed a zinc-deficient diet for 5 days compared with pair-fed control rats.

Sandstead et al. (191,192) have investigated some effects of zinc deficiency on liver composition, RNA synthesis, and polyribosomal profiles in rats fed a zinc-deficient diet for 21 days compared with pair-fed controls. Zinc deficiency resulted in lowered amounts of RNA and protein in the livers in comparison with livers of rats fed the control diet. Depressed levels of RNA polymerase activity and decreased incorporation of labeled uridine into RNA of heavier polyribosomes were reported in the livers of zinc-deficient animals when compared with values from zinc-adequate controls (191). Furthermore, zinc deficiency also causes an increase in the monomer-dimer fraction with a corresponding decrease in the heavier polyribosomal material (192). A number of factors may be implicated as causes of these alterations resulting from zinc deficiency. For example, free amino acid pools have been shown to influence polyribosomal dynamics. Indeed, alterations in essential free amino acids in the plasma, such as leucine and phenylalanine, have been observed to occur with zinc deficiency (193). The levels of hepatic RNase activity were unaffected by zinc deficiency in this study, and therefore such an alteration does not account for the polyribosomal disaggregation. These studies suggest that decreased dietary intake as demonstrated in pair-fed controls may account for some of the effects on protein and RNA synthesis caused by the zinc-deficient diet. However studies with both pair-fed and ad libitum-fed control and experimental rats indicate that the changes in the hepatic polyribosomal profiles appear to be altered more in response to zinc deficiency than by inanition alone. Currently available information does not elucidate the mechanism(s) responsible for the effects of zinc-deficiency on protein synthesis in liver.

2. *Vitamins*

Vitamins are essential dietary constituents and their deficiencies, which result in a variety of pathological states, have been well recognized. However very little is known concerning the effects of dietary vitamin deficiencies on liver protein synthesis.

The absence of vitamin A from the diet of animals has been reported to cause retardation of growth and development (194). Therefore it is reasonable to expect that vitamin A deficiency may affect RNA and protein biosynthesis. Johnson et al. (195) have investigated the effect on liver RNA synthesis of vitamin A deficiency and of vitamin A administration to vitamin A-deficient animals. Nuclear RNA synthesis was inhibited in the livers of vitamin A-deficient rats, and the administration of vitamin A to vitamin A-deficient rats was shown to stimulate the rate of synthesis of a liver nuclear RNA fraction with messenger properties. Altered messenger RNA synthesis was also reported in the livers of vitamin A-deficient rats (196). However Tryfiates and Krause (197) observed that the in vitro capacity of liver ribosomes to synthesize protein was enhanced in rats made vitamin A deficient. Further studies demonstrated that this stimulatory effect of vitamin A deficiency was due to alterations in the activities of aminoacyl-tRNA synthetase enzymes in the cytosol fraction (198). Ganguly and his co-workers (199,200) found no changes in the contents of rRNA, tRNA, and poly(A), in

tRNA acceptor activity, and in incorporation of ^{32}P into rRNA and tRNA of the livers of rats fed a vitamin A-deprived diet in comparison with rats fed a normal diet.

The effect of biotin deficiency on liver protein biosynthesis was investigated by Dakshiamurti et al. (201,202). Amino acid incorporation into proteins, both in vivo and in vitro, was significantly decreased with biotin deficiency. Biotin treatment of biotin-deficient rats resulted in stimulation (more than twofold) of amino acid incorporation into protein. Analysis of the products of amino acid incorporation into liver protein revealed that the synthesis of only some proteins were stimulated, indicating a specificity in the stimulation of protein synthesis mediated by administration of the biotin.

These described studies, although in preliminary stages of experimentation, suggest that some vitamins may exert an influence at the transcriptional or translational, or both, levels of control of protein biosynthesis.

IV. CONCLUSIONS

This review has dealt with numerous examples whereby nutritional components were found to act in controlling or regulating hepatic protein synthesis, particularly the role of amino acids. In addition, the influences of selected dietary components, other than amino acids, have been cited. It is now evident that altered dietary intake may affect protein synthesis in the liver via transcriptional, posttranscriptional, and translational controls. Some experimental studies clearly implicate one or more of these mechanisms. Other studies suggest that such alterations are involved, and further experimentation is needed for elucidation of the specific mechanisms involved. As our understanding becomes more complete concerning how nutrition may affect protein synthesis in a specific organ, such as the liver, rational utilization of dietary intake may play an important role in the treatment and prevention of many disease states. Such is indeed the goal of experimentation in nutritional pathology.

ACKNOWLEDGMENTS

Many of the experimental studies reported in this chapter were supported by U.S. Public Health Service Research Grants AM-27339 from the National Institute of Arthritis, Diabetes and Digestive and Kidney Diseases and CA-26557 from the National Cancer Institute.

REFERENCES

1. H. N. Munro, Role of amino acid supply in regulating ribosome function. *Fed. Proc.* 27:1231 (1968).
2. H. N. Munro, A general survey of mechanisms regulating protein metabolism in mammals, in *Mammalian Protein Metabolism*, Vol. 4 (H. N. Munro, ed.). Academic Press, New York, 1970, p. 3.
3. H. Sidransky, Chemical and cellular pathology of experimental acute amino acid deficiency. *Methods Achiev. Exp. Pathol.* 6:1 (1972).
4. H. Sidransky, Regulatory effect of amino acids on polyribosomes and protein synthesis of liver. in *Prog. Liver Dis.* 4:31 (1972).
5. M. A. Rothschild, M. Oratz, and S. S. Schreiber, *Alcohol and Abnormal Protein Synthesis, Biochemical and Clinical.* Pergamon Press, New York, 1975.
6. H. Sidransky, Nutritional disturbances of protein metabolism in the liver. *Am. J. Pathol.* 84:649 (1976).

7. M. A. Zern, D. A. Shafritz, and D. Shields, Hepatic protein synthesis and its regulation, in *The Liver: Biology and Pathobiology* (I. Arias, H. Popper, D. Schachter, and D. A. Shafritz, eds.). Raven Press, New York, 1982, p. 103.

8. S. A. Austin and M. J. Clemens, the regulation of protein synthesis in mammalian cells by amino acid supply. *Biosci. Rep. 1:* 35 (1981).

9. J. M. Kinney and D. H. Elwyn, Protein metabolism and injury. *Ann. Rev. Nutr. 3:*433 (1983).

10. H. Lodish, Translational control of protein synthesis. *Ann. Rev., Biochem. 49:*39 (1976).

11. M. Revel, and Y. Groner, Post-transcriptional and translational controls of gene expression in eukaryotes. *Ann. Rev. Biochem.* 47:1079 (1978).

12. S. Ochoa, and C. de Haro, Regulation of protein synthesis in eukaryotes. *Ann. Rev. Biochem. 48:*549 (1979).

13. F. H. C. Crick, Central dogma of molecular biology. *Nature 227:*561 (1970).

14. J. E. Darnell, Transcriptional units of mRNA production in eukaryotic cells and their DNA viruses. *Prog. Nucleic Acid Res. Mol. Biol. 22:*327 (1979).

15. U. Z. Littauer and H. Soreq, The regulatory role of poly(A) and adjacent 3' sequences in translated RNA. *Prog. Nucleic Acid Res. Mol. Biol. 27:*53 (1982).

16. H. Schwartz and J. E. Darnell, The association of protein with the polyadenylic acid of HeLa cell mRNA: Evidence for a "transport" role of a 75,000 molecular weight polypeptide. *J. Mol. Biol. 104:*833 (1976).

17. H. Noll, T. Staehelin, and F. O. Wettstein, Ribosomal aggregates engaged in protein synthesis: Ergosome breakdown and messenger ribonucleic acid transport. *Nature 198:*632 (1963).

18. P. Siekevitz and G. E. Palade, A cytochemical study on the pancreas of the guinea pig. II. Functional variations in the enzymatic activity of microsomes. *J. Biophys. Biochem. Cytol. 4:* 309 (1958).

19. M. S. C. Birbeck, and E. H. Mercer, Cytology of cells which synthesize protein. *Nature 189:*558 (1961).

20. C. N. Murty and H. Sidransky, Studies on the turnover of mRNA in free and membrane-bound polyribosomes in rat liver. *Biochim. Biophys. Acta 281:*69 (1972).

21. C. N. Murty and H. Sidransky, Studies on the metabolism of ribosomal RNA and protein of free and membrane-bound polyribosomes of rat liver. *Biochim. Biophys. Acta 335:*226 (1974).

22. T. Hallinan, C. N. Murty, and J. H. Grant, Early labeling with glucosamine-^{14}C of granular and agranular endoplasmic reticulum

and free ribosomes from rat liver. *Arch. Biochem. Biophys. 125:* 715 (1968).

23. C. M. Redman, Biosynthesis of serum proteins by free and attached ribosomes of rat liver. *J. Biol. Chem. 244:*4308 (1969).

24. S. J. Hicks, J. W. Drysdale, and H. N. Munro, Preferential synthesis of ferritin and albumin by different populations of liver polysomes. *Science 164:*584 (1969).

25. M. C. Ganoza and C. A. Williams, In vitro synthesis of different categories of specific protein by membrane-bound and free ribosomes. *Proc. Nat. Acad. Sci. USA 63:*1370 (1969).

26. K. S. Bhat and G. Padmanaban, Site of synthesis of cytochrome P-450 in liver. *Biochem. Biophys. Res. Commun. 84:*1 (1978).

27. Y. Fujii-Kuriyama, M. Negishi, R. Mikawa, and Y. Tashiro, Biosynthesis of cytochrome P-450 on membrane-bound ribosomes and its subsequent incorporation into rough and smooth microsomes in rat hepatocytes. *J. Cell. Biol. 81:*510 (1979).

28. A. Bar-Nan, G. Kreibich, M. Adesmik, L. Alterman, M. Negishi, and D. D. Sabatini, Synthesis and insertion of cytochrome P-450 into endoplasmic reticulum membranes. *Proc. Nat. Acad. Sci. USA* 77:965 (1980).

29. D. G. Puro and G. W.Richter, Ferritin synthesis by free and membrane-bound (poly)ribosomes of rat liver. *Proc. Soc. Exp. Biol. Med. 138:*399 (1971).

30. V. Maitra, E. A. Stringer and A. Chandhuri, Initiation factors in protein synthesis. *Ann. Rev. Biochem. 51:*869 (1982).

31. T. E. Webb, D. E. Schumm, and T. Palayoor, Nucleocytoplasmic transport of mRNA. *Cell Nucleus 9:*199 (1981).

32. D. E. Schumm and T. E. Webb, The in vivo equivalence of a cell-free system for RNA processing and transport. *Biochem. Biophys. Res. Commun. 58:*354 (1974).

33. P. S. Agutter, B. McCaldin, and H. J. McArdle, Importances of mammalian nuclear-envelope nucleoside triphosphatase in nucleocytoplasmic transport of ribonucleoproteins. *Biochem. J. 182:*811 (1979).

34. J. R. McDonald and P. S. Agutter, The relationship between polyribonucleotide binding and the phosphorylation and dephosphorylation of nuclear envelope protein. *FEBS Lett. 116:*145 (1980).

35. R. B. Moffett and T. E. Webb, Regulated transport of messenger ribonucleic acid from isolated liver nuclei by nucleic acid binding proteins. *Biochemistry 20:*3253 (1981).

36. D. E. Schumm and T. E. Webb, Effect of adenosine 3':5-monophosphate and guanosine 3':5-monophosphate on RNA release from isolated nuclei. *J. Biol. Chem. 253:*8513 (1978).

37. J. B. Gurdon, Nuclear transplantation and control of gene activity in animal development. *Proc. R. Soc. Lond. B Biol. Sci. 176:*303 (1970).

38. P. L. Paine, Mechanisms of nuclear protein concentration, in *The Nuclear Envelope and the Nuclear Matrix*, Vol. 2 (G. G. Maul, ed.). A. R. Liss, New York, 1982, pp. 75-83.

39. H. N. Munro and C. M. Clark, The incorporation of ^{14}C-labeled precursors into the ribonucleic acid of rat liver following administration of a nutritionally incomplete amino acid mixture. *Biochim. Biophys. Acta 33*:551 (1959).

40. A. Fleck, J. Shepherd, and H. N. Munro, Protein synthesis in rat liver. Influence of amino acids in diet on microsomes and polysomes. *Science 150*:628 (1965).

41. W. H. Wunner, J. Bell, and H. N. Munro, The effect of feeding with a tryptophan-free amino acid mixture on rat liver polysomes and ribosomal ribonucleic acid. *Biochem. J. 101*:417 (1966).

42. T. Staehelin, E. Verney, and H. Sidransky, The influence of nutritional change on polyribosomes of the liver. *Biochem. Biophys. Acta 145*:105 (1967).

43. H. Sidransky, D. S. R. Sarma, M. Bongiorno, and E. Verney, Effect of dietary tryptophan on hepatic polyribosomes and protein synthesis in fasted mice. *J. Biol. Chem. 243*:1123 (1968).

44. A. W. Pronezuk, B. S. Baliga, J. W. Triant, and H. N. Munro, Comparison of the effect of amino acid supply on hepatic polysome profiles in vivo and in vitro. *Biochim. Biophys. Acta 157*: 204 (1968).

45. H. Sidransky, E. Verney, and D. S. R. Sarma, Effect of tryptophan on polyribosomes and protein synthesis in liver. *Am. J. Clin. Nutr. 24*:779 (1971).

46. P. Cammarano, G. Chinali, S. Gaetani, and M. A. Spadoni, Involvement of adrenal steroids in the changes of polysome organization during feeding of imbalanced amino acid diets. *Biochim. Biophys. Acta 155*:302 (1968).

47. M. Oravec and T. L. Sourkes, Inhibition of hepatic protein synthesis by α-methyl-DL-tryptophan in vivo: Further studies on the glyconeogenic action of α-methyl-tryptophan. *Biochemistry 9*:4458 (1970).

48. O. J. Park, L. M. Henderson, and P. B. Swan, Effects of the administration of single amino acids on ribosome aggregation in rat liver. *Proc. Soc. Exp. Biol. Med. 142*:1023 (1973).

49. A. J. F. Jorgensen and A. P. N. Majumdar, Bilateral adrenalectomy: Effect of a single tube-feeding of tryptophan on amino acid incorporation into plasma albumin and fibrinogen in vivo. *Biochem. Med. 13*:231 (1975).

50. C. N. Murty and H. Sidransky, The effect of tryptophan on mRNA in the livers of fasted mice. *Biochim. Biophys. Acta 262*:328 (1972).

51. C. N. Murty, E. Verney, and H. Sidransky, Effect of tryptophan on polyriboadenylic acid and polyadenylic acid-messenger ribonucleic acid in rat liver. *Lab. Invest. 34*:77 (1976).

52. C. N. Murty, E. Verney, and H. Sidransky, The effect of tryptophan on nucleocytoplasmic translocation of RNA in rat liver. *Biochim. Biophys. Acta 474*:117 (1977).
53. C. N. Murty, E. Verney, H. Sidransky, In vivo and in vitro studies on the effect of tryptophan on translocation of RNA from nuclei of rat liver. *Biochem. Med. 22*:98 (1979).
54. S. H. Yap, R. K. Strain, and D. A. Shafritz, Identification of albumin mRPNs in the cytosol of fasting rat liver and influence of tryptophan or a mixture of amino acids. *Biochem. Biophys. Res. Commun. 83*:427 (1978).
55. C. T. Garrett, V. Cairns, C. N. Murty, E. Verney, and H. Sidransky, Effect of tryptophan on informosomal and polyribosome-associated messenger RNA in rat liver. *J. Nutr. 114*:50 (1984).
56. C. N. Murty, E. Verney, and H. Sidransky, Effect of tryptophan on nuclear envelope nucleoside triphosphatase activity in rat liver. *Proc. Soc. Exp. Biol. Med. 163*:155 (1980).
57. C. N. Murty, R. Hornseth, E. Verney, and H. Sidransky, Effect of tryptophan on enzymes and proteins of hepatic nuclear envelopes of rats. *Lab. Invest. 48*:256 (1983).
58. W. H. Wunner, The time sequence of RNA and protein synthesis in cellular compartments following an acute dietary challenge with amino acids mixtures. *Proc. Nutr. Soc. 27*:153 (1967).
59. J. Vesley and A. Cihak, Enhanced DNA-dependent RNA polymerase and RNA synthesis in rat liver nuclei after administration of L-tryptophan. *Biochim. Biophys. Acta 204*:610 (1970).
60. A. R. Henderson, The effect of feeding with a tryptophan-free amino acid mixture on rat liver magnesium ion-activated deoxyribonucleic acid-dependent ribonucleic acid polymerase. *Biochem. J. 120*:205 (1970).
61. M. Oravec and A. Korner, Stimulation of ribosomal and DNA-like RNA synthesis by tryptophan. *Biochim. Biophys. Acta 247*:404 (1971).
62. A. P. N. Majumdar, Effect of tryptophan on hepatic nuclear free and engaged RNA polymerases in young and adult rats. *Experientia 34*:1258 (1978).
63. F. B. Adamstone and H. Spector, Tryptophan deficiency in the rat: Histologic changes induced by force-feeding of an acid hydrolyzed casein diet. *Arch. Pathol. 49*:173 (1950).
64. H. Sidransky and E. Farber, Chemical pathology of acute amino acid deficiencies. I. Morphological changes in immature rats fed threonine-, methionine-, or histidine-devoid diets. *Arch. Pathol. 66*:119 (1958).
65. H. Sidransky and E. Farber, Chemical pathology of acute amino acid deficiencies. II. Biochemical changes in rats fed threonine- or methionine-devoid diets. *Arch. Pathol. 66*:135 (1958).

66. H. Sidransky and T. Baba, Chemical pathology of acute amino acid deficiencies. III. Morphologic and biochemical changes in young rats fed valine- or lysine-devoid diets. *J. Nutr.* 70:463 (1960).

67. H. Sidransky and E. Farber, Sex differences in induction of periportal fatty liver by methionine deficiency in the rat. *Proc. Soc. Exp. Biol. Med.* 98:293 (1958).

68. H. Sidransky and E. Verney, Chemical pathology of acute amino acid deficiencies. VII. Morphologic and biochemical changes in young rats force-fed arginine-, leucine-, isoleucine-, or phenyl-alanine-devoid diets. *Arch. Pathol.* 78:134 (1964).

69. H. Sidransky and E. Verney, Decreased protein synthesis in the skeletal muscle of rats force-fed a threonine diet. *Biochim. Biophys. Acta.* 138:426 (1967).

70. H. Sidransky, T. Staehelin, and E. Verney, Protein synthesis enhanced in the livers of rats force-fed a threonine-devoid diet. *Science* 146:766 (1964).

71. H. Sidransky and E. Verney, Studies on hepatic protein synthesis in rats force-fed a threonine-devoid diet. *Exp. Mol. Pathol.* 13:12 (1970).

72. H. Sidransky, E. Verney, and H. Shinozuka, Alterations in distribution of free and membrane-bound ribosomes in the livers of rats force-fed a threonine-devoid diet. *Exp. Cell Res.* 54:37 (1969).

73. C. N. Murty, E. Verney, and H. Sidransky, Initiation factors in protein synthesis of livers and skeletal muscle of rats force-fed a threonine-devoid diet. *Proc. Soc. Exp. Biol. Med.* 145:74 (1974).

74. H. Sidransky, E. Verney, and C. N. Murty, Studies dealing with hepatic RNA metabolism in rats force-fed a threonine-devoid diet. *J. Nutr.* 110:2514 (1980).

75. H. Sidransky, S. M. Epstein, E. Verney, and R. S. Verbin, The effect of cycloheximide on hepatic RNA synthesis and nucleolar size in rats force-fed a threonine-devoid diet. *J. Nutr.* 106:930 (1976).

76. H. Sidransky, E. Verney, and C. N. Murty, Studies on mRNA of the livers of rats force-fed a threonine-devoid diet or complete diet. *J. Nutr.* 104:726 (1974).

77. H. Sidransky and E. Verney, Skeletal muscle protein metabolism changes in rats force-fed a diet inducing an experimental kwashiorkor-like model. *Am. J. Clin. Nutr.* 23:1154 (1970).

78. E. Verney and H. Sidransky, Chemical pathology of acute amino acid deficiencies: Reversibility of the pathological changes in rats force-fed a threonine-devoid diet. *J. Nutr.* 101:1727 (1971).

79. H. Sidransky and E. Verney, Influence of quantity of diet on

protein synthesis of rats force-fed a threonine-devoid diet. *Lab. Invest. 32*:65 (1975).

80. H. Sidransky and E. Verney, Enhanced hepatic protein synthesis in rats force-fed a tryptophan-devoid diet. *Proc. Soc. Exp. Biol. Med. 135*:618 (1970).
81. H. Sidransky and E. Verney, Chemical pathology of acute amino acid deficiencies. Effects of single and multiple amino acid deficiencies in the rat. *Nutr. Rep. Int. 13*:367 (1976).
82. H. Patrick and L. K. Bennington, Biochemical changes in the rat produced by feeding a tryptophan-deficient ration. *Proc. W. Va. Acad. Sci. 41*:161 (1969).
83. M. E. Nimni and L. A. Bavetta, Dietary composition and tissue protein synthesis. I. Effect of tryptophan deficiency. *Proc. Soc. Exp. Biol. Med. 108*:38 (1961).
84. R. Bocker, I. K. Jones, and W. Kersten, Metabolism of protein and RNA in liver of rats deprived of tryptophan. *J. Nutr. 107*: 1737 (1977).
85. A. Yoshida, P. M. B. Leung, Q. R. Rogers, and A. E. Harper, Effect of amino acid imbalance on the fate of the limiting amino acid. *J. Nutr. 89*:80 (1966).
86. N. J. Benevenga, A. E. Harper, and Q. R. Rogers, Effect of an amino acid imbalance on the metabolism of the most-limiting amino acid in the rat. *J. Nutr. 95*:434 (1968).
87. A. W. Pronczuk, Q. R. Rogers, and H. N. Munro, Liver polysome patterns of rats fed amino acid imbalanced diets. *J. Nutr. 100*:1249 (1970).
88. C. C. Y. Ip and A. E. Harper, Liver polyribosome profile and protein synthesis in rats fed a threonine-imbalanced diet. *J. Nutr. 104*:252 (1974).
89. C. Shaw and L. C. Fillios, RNA polymerase activities and other aspects of hepatic protein synthesis during early protein depletion in the rat. *J. Nutr. 96*:327 (1968).
90. R. W. Wannemacher, Jr., C. F. Wannemacher, and M. B. Yatvin, Amino acid regulation of synthesis of ribonucleic acid and protein in the liver of rats. *Biochem. J. 124*:385 (1971).
91. R. Shields and A. Korner, Regulation of mammalian ribosome synthesis by amino acids. *Biochim. Biophys. Acta 204*:521 (1970).
92. A. J. Clark and M. Jacob, Effect of diet on RNA polymerase activity in rats. *Life Sci. 11*:1147 (1972).
93. P. Mandel and C. Quirin-Stricker, Effect of protein deprivation on soluble and aggregate RNA polymerase in rat liver. *Life Sci. 6*:1299 (1967).
94. A. von der Decken and G. M. Andersson, Effect of protein intake on DNA-dependent RNA polymerase activity and protein synthesis in vitro in rat liver and brain. *Nutr. Rep. Int. 5*: 413 (1972).

95. G. M. Andersson and A. von der Decken, Deoxyribonucleic acid-dependent ribonucleic acid polymerase activity in rat liver after protein restriction. *Biochem. J. 148*:49 (1975).

96. C. G. Lewis and M. Winick, Studies on ribosomal RNA synthesis in vivo in rat liver during short-term protein malnutrition. *J. Nutr. 108*:329 (1978).

97. C. O. Enwonwu and H. N. Munro, Rate of RNA turnover in rat liver in relation to intake of protein. *Arch. Biochem. Biophys. 138*:532 (1970).

98. H. Sidransky and E. Verney, Studies on hepatic polyribosomes and protein synthesis in rats forced-fed a protein-free diet. *J. Nutr. 101*:1153 (1971).

99. A. Sato, K. Noda, and Y. Natori, The effect of protein depletion on the rate of protein synthesis in rat liver. *Biochim. Biophys. Acta 561*:475 (1979).

100. H. Yokogoshi and A. Yoshida, Effect of supplementation of methionine and threonine on hepatic polyribosome profile in rats meal-fed by a protein-free diet. *J. Nutr. 109*:148 (1979).

101. P. Quartey-Papafio, P. J. Garlic, and V. M. Pain, Effect of dietary protein on liver protein synthesis. *Biochem. Soc. Trans. 8*:357 (1980).

102. V. M. Pain, M. J. Clemens, and P. J. Garlic, The effect of dietary protein deficiency on albumin synthesis and on the concentration of active albumin messenger ribonucleic acid. *Biochem. J. 172*:129 (1978).

103. M. A. McNurlan and P. J. Garlic, Protein synthesis in liver and small intestine in protein deprivation and diabetes. *Am. J. Physiol. 241*:E238 (1981).

104. E. H. Morgan and T. Peters, Jr., The biosynthesis of rat serum albumin. V. Effect of protein depletion and refeeding on albumin and transferrin synthesis. *J. Biol. Chem. 246*:3500 (1971).

105. U. Stenram, Interferometric determination of the ribose nucleic acid concentration in liver nucleoli of protein-fed and protein-deprived rats. *Exp. Cell. Res. 15*:174 (1958).

106. T. Kawada, T. Fujisawa, K. Imai, and K. Ogato, Effects of protein deficiency on the biosynthesis and degradation of ribosomal RNA in rat liver. *J. Biochem. 81*:143 (1977).

107. R. P. Bailey, M. J. Vrooman, Y. Sawai, K. Tsukada, J. Short, and I. Lieberman, Amino acids and control of nucleolar size, the activity of RNA polymerase I and DNA synthesis in liver. *Proc. Nat. Acad. Sci. USA 73*:3201 (1976).

108. H. Nordgren and U. Stenram, Decreased half-life of the RNA of free and membrane-bound ribosomes. *Hoppe-Seyler's Z. Physiol. Chem. 353*:1832 (1972).

109. P. Feigelson, M. Feigelson, and O. Greengard, Comparison of the mechanisms of hormonal and substrate induction of rat liver tryptophan pyrrolase. *Recent Prog. Horm. Res. 18*:491 (1962).

110. S. R. Wagle, The influence of growth hormone, cortisol and insulin on the incorporation of amino acids into protein. *Arch. Biochem.* 102:373 (1963).

111. H. A. Leon, Early effects of corticosterone on amino acid incorporation by rat liver systems subsequent to in vivo injection. *Endocrinology* 78:481 (1966).

112. H. Sidransky and D. S. Wagle, Hepatic protein synthesis in rats force-fed a threonine-devoid diet and treated with cortisone acetate or threonine, *Lab. Invest.* 20:364 (1969).

113. C. O. Enwonwu and H. N. Munro, Changes in liver polysome patterns following administration of hydrocortisone and actinomycin D. *Biochim. Biophys. Acta* 238:264 (1971).

114. E. Verney and H. Sidransky, Further enhancement by tryptophan of hepatic protein synthesis stimulated by phenobarbital or cortisone acetate. *Proc. Soc. Exp. Biol. Med.* 158:245 (1978).

115. A. Korner, The effect of hypophysectomy of the rat and of treatment with growth hormone on the incorporation of amino acids into rat liver proteins in a cell-free system. *Biochem. J.* 73:61 (1959).

116. A. Korner, Regulation of the rate of synthesis of messenger ribonucleic acid by growth hormones. *Biochem. J.* 92:449 (1964).

117. A. K. Roy and D. J. Dowbenko, Role of growth hormone in the multihormonal regulation of messenger RNA for α_{2u} globulin in the liver of hypophysectomized rats. *Biochemistry* 16:3918 (1977).

118. G. H. Keller and J. M. Taylor, Effect of hypophysectomy and growth hormone treatment on albumin mRNA levels in the rat liver. *J. Biol. Chem.* 254:276 (1978).

119. W. D. Wicks, Regulation of protein synthesis by cAMP. *Adv. Cyclic Nucleotide Res.* 4:335 (1974).

120. M. S. Ayuso-Parrilla, A. Martin-Requero, J. P. Diaz, and R. Parrilla, Role of glucagon on the control of hepatic protein synthesis and degradation in the rat in vivo. *J. Biol. Chem.* 251:7785 (1976).

121. D. E. Peavy, J. M. Taylor, and L. S. Jefferson, Correlation of albumin production rates and albumin mRNA levels in livers of normal, diabetic and insulin-treated diabetic rats. *Proc. Nat. Acad. Sci. USA* 75:5879 (1978).

122. L. S. Jefferson, Role of insulin in the regulation of protein synthesis. *Diabetes* 29:487 (1980).

123. J. R. Tata and C. C. Widnell, Ribonucleic synthesis during the early action of thyroid hormones. *Biochem. J.* 98:604 (1966).

124. R. W. Mathews, W. A. Oronsky, and A. E. V. Haschemeyer, Effect of thyroid hormone on polypeptide chain assembly kinetics in liver protein synthesis in vivo. *J. Biol. Chem.* 248:1329 (1973).

125. W. J. Carter, F. H. Faas, and J. O. Wynn, Stimulation of peptide elongation by thyroxine. *Biochem. J. 156*:713 (1976).

126. W. J. Carter and F. H. Faas, Early stimulation of rat liver microsomal protein synthesis after tri-iodothyronine injection in vivo.
Biochem. J. 182:651 (1979).

127. D. T. Kurtz, A. E. Sippel, R. Ansah-Yiadon, and P. Feigelson,
Effect of sex hormones on the level of messenger RNA for the rat
hepatic protein α_{2u} globulin. *J. Biol. Chem. 251*:3594 (1974).

128. H. N. Munro and D. Mukerji, The mechanism by which administration of individual amino acids cause changes in ribonucleic
acid metabolism of the liver. *Biochem. J. 82*:520 (1962).

129. J. C. Floyd, S. S. Fajans, J. W. Conn, R. P. Knopf, and
J. Rull, Stimulation of insulin secretion by amino acids. *J.
Clin. Invest. 45*:1487 (1966).

130. J. K. Tews, N. A. Woodcock, and A. E. Harper, Effect of
protein intake on amino acid transport and adenosine 3',5'-
monophosphate content in rat liver. *J. Nutr. 102*:409 (1972).

131. A. J. F. Jorgensen and A. P. N. Majumdar, Bilateral adrenalectomy: Effect of tryptophan force-feeding on amino acid incorporation into ferritin, transferrin, and mixed proteins of liver,
brain and kidneys in vivo. *Biochem. Med. 16*:37 (1976).

132. J. Vesely and A. Cihak, Effect of L-tryptophan on DNA-dependent RNA polymerase in rat liver nuclei. *Collect. Czech.
Chem. Commun. 35*:1892 (1970).

133. H. Sidransky, Regulatory effect of amino acids on polyribosomes
and protein synthesis of liver. *Prog. Liver Dis. 4*:31 (1972).

134. D. S. Wagle and H. Sidransky, Effect of adrenalectomy on
pathologic changes in rats force-fed a threonine devoid diet.
Metabolism. 14:932 (1965).

135. H. Sidransky and E. Verney, Effect of hypophysectomy on
pathologic changes in rats force-fed a threonine devoid diet.
J. Nutr. 96:28 (1968).

136. C. C. Y. Ip and A. E. Harper, Effect of threonine supplementation on hepatic polyribosome pattern and protein synthesis of
rats fed a threonine-deficient diet. *Biochim. Biophys. Acta 331*:
251 (1973).

137. R. C. Feddhoff, J. M. Taylor, and L. S. Jefferson, Synthesis
and secretion of rat albumin in vivo, in perfused liver, and in
isolated hepatocytes. *J. Biol. Chem. 252*:3611 (1977).

138. E. H. Morgan and T. Peters, Jr., The biosynthesis of rat
serum albumin. V. Effect of protein depletion and refeeding
on albumin and transferrin synthesis. *J. Biol. Chem. 246*:
3500 (1971).

139. L. S. Jefferson and A. Korner, Influence of amino acid supply
on ribosomes and protein synthesis of perfused rat liver.
Biochem. J. 111:703 (1969).

140. M. VanDenBorre and T. E. Webb, Perfusate composition and stability of polyribosomes in perfused liver. *Life Sci.* *11*:347 (1972).

141. L. Kelman, S. J. Saunders, S. Wicht, L. Frith, A. Corrigall, R. E. Kirsch, and J. Terblanche, The effects of amino acids on albumin synthesis by the perfused rat liver. *Biochem. J.* *129*:805 (1972).

142. K. H. Woodside and G. E. Mortimore, Suppression of protein turnover by amino acids in the perfused rat liver. *J. Biol. Chem.* *247*:6474 (1972).

143. A. S. Tavill, A. G. East, E. G. Black, D. Nadkarni, and R. Hoffenberg, Regulatory factors in the synthesis of plasma proteins by the isolated perfused rat liver. *Ciba Found. Symp.* *9*: 155 (1973).

144. D. E. Peavy and R. J. Hansen, Lack of effect of amino acid concentration on protein synthesis in the perfused rat liver. *Biochem. J.* *160*:797 (1976).

145. K. E. Flaim, D. E. Peavy, W. V. Everson, and L. S. Jefferson, The role of amino acids in the regulation of protein synthesis in perfused rat liver. I. Reduction in rates of synthesis resulting from amino acid deprivation and recovery during flow through perfusion. *J. Biol. Chem.* *257*:2932 (1982).

146. K. E. Flaim, W. S. Liao, D. E. Peavy, J. M. Taylor, and L. S. Jefferson, The role of amino acids in the regulation of protein synthesis in perfused rat liver. II. Effects of amino acid deficiency on peptide chain initiation, polysomal aggregation, and distribution of albumin mRNA. *J. Biol. Chem.* *257*:2939 (1982).

147. G. E. Mortimore and C. E. London, Inhibition by insulin of valine turnover in liver. Evidence for a general control of proteolysis. *J. Biol. Chem.* *245*:2375 (1970).

148. C. M. Schworer and G. E. Mortimore, Glucagon-induced autophagy and proteolysis in rat liver: Mediation by selective deprivation of intracellular amino acids. *Proc. Nat. Acad. Sci. USA* *76*:3169 (1979).

149. E. McGown, A. G. Richardson, L. M. Henderson, and P. B. Swan, Effect of amino acids on ribosome aggregation and protein synthesis in perfused rat liver. *J. Nutr.* *103*:109 (1973).

150. L. L. Miller and E. E. Griffin, Effects on net plasma protein synthesis of removal of L-tryptophan or L-threonine from a complete amino acid mixture: Studies in the isolated perfused rat liver. *Am. J. Clin. Nutr.* *24*:718 (1971).

151. M. A. Rothschild, M. Oratz, J. Mongelli, L. Fishman, and S. S. Schreiber, Amino acid regulation of albumin synthesis. *J. Nutr.* *98*:395 (1969).

152. M. Oratz, M. A. Rothschild, and S. S. Schreiber, Alcohol, amino acids, and albumin synthesis. II. Alcohol inhibition of

albumin synthesis reversed by arginine and spermine. *Gastro-enterology 71*:123 (1976).

153. R. B. Howard, A. K. Christensen, F. A. Gibbs, and L. A. Persch, The enzymatic preparation of isolated intact parenchymal cells from rat liver. *J. Cell Biol. 35*:675 (1967).

154. M. N. Berry and D. S. Friend, High yield preparation of isolated rat liver parenchymal cells. *J. Cell Biol. 43*:506 (1969).

155. P. O. Seglen, Preparation of rat liver cells. *Meth. Cell Biol. 13*:29 (1976)

156. E. M. Suolinna, Isolation and culture of liver cells and their use in the biochemical research of xenobiotics. *Med. Biol. 60*: 237 (1982).

157. G. Schreiber and M. Schreiber, Protein synthesis in single cell suspensions from rat liver. *J. Biol. Chem. 247*:6340 (1972).

158. A. G. Grant and E. G. Black, Polyribosome aggregation in isolated rat-liver cells. *Eur. J. Biochem. 47*:397 (1974).

159. K. Weigland, H. Wernze, and C. Falge, Synthesis of angiotensinogen by isolated rat liver cells and its regulation in comparison to serum albumin. *Biochem. Biophys. Res. Commun. 75*: 102 (1977).

160. C. F. A. Van Bezooijen, T. Grell, and D. L. Knook, The effect of age on protein synthesis by isolated liver parenchymal cells. *Mech. Ageing Dev. 6*:293 (1977).

161. P. O. Seglen, Incorporation of radioactive amino acids into protein in isolated rat hepatocytes. *Biochim. Biophys. Acta 442*: 391 (1976).

162. P. O. Seglen and A. E. Solheim, Effects of aminooxyacetate, alanine and other amino acids on protein synthesis in isolated rat hepatocytes. *Biochim. Biophys. Acta 520*:630 (1978).

163. P. O. Seglen, A. E. Solheim, B. Grinde, P. B. Gordon, P. E. Schwarze, R. Gjessing, and A. Poli, Amino acid control of protein synthesis and degradation in isolated rat hepatocytes. *Ann. N.Y. Acad. Sci. 349*:1 (1980).

164. P. E. Schwarze, A. E. Solheim and P. O. Seglen, Amino acid energy requirement for rat hepatocytes in primary culture. *In Vitro 18*:43 (1982).

165. H. E. S. J. Hensgens and A. J. Meijer, The interrelationships between ureogenesis and protein synthesis in isolated rat-liver cells. *Biochim. Biophys. Acta 582*:525 (1979).

166. T. Girbes, A. Susin, M. S. Ayuso, and R. Parilla, Acute effects of ethanol in the control of protein synthesis in isolated rat liver cells. *Arch. Biochem. Biophys. 226*:37 (1983).

167. M. S. Ayuso-Parrilla, A. Martin-Requero, J. Perez-Diaz, and R. Parrilla, Role of glucagon on the control of hepatic protein synthesis in the rat in vivo. *J. Biol. Chem. 251*:7785 (1976).

168. J. Hoshino, V. Kuhne, G. Studinger, and H. Kroger, In vitro protein synthesizing activity of rat liver as influenced by a

physiological dose of cortisone and dibutyrl cAMP. *Biochem. Biophys. Res. Commun. 74*:663 (1977).

169. S. R. Wagle and L. Sampson, Studies on the differential response to insulin on the stimulation of amino acid incorporation into protein in isolated hepatocytes containing different levels of glycogen. *Biochem. Biophys. Res. Commun. 64*:72 (1975).

170. A. Klaipongpan, D. Bloxham, and M. Akhtar, Enhancement of the anti-anabolic response to adenosine 3':5'-cyclic monophosphate during development. The inhibition of hepatic protein synthesis. *Biochem. J. 168*:271 (1977).

171. L. J. Crane and D. L. Miller, Plasma protein synthesis by isolated rat hepatocytes. *J. Cell. Biol. 72*:11 (1977).

172. T. E. Webb, G. Blobel, and V. R. Potter, Polyribosomes in rat tissues. III. The response of the polyribosome pattern of rat liver to physiologic stress. *Cancer Res. 26*:253 (1966).

173. J. S. Wittman, K. L. Lee, and O. N. Miller, Dietary and hormonal influences on rat liver polysome profiles: Fat, glucose, and insulin. *Biochim. Biophys. Acta 174*:536 (1969).

174. J. Dich and I. C. Tonneson, Effects of ethanol, nutritional status, and composition of the incubation medium on protein synthesis in isolated rat liver parenchymal cells. *Arch. Biochem. Biophys. 204*:640 (1980).

175. H. A. Krebs, N. W. Cornell, P. Lund, and R. Hems, Isolated liver cells as experimental material, in *Regulation of Hepatic Metabolism* (F. Lundquist and N. Tystrup, eds.), Academic Press, New York, 1974, pp. 726-750.

176. K. E. Crow, N. W. Cornell, and R. L. Veech, Lactate-stimulated ethanol oxidation in isolated rat hepatocytes. *Biochem. J. 172*: 29 (1978).

177. N. B. Ruderman, K. C. Richards, V. Valles de Bourges, and A. L. Jones, Regulation of production and release of lipoprotein by the perfused rat liver. *J. Lipid Res. 9*:613 (1968).

178. L. G. Alaindor, R. Infante, C. Solar-Argilaga, A. Raisonner, J. Polonovski, and J. Caroli, Induction of hepatic synthesis of β-lipoproteins by high concentrations of fatty acids. Effect of actinomycin D. *Biochim. Biophys. Acta 210*:483 (1970).

179. A. Raisonnier, M. E. Bouma, C. Salvat, and R. Infante, Ribonucleic acid synthesis in rat liver during fatty acid stimulated secretion of very low density lipoproteins. *Biochimie 60*:743 (1978).

180. G. D. Pfeifer and B. Szepesi, Changes in hepatic RNA synthesis in the starve-refeed response of the rat. *J. Nutr. 104*:1178 (1974).

181. Z. Glick, M. A. McNurlan, and P. J. Garlick, Protein synthesis rate in liver and muscle of rats following four days of overfeeding. *J. Nutr. 112*:391 (1982).

182. E. Verney and H. Sidransky, Protein synthesis in rats forcefed for one day purified diets containing complete, threonine-

devoid or no amino acids and adequate or low carbohydrate. *Exp. Mol. Pathol. 22*:85 (1975).

183.   T. K. Li and B. L. Vallee, The biochemical and nutritional roles of other trace elements, in *Modern Nutrition in Health and Disease,* 6th ed. (R. S. Goodhart and M. E. Shils, eds.), 1980, pp. 409-441.

184.   S. K. Czarnecki and D. Kritchevsky, Trace elements, in *Human Nutrition, Nutrition and the Adult: Micronutrients.* Vol. 3B (R. B. Alfin-Slater and D. Kritchevsky, eds.), 1980, pp. 319-350.

185.   P. Lengyel, On peptide chain initiation. *Mol. Genet. 2*:193 (1967).

186.   M. Jost, N. Shoemaker, and H. Noll, Stepwise reconstruction of a ternary complex in protein synthesis. *Nature 218*:1217 (1968).

187.   V. Weser, S. Seeber, and P. Warnecke, Reactivity of Zn^{2+} on nuclear DNA and RNA biosynthesis of regenerating liver. *Biochim. Biophys. Acta 179*:422 (1969).

188.   R. Schwartz, N. A. Woodcock, J. Blakely, F. L. Wang, and E. A. Khairallah, Effect of magnesium deficiency in growing rats on synthesis of liver proteins and serum albumin. *J. Nutr. 100*:123 (1970).

189.   F. J. Zieve, K. A. Freude, and L. Zieve, Effect of magnesium deficiency on protein and nucleic acid synthesis in vivo. *J. Nutr. 107*:2178 (1977).

190.   R. B. Williams and J. K. Chesters, The effects of early zinc deficiency on DNA and protein synthesis, in the rat. *Br. J. Nutr. 24*:1053 (1970).

191.   M. W. Terhune and H. H. Sandstead, Decrease RNA polymerase activity in mammalian zinc deficiency. *Science 177*:68 (1972).

192.   G. J. Fosmire, M. N. Fosmire, and H. H. Sandstead, Zinc deficiency in the weanling rat: Effects on liver composition and polyribosomal profiles. *J. Nutr. 106*:1152 (1976).

193.   P. R. Griffith and J. C. Alexander, Effect of zinc deficiency on amino acid metabolism of the rat. *Nutr. Rep. Int. 6*:9 (1972).

194.   K. C. Hays, On the pathophysiology of vitamin A deficiency. *Nutr. Rev. 29*:3 (1971).

195.   C. Johnson, M. Kennedy, and N. Chiba, Vitamin A and nuclear RNA synthesis. *Am. J. Clin. Nutr. 22*:1048 (1969).

196.   G. P. Tryfiates and R. F. Krause, Altered messenger RNA synthesis in vitamin A deficient rat liver. *Life. Sci. 10*:1097 (1971).

197.   G. P. Tryfiates and R. F. Krause, Effect of vitamin A deficiency on the protein synthetic activity of rat liver ribosomes. *Proc. Soc. Exp. Biol. Med. 136*:946 (1971).

198. G. P. Tryfiates, R. F. Krause, and J. K. Shuler, Transfer ribonucleic acid of vitamin A-deficient rats. *Am. J. Clin. Nutr. 26*: 41 (1973).

199. M. Jayaram and J. Ganguly, Effect of vitamin A nutritional status on the ribonucleic acids of liver, intestinal mucosa and testes of rats. *Biochem. J. 166*:339 (1977).

200. J. Ganguly, M. R. S. Rao, S. K. Murty, and K. Sarada, Systemic mode of action of vitamin A. *Vit. Horm. 38*:1 (1980).

201. K. Dakshinamurti and S. Litvak, Biotin and protein synthesis in rat liver. *J. Biol. Chem. 245*:5600 (1970).

202. R. L. Boeckx and K. Dakshinamurti, Biotin-mediated protein biosynthesis. *Biochem. J. 140*:549 (1974).

3

Nutritionally Induced Pancreatic Disease

Daniel S. Longnecker

Dartmouth Medical School
Hanover, New Hampshire

I. INTRODUCTION

The pancreas plays a central role in nutrition and metabolism by
virtue of the production of digestive enzymes by the exocrine tissue
and the secretion of polypeptide hormones by the islets. Pancreatic
digestive enzymes include trypsin, chymotrypsin, carboxypeptidase,
elastase, lipase, RNase, DNase, and amylase (1). A deficiency of
these enzymes in the small intestine results in digestive failure and
malabsorption, which can lead to systemic nutritional disturbances affecting the growth and function of other organs and tissues. The
effect of diseases of the pancreas on absorption and nutrition has
been fully reviewed elsewhere (2), and a discussion of those malabsorption states that are secondary to pancreatic diseases such as
cystic fibrosis and chronic pancreatitis will not be included here.
Many abnormalities of the pancreas remain clinically "silent" because
of the remote internal location of the pancreas and because the organ
possesses a large reserve capacity with regard to both exocrine and
endocrine functions (3). Secretion by the exocrine pancreas must be
reduced by more than 90% before there is clinical evidence of malabsorption, and in subtotal pancreatectomy, preservation of about 10%
of the gland prevents diabetes.
   The teleology of allowing an organ that is influential in maintaining
normal nutrition to remain relatively immune to nutritional deficiencies
is obvious. Any other arrangement would place the maintenance of

normal nutrition in a state of double jeopardy. Thus with the exception of kwashiorkor, there are few clinically recognized syndromes of pancreatic dysfunction attributable to nutritional factors in humans. The effects of nutritional factors on the pancreas are more easily recognized in the laboratory where attention can be focused on a single factor under controlled conditions. The effects of selenium and copper deficiencies on the pancreas were found by this approach, but there are no clinically recognized counterparts in humans.

II. EXOCRINE PANCREAS

A. Protein-Calorie Malnutrition

The recycling of amino acids probably represents a subtle but important function of the pancreas. Digestive enzymes are synthesized and secreted even during periods of fasting. It is reasonable to assume that these enzyme proteins are digested in the small intestine and their amino acids reabsorbed. The amount of zymogen stored in the pancreas varies greatly with cycles of feeding and fasting. This seems to represent an ideal system for short-term storage and release of amino acids. The normal pancreas constitutes about 1% of the body's total weight. More than 80% of its mass is acinar tissue, which provides the basis for storage of an estimated 20 to 30 g of zymogen protein in the adult. The impact of the loss of exocrine pancreatic function on normal amino acid metabolism is not specifically defined.

The impact of protein deprivation (4) and single amino acid deficiency (5) on the pancreas has been demonstrated in experimental animals, thus providing experimental models for kwashiorkor (6). Pancreatic atrophy has been described in rats fed diets that were deficient in histidine, isoleucine, leucine, lysine, phenylalanine, tryptophan, threonine, or valine (5). Most of these studies were for 1 week or less, and the principal histological change was a reduction in the size of acinar cells, showing a reduced zymogen content. The pathological changes were similar regardless of which amino acid was restricted. Atrophy did not develop with deficiencies of methionine or arginine (5). Similar changes were noted in rats that were force-fed several plant proteins of poor quality but not in rats that were allowed to eat these diets ad libitum (7). This difference between forced-feeding and ad libitum consumption was also observed in studies of rats fed diets that were deficient in single amino acids (5). In several instances edema was noted in the pancreas of rats force-fed amino acid-deficient diets. This finding suggests that there was mild pancreatitis, although infiltration by leukocytes was not described. Studies of longer duration have utilized ad libitum feeding. Ultrastructural studies in lysine-deficient rats showed reduced zymogen, increased intracellular lipid deposits, and mitochondrial abnormalities during an experiment lasting 95 days (8). In the aggregate, these studies demonstrate the sensitivity of the acinar cell to protein deficiency and amino acid imbalances.

1. The Pancreas in Kwashiorkor

The clinical syndrome of kwashiorkor is complex and reflects the systemic effects of prolonged consumption of a protein-deficient diet (9), although it has been pointed out that the dietary deficiency is more correctly described as protein deficiency with a relatively high-carbohydrate intake (6), as contrasted with complete starvation. The pancreas is affected early and the loss of exocrine pancreatic function contributes to malabsorption, which further compounds the nutritional problem. Scrimshaw and Béhar state that lipase, trypsin, and amylase activities in the duodenal secretions are lowered almost to zero in the acute stage of kwashiorkor (10). Veghelyi et al., in describing observations of infants fed diets lacking animal protein, report that atrophic changes were present in pancreatic acinar cells as early as 17 days after the beginning of the deficient diet (11). Patients with kwashiorkor suffer from steatorrhea, with stools that are bulky and soft and contain considerable quantities of undigested food (10). The activities of the pancreatic enzymes apparently return rapidly to normal when an adequate diet is given to patients in an acute phase of kwashiorkor; excellent recovery was reported in adults with a history of prolonged protein-calorie malnutrition when a high-protein diet was consumed for 12 to 14 weeks (12). This sequence in the clinical picture reflects the dependence of the pancreas on dietary amino acids to maintain the high rate of protein synthesis required for production of the pancreatic exocrine digestive enzymes.

Minor differences in the clinical syndrome have been described in various geographical locations. It has been suggested that this might reflect relatively greater degrees of deficiency of specific amino acids in the diets in various regions and population groups, but the possibility that there are deficiencies of other trace elements should be considered as well. The disease develops following weaning and has its peak prevalence between the ages of 1 and 5; it can also develop in children of all ages and in adults who consume an inadequate amount of dietary protein.

The pathological changes in the pancreas include atrophy of acinar cells with a striking decrease in their zymogen content at an early stage and reflect the functional changes previously described. A single ultrastructural study has been carried out in infants dying of kwashiorkor, and in this study autophagy was a conspicuous process in the acinar cells during the acute phase (13). After 1 month of protein deficiency, the acini are observed to be markedly dilated and lined by flat epithelium. When kwashiorkor is prolonged, there is perilobular and periductal fibrosis. In advanced cases, the acinar tissue virtually disappears and

may be replaced by fibrous tissue. Davies and Brist describe an advanced case in a 24-year-old man in whom the pancreas had been almost
completely replaced by fibrous tissue (14). Islets may survive for
several months up to several years, but the number of islets is reported
to be decreased in late cases. Calcification of duct contents has been
found in some cases (fewer than half)—a change regarded as a characteristic part of the process (15).

2. *Nutritional Pancreatitis*

A subset of patients with protein-calorie malnutrition has been identified
by Pitchumoni and others under the diagnosis of "nutritional pancreatitis"
(16). The group described by Shaper (15) has been included in this
category. The characteristic difference between this group and patients
with typical kwashiorkor is the presence of extensive intrapancreatic
calcification (calculi) and diabetes mellitus. Pancreatic atrophy and fibrosis were also present. More than 400 such patients have been studied
in the state of Kerala, India, whereas only a few cases have been reported from other parts of India. The peak prevalence occurred between the ages of 16 and 20, usually in patients with a history of epi-
sodes of upper-abdominal pain during childhood. Chronic alcoholism
was excluded as a cause in the adolescent patients, but the disease was
not attributed solely to protein malnutrition. It was suggested that some
toxin, perhaps of dietary origin, initiates the process, based on the
view that pancreatic injury by toxic agents may be accentuated by protein deficiency. While there is insufficient data regarding the etiology
and pathogenesis of nutritional pancreatitis to ascertain that this disease
is distinct from severe, advanced-stage kwashiorkor complicated by
diabetes, the focused geographical distribution and characteristic clinical picture are consistent with the view that it may be a separate, but
related, disease.

B. Selenium Deficiency

Pancreatic atrophy and fibrosis have been described in chicks maintained
on a selenium-deficient diet (17). Whereas several lesions noted in
chicks maintained on such diets were attributable to vitamin E deficiency
resulting from fat malabsorption, pancreatic lesions developed even
when the model was manipulated to prevent this vitamin deficiency.
Selenomethionine was highly effective in preventing such pancreatic
atrophy, presumably because of its great affinity for the pancreas.
    A functional counterpart consisted of progressive loss of exocrine
pancreatic secretion with decreased production of lipase, trypsin, and
chymotrypsin. Although food intake and body weight both decreased,
forced-feeding experiments have supported the view that this loss of
pancreatic function is specifically the result of selenium deficiency.
The biochemical mechanism of the essential role of selenium in the chick

pancreas remains to be elucidated. No corresponding clinical deficiency state is known in man, but it is of interest that low blood selenium levels were reported in children with kwashiorkor (18). This raises the possibility that selenium deficiency may play a role in at least some cases of kwashiorkor, because of similarities in the pancreatic lesions.

C. Copper Deficiency

Atrophy of acinar tissue and its replacement by fat has been reported in rats that were maintained on copper-free diets for prolonged periods. The rate of development of clinically significant copper deficiency was accelerated by feeding a chelating agent such as penicillamine to speed the depletion of tissue stores (19). Islet and ductal tissues were little affected, and the atrophic changes were not accompanied by significant inflammation or fibrosis. The periinsular acinar cells were relatively resistent to the injury and survived longer than the bulk of acinar cells. The biochemical basis of the effect of copper deficiency in the pancreas is not known, but deficient activity of one or more metallo-enzymes seems likely. Severe depletion of cytochrome oxidase activity, a copper-dependent enzyme, was reported in the acinar tissue of copper-deficient rats. Ultrastructural studies in these rats revealed enlargement, abnormal configuration, and degenerative changes in the mitochondria in acinar cells. Vesiculation of rough endoplasmic reticulum, autophagy, and reduced zymogen content were also noted (20).

D. Other Trace Elements and Vitamins

The effect of dietary zinc deficiency on the activity of pancreatic carboxypeptidase, an enzyme which contains zinc, has been investigated in rats. Carboxypeptidase activity was decreased to about 50% of control levels in the pancreas of deficient rats while trypsin and chymotrypsin activity remained normal (21).

Although it would seem inevitable that vitamin deficiencies affect pancreatic function, there are few reports of pancreatic dysfunction or lesions associated with specific vitamin deficiencies. Squamous metaplasia of the interlobular pancreatic ductal epithelium has been described in rats maintained on a vitamin A-free diet for several months (22); squamous metaplasia also has been found in the pancreatic ducts of patients with cystic fibrosis, some of whom were known to be deficient in vitamin A (23).

E. Choline Deficiency

A recently developed model of acute hemorrhagic pancreatitis in mice implicates choline deficiency as an important factor in predisposing the pancreas to catastrophic injury (24). It has long been known that ethionine is toxic for the pancreas in rats. While injections of large doses (1 g/kg) induced pancreatitis (25), injections or feeding of lower levels caused severe, albeit reversible, pancreatic atrophy [26, 27,28]. A conspicuous degeneration of acinar cells occurred in such rats. Subsequently, Lombardi has shown that female mice fed a choline-deficient semipurified diet containing ethionine develop a fatal acute hemorrhagic pancreatitis over a 4- to 5-day course, while mice fed a choline-supplemented version of the diet do not develop pancreatitis (24). A series of studies concerned with the pathogenesis of the disease in this model has implicated the intracellular activation of digestive enzymes with autodigestion of the pancreatic tissues. Increased cathepsin B_1 activity in the pancreas, and the presence of this enzyme in the cytosol have been demonstrated (29). It is known that cathepsin B_1 can activate trypsinogen. The hypothesis has been advanced that an alteration of intracellular membrane permeability or stability resulting from altered phospholipid metabolism in the choline-deficient, ethionine-fed mice is important in the pathogenesis of the acute autodestruction of the acinar cells (24). Such changes in the membrane give the lysosomal cathepsin access to the normally segregated zymogens.

Integrity of intracellular compartmentation is thought to be influential in preventing the intracellular activation of pancreatic enzymes (3). The significance of choline in phospholipid metabolism, and thus in membrane synthesis, is established. Ethionine seems to play a major pancreatoxic role in this model, since mice fed the choline-deficient diet do not develop pancreatitis, nor has pancreatitis been a feature of experimental choline deficiency in other species. The delicacy of the pathogenetic mechanism is emphasized by the fact that male mice fed the choline-deficient, ethionine-containing diet do not develop acute pancreatitis unless they are also treated with estrogen (30). The importance of choline deficiency as a predisposing factor for pancreatitis in humans is unknown.

One other interesting observation on the effect of choline deficiency on experimentally induced pancreatic disease is that there were fewer carcinogen-induced lesions in the pancreas of azaserine-treated rats fed deficient diets than in rats fed a choline-supplemented diet (31). This study lasted 6 months, and no carcinomas were observed in either group, but the observation suggests that choline-deficiency could alter the process of carcinogenesis in the pancreas in the azaserine-rat model.

E. Chronic Alcoholism

Although the etiology of chronic pancreatitis, one of the major diseases of the pancreas, is often unknown, chronic alcoholism is one of the major known causes of this disease and one of the apparent causes of acute pancreatitis. The characteristic pathological changes in chronic pancreatitis are atrophy and fibrosis of the exocrine pancreas. Fibrous and interstitial tissue may contain a lymphocytic infiltrate.

The pancreatic disease of chronic alcoholism may be more correctly classed as an example of toxic cellular injury than as a nutritional disease. Pitchumoni et al. examined the nutritional history of chronic alcoholic patients with chronic pancreatitis and concluded "...that patients with alcoholic pancreatitis in New York City consume less protein, fat, and total calories than the national norms...," but also found that the signs and symptoms of malnutrition, e.g., anemia, macrocytosis, and folate deficiency were more common among alcoholics with cirrhosis than among those with chronic pancreatitis (32). Acinar and ductal cell function is altered in chronic alcoholism to the extent that protein secretion is increased while the secretion of bicarbonate and water is diminished. Sarles and Laugier have suggested that intralobular ductal and ductular obstruction may be a factor in the pathogenesis as a result of altered protein content and concentration in the pancreatic juice (33). They envision the obstruction as initially resulting from protein precipitates or "plugs" within the duct system. Later, these plugs may become calcified. The importance of duct obstruction as a cause of pancreatitis and atrophy of acinar tissue is recognized. The plugs and their calcified derivatives are demonstrable in the pancreas of alcoholics, but whether or not they play a primary pathogenetic role in causing duct obstruction and pancreatitis, as opposed to having the status of a secondary change reflecting the dysfunction of injured cells, is still a matter of debate. Recent reviews provide discussion of these issues (3,33).

F. Trypsin Inhibitors

A recent fascinating and potentially clinically important development is the recognition that nutritional factors may influence pancreatic carcinogenesis by indirect but definable mechanisms. The fact that the ratio of digestive enzymes produced by the pancreas changes in response to long-term changes in the composition of the diet is well established (34). Feeding a raw soy flour diet to rats induces hyperplasia of the pancreas. Diets containing several other natural or chemical trypsin inhibitors have a similar effect. It seems likely that the mechanisms by which pancreatic enzyme production is adapted to the composition of the diet are similar to those that mediate the effects of diet on pancreatic hypertrophy and hyperplasia. These effects can be regarded as adaptive physiological responses rather than pathological, and they will not be considered in detail here except in the context of carcinogenesis. A hypothesis has been formulated to explain and relate these observations. Pancreatic enzyme secretion is stimulated by cholecystokinin (CCK), which is made in the mucosa of the proximal small intestine and released when food enters this segment. The release of CCK appears, at least in part, to be controlled by a negative-feedback mechanism such that the presence of free trypsin in the intestine suppresses CCK release. Dietary substrates of trypsin, i.e., proteins, bind the enzyme in the lumen leading to CCK secretion which in turn stimulates further pancreatic secretion of trypsin and other enzymes. Trypsin inhibitors, such as the one contained in raw soy flour (soy bean trypsin inhibitor, SBTI) also bind trypsin in the bowel lumen and cause CCK release.

Cholecystokinin is a trophic hormone for the pancreas. Prolonged administration of exogenous CCK, or of closely related peptides such as cerulein, that stimulate pancreatic enzyme secretion, has been shown to stimulate pancreatic growth (34). Both hypertrophy and hyperplasia of the pancreatic tissue have been reported. The hypothesis that consumption of a diet containing a trypsin inhibitor, such as SBTI, leads to hyperplasia because of sustained CCK secretion followed logically and led to evaluation of a derivative hypothesis that such hyperplasia might enhance the process of carcinogenesis in the pancreas. It was first shown that rats fed a raw soy flour diet during exposure to a chemical carcinogen, azaserine, developed a higher frequency of pancreatic carcinomas than a control group that was fed heated soy flour (35). Heating inactivates SBTI. In a subsequent study, the same trend was shown in rats treated with another carcinogen, N-nitrosobis(2-hydroxypropyl)amine (36). During the course of these studies, an increased prevalence of hyperplastic foci and adenomas in the pancreas was observed in control groups that were fed raw soy flour but not treated with carcinogen. Finally, in long-term studies, a low prevalence of carcinomas has been observed in groups of rats fed raw soy flour diet without prior carcinogen treatments (37).

The significance of these observations for human nutrition is unknown, and the fact that SBTI is largely inactivated by heating is emphasized. No epidemiological data has been presented that implicates consumption of diets containing trypsin inhibitors as a risk factor for pancreatic carcinoma in humans. On the other hand, the causes of pancreatic carcinoma in man are still largely unknown.

G. High-Fat Diets

The pancreas has been shown to increase the synthesis and secretion of lipase in response to high dietary levels of fat. This mechanism, at least in part, appears to be a response to circulating levels of fatty acids and monoglycerides (34). There is no clear evidence that high-fat diets cause pancreatic hyperplasia.

Epidemiological studies have shown a general correlation of national levels of dietary fat consumption with the prevalence of carcinoma of the breast (38) and of the pancreas (39). In experimental studies, high-fat diets enhanced breast carcinogenesis (40). This finding has stimulated similar studies with experimental models of pancreatic carcinogenesis. Enhanced progression of pancreatic carcinomas has been demonstrated in azaserine-treated rats that were fed diets containing 20% corn oil or 20% safflower oil (41), and in N-nitrosobis(2-oxopropyl)-amine-treated hamsters fed a high corn oil diet during the postinitiation phase of carcinogenesis (42). It is of interest that the prevalence of pancreatic neoplasms was similar in a group of rats fed a diet containing 18% coconut oil plus 2% corn oil and in a group fed a diet containing 5% corn oil (41), and that the prevalence in both of these groups was significantly lower than in groups fed diets containing 20%-unsaturated fat. A more complex hypothesis is needed to accommodate these observations than the one regarding raw soy flour diets, but the implications for human nutrition may be clearer.

III. ENDOCRINE PANCREAS

The effect of nutritional factors on the islet cells is less clear than are the nutritional relationships in the exocrine pancreas. The responsiveness of the B cells to blood glucose levels is well established. It has been postulated that persistent hyperglycemia in the mild-moderate type 2 diabetic, "stresses" the B cells and leads to a progressive decline in the functional capacity to secrete insulin. This somewhat mystical pathogenetic scheme seems to neglect the more important consideration of the underlying cause of the islet dysfunction and diabetes.

Trace elements, e.g., zinc, are known to complex with insulin in vitro, but deficiency states are not recognized as clinically affecting islet function. Dietary chromium deficiency has been identified as a cause or contributing factor in the glucose intolerance of infants with protein-calorie malnutrition, of adults with maturity-onset diabetes, and of certain other patient groups (43). Impaired glucose tolerance has been reported in chromium-deficient rats. These observations suggest that chromium may be an essential trace element for normal B cell function.

Islet function may suffer as a result of diseases of the exocrine tissue, e.g., in advanced stages of chronic pancreatitis and the pancreatic atrophy of kwashiorkor. Rats fed a protein-deficient diet for 8 weeks were reported to have reduced levels of plasma insulin, although their glucagon levels were similar to those in control rats (44). Thus nutritional deficits may be an indirect cause of diabetes under special circumstances.

IV. CONCLUSION

Although kwashiorkor is the only clinically important pancreatic disease known to be due to nutritional deficiency, the experimental observations noted earlier require a caveat. Even though the pancreatic lesion of kwashiorkor is usually attributed entirely to nutritional imbalance with deficient dietary protein, the possible contribution of some trace element deficiency should be recognized. Furthermore, the etiology of many cases of pancreatitis, pancreatic carcinoma, and diabetes remains unknown (45). Until the causes of these diseases are better understood, the possibility that nutritional factors play an etiological role in at least some of the cases should merit consideration.

REFERENCES

1. H. W. Davenport, *Physiology of the Digestive Tract.* Year Book Medical Publishers, Inc., Chicago, 1977.
2. P. Dhar, N. Zamcheck, and S. A. Broitman, Nutrition in diseases of the pancreas, in *Modern Nutrition in Health and Disease,* 6th ed. (R. S. Goodhart and M. E. Shils, eds.). Lea & Febiger, Philadelphia, 1980, p. 953.
3. D. S. Longnecker, Pathology and pathogenesis of the diseases of the pancreas. *Am. J. Pathol. 107*:99 (1982).
4. B. Weisblum, L. Herman, and P. J. Fitzgerald, Changes in pancreatic acinar cells during protein deprivation. *J. Cell Biol. 12*:313 (1962).
5. H. Sidransky, Chemical and cellular pathology of experimental acute amino acid deficiency. *Methods Achiev. Exp. Pathol. 6*:1 (1972).
6. H. Sidransky, Nutritional disturbances of protein metabolism in the liver. *Am. J. Pathol. 84*:649 (1976).
7. H. Sidransky, Chemical pathology of nutritional deficiency induced by certain plant proteins. *J. Nutr. 71*:387 (1960).
8. E. B. Scott, Histopathology of amino acid deficiencies. *Arch. Pathol. 82*:119 (1966).
9. F. E. Viteri and B. Torun, Protein-calorie malnutrition, in *Modern Nutrition in Health and Disease,* 6th Ed. (R. S. Goodhart and M. E. Shils, eds.), Lea & Febiger, Philadelphia, 1980, p. 697.
10. N. S. Scrimshaw and M. Béhar, Protein malnutrition in young children. *Science 133*:2039 (1961).
11. P. V. Veghelyi, T. T. Kemény, J. Pozsonyi, and J. Sos, Dietary lesions of the pancreas. *Am. J. Dis. Child 79*:658 (1950).
12. B. N. Tandon, P. A. Banks, P. K. George, S. K. Sama, K. Ramachandran, and P. C. Gandhi, Recovery of exocrine pancreatic function in adult protein-calorie malnutrition. *Gastroenterology 58*:358 (1970).
13. W. R. Blackburn and K. Vinijchaikul, The pancreas in kwashiorkor; an electron microscopy study. *Lab. Invest. 20*:305 (1969).
14. J. N. P. Davies and M. B. Brist, The essential pathology of kwashiorkor. *Lancet 1*:317 (1948).
15. A. G. Shaper, Chronic pancreatic disease and protein malnutrition. *Lancet 1*:1223 (1960).
16. C. S. Pitchumoni, Pancreas in primary malnutrition disorders. *Am. J. Clin. Nutr. 26*:374 (1973).

17. G. F. Combs, Jr. and M. J. Bunk, The role of selenium in pancreatic function, in *Selenium in Biology and Medicine*, (J. E. Spallholz, J. L. Martin, and H. E. Ganther, eds.). Avi Publishing Co., Inc., Westport, Conn. 1981, p. 70.

18. R. J. Levine and R. E Olson, Blood selenium in Thai children with protein-calorie malnutrition. *Proc. Soc. Exp. Med. 134*: 1030 (1970).

19. U. R. Fölsch and W. Creutzfeldt, Pancreatic duct cells in rats: Secretory studies in response to secretin, cholecystokinin-pancreozymin and gastrin in vivo. *Gastroenterology 73*:1053 (1977).

20. B. F. Fell, T. P. King, and N. T. Davies, Pancreatic atrophy in copper-deficient rats: Histochemical and ultrastructural evidence of a selective effect on acinar cells. *Histochem. J. 14*: 665 (1982).

21. C. F. Mills, J. Quarterman, R. B. Williams, and A. C. Dalgarno, The effects of zinc deficiency on pancreatic carboxypeptidase activity and protein digestion and absorption in the rat. *Biochem. J. 102*:712 (1967).

22. N. Raica, Jr., M. A. Stedham, Y. F. Herman, and H. E. Sauberlich, Vitamin A deficiency in germ-free rats, in *The Fat-Soluble Vitamins*, (H. F. DeLuca and J. W. Suttie, eds.), University of Wisconsin Press, Madison, Wis., 1969, p. 283.

23. K. D. Blackfan and C. D. May, Inspissation of secretion, dilatation of the ducts and acini, atrophy and fibrosis of the pancreas in infants. *J. Pediatr. 13*:627 (1938).

24. B. Lombardi, Influence of dietary factors on the pancreatotoxicity of ethionine. *Am. J. Pathol. 84*:633 (1976).

25. E. Farber and H. Popper, Production of acute pancreatitis with ethionine and its prevention by methionine. *Proc. Exp. Biol. Med. 74*:838 (1950).

26. R. C. Goldberg, I. L. Chaikoff, A. H. Dodge, Destruction of pancreatic acinar tissue by DL-ethionine. *Proc. Soc. Exp. Biol. Med. 74*:869 (1950).

27. L. Herman and P. J. Fitzgerald, The degenerative changes in pancreatic acinar cells caused by DL-ethionine. *J. Cell Biol. 12*: 277 (1962).

28. L. Herman and P. J. Fitzgerald, Restitution of pancreatic acinar cells following ethionine. *J. Cell Biol. 12*:297 (1962).

29. K. N. Rao, M. F. Zuretti, F. M. Baccino, and B. Lombardi, Acute hemorrhagic pancreatic necrosis in mice: The activity of lysosomal enzymes in the pancreas and the liver. *Am. J. Pathol. 98*:45 (1980).

30. K. N. Rao, P. K. Eagon, K. Okamura, D. H. Van Thiel, J. S. Gavaler, R. H. Kelley, and B. Lombardi, Acute hemorrhagic pancreatic necrosis in mice. Induction in male mice treated with estradiol. *Am. J. Pathol. 109*:8 (1982).

31. H. Shinozuka, S. L. Katyal, and B. Lombardi, Azaserine carcinogenesis: Organ susceptibility change in rats fed a diet devoid of choline. *Int. J. Cancer* 22:36 (1978).
32. C. S. Pitchumoni, M. Sonnenshein, F. M. Candido, M. D. Panchaharam, and J. M. Cooperman, Nutrition in the pathogenesis of alcoholic pancreatitis. *Am. J. Clin. Nutr.* 33:631 (1980).
33. H. Sarles and R. Laugier, Alcoholic pancreatitis. *Clin. Gastroenterol.* 10:401 (1981).
34. T. E. Solomon, Regulation of exocrine pancreatic cell proliferation and enzyme synthesis, in *Physiology of the Gastrointestinal Tract, Vol. 2.* (L. R. Johnson, ed.). Raven Press, New York, 1981, p. 873.
35. R. G. H. Morgan, D. A. Levinson, D. Hopwood, J. H. B. Saunders, and K. G. Wormsley, Potentiation of the action of azaserine on the rat pancreas by raw soya bean flour. *Cancer Lett.* 3:87 (1977).
36. D. A. Levison, R. G. H. Morgan, J. S. Brimacombe, D. Hopwood, G. Coghill, and K. G. Wormsley, Carcinogenic effects of di(2-hydroxypropyl)nitrosamine (DHPN) in male Wistar rats: Promotion of pancreatic cancer by a raw soya flour diet. *Scand. J. Gastroenterol.* 14:217 (1979).
37. E. E. McGuinness, R. G. H. Morgan, D. A. Levison, D. L. Frape, D. Hopwood, and K. G. Wormsley, The effects of long-term feeding of soya flour on the rat pancreas. *Scand. J. Gastroenterol.* 15:497 (1980).
38. K. K. Carroll and H. T. Khor, Dietary fat in relation to tumorigenesis. *Prog. Biochem. Pharmacol.* 10:30 (1975).
39. E. L. Wynder, An epidemiological evaluation of the causes of cancer of the pancreas. *Cancer Res.* 35:2228 (1975).
40. P.-C. Chan, K. A. Ferguson, and T. L. Dao, Effects of different dietary fats on mammary carcinogenesis. *Cancer Res.* 43:1079 (1983).
41. B. D. Roebuck, J. D. Yager Jr., D. S. Longnecker, and S. A. Wilpone, Promotion by unsaturated fat of azaserine-induced pancreatic carcinogenesis in the rat. *Cancer Res.* 41:3961 (1981).
42. D. F. Birt, S. Salmasi, and P. M. Pour, Enhancement of experimental pancreatic cancer in Syrian golden hamsters by dietary fat. *J. Nat. Cancer Inst.* 67:1327 (1981).
43. K. M. Hambidge, Chromium nutrition in man. *Am. J. Clin. Nutr.* 27:505 (1974).
44. L. E. Anthony and G. R. Faloona, Plasma insulin and glucagon levels in protein-malnourished rats. *Metabolism* 23:303 (1974).
45. D. S. Longnecker, Environmental factors and diseases of pancreas, *Environ. Health Perspect.* 20:105 (1977).

4

Nutrition and Cardiovascular Disease

David Kritchevsky

The Wistar Institute of Anatomy and Biology
Philadelphia, Pennsylvania

I. INTRODUCTION

Atherosclerotic heart disease (ASHD) is a multifactorial disease. Its manifestation is a composite of genetic factors; physiological factors, including metabolism of the arteries and coronary vessels; humoral factors, including the presence of lipids and clotting factors in the circulation; psychological factors, including the effects of stress; and ecological factors, including cigarette smoking and components of the diet. Davignon (1) has attempted a graphic integration of these factors and has succeeded in demonstrating the intricate interconnections among the many risk factors.

In the absence of a simple test for certainty of a coronary attack, we must be satisfied with medical or behavioral factors that lead to increased risk of susceptibility to ASHD. A recent report (2) listed five factors that increase the danger of coronary heart disease (CHD). These are: (a) elevated plasma cholesterol, (b) elevated blood pressure, (c) smoking, (d) diabetes, and (e) obesity. These five factors are unquestionably the major risk factors for most people, but compendia of risk factors (3,4) have named as many as 246 elements of risk. Hopkins and Williams (4) classified the risk factors as initiators, promoters, potentiators, and precipitators. There were 21 positive factors associated with dietary excess.

The factor that most lends itself to intervention is diet. It can be altered to conform to specific dietary theories of atherogenesis, and results can be monitored by analysis of plasma lipids in man or by actual examination of the extent of atherosclerosis in experimental animals. Plasma cholesterol levels have been shown to correlate with risk of ASHD (5), as have triglycerides (6). Since dietary cholesterol is a source of plasma cholesterol and can induce atherosclerosis in susceptible animal species, it is logical to study the impact of dietary cholesterol and fat on this disease. However other dietary components can also influence lipemia (7), and not all animal species show a similar response to diet (8,9). Thus we return to consideration of the multifaceted nature of this disease. The purpose of this exposition is to consider how nutritional factors influence ASHD.

II. CHOLESTEROL AND LIPOPROTEINS

Aspects of plasma and dietary cholesterol have been at the center of atherosclerosis research for over 70 years, or since it was shown that rabbits fed cholesterol developed atherosclerosis (10,11). The logical belief being that if the cholesterol content of the diet were reduced, less would appear in the circulation.

Cholesterol found in the intestine is derived principally from three sources—diet, bile, and intestinal secretion. The diet contributes about 500 mg of cholesterol in the average subject; the biliary contribution is between 750 and 1250 mg/day (12); the intestinal contribution is small. Cholesterol is solubilized through the formation of mixed micelles containing bile acids, monoglycerides, and free fatty acids. The amount of cholesterol absorbed ranges between 30 and 60% (13,14), but the capacity for absorption can be increased with increasing levels of dietary cholesterol (15). Biliary cholesterol is derived from hepatic synthesis (9-13 mg/kg/day), and absorption is affected by synthesis (16). The endogenous synthesis of cholesterol and its central role in metabolism (as a precursor of bile acids and steroid hormones) make it difficult to regard it simply as a noxious substance.

There have been numerous studies of the effect of dietary cholesterol on levels of plasma cholesterol. A number of studies, usually conducted under clinical conditions, have shown that the level of dietary cholesterol is a determinant in raising cholesterol levels (17-20). However the findings are by no means unaminous. Gertler et al. (21) compared serum cholesterol levels and cholesterol intake in groups of subjects with and without CHD. In every category subjects with CHD exhibited significantly higher levels of cholesterol, but there was little relation to cholesterol ingestion. In the CHD group, men who ate 5.67 g of cholesterol per week had serum cholesterol levels of 288 mg/dl, and those who ate only 1.34 g/wk had cholesterol levels of 271 mg/dl. In the control group, subjects ingesting the highest and lowest amounts of cholesterol (6.98 g/wk or 1.37 g/wk) had cholesterol levels of 213 mg/dl and 222 mg/dl. respectively. In both the CHD and control groups, the subgroups with highest serum cholesterol levels (416 ± 17 mg/dl in CHD group and 313 ± 3 mg/dl in normals, $p < 0.001$) ate 4.12 and 4.26 g/wk of cholesterol, respectively. Several studies (22-24) have found no relationship between egg intake and cholesterolemia. In the Framingham study (25) there seemed to be no relationship between dietary components and levels of cholesterol. Nichols et al. (26) found no effect of diet on lipidemia in over 4000 subjects studied in Tecumseh, Michigan. A comparison of diet and CHD in three major prospective heart studies (27) found correlations with alcohol and carbohydrate intake but none with any other aspect of the diet (Table 1).

Cholesterol and other lipids are transported in the plasma as part of a continuum of lipid and protein complexes whose hydrated density is less than 1.210 g/ml. Other plasma proteins, which have minimal lipid transport functions, have a density of about 1.33 g/ml. The lipoproteins can be separated by electrophoresis or by their rates of flotation in the ultracentrifuge. Gofman et al. (28) devised the method of flotation in the ultracentrifuge in salt solutions, whose density was 1.063 g/ml at 26°C. The various lipoprotein fractions were defined by their flotation rates in S_f units (for Svedberg units of flotation). The classification of the plasma lipoproteins is summarized in Table 2. There is now intensive research on the structure and function of the apolipoproteins and in their roles as predictors of ASHD risk.

The primary hyperlipoproteinemias (HLP) are described and can be diagnosed by lipoprotein type. In type I HLP, cholesterol is normal or slightly elevated and triglycerides are very high. Chylomicron levels are also high. Type IIA HLP is characterized by elevated low-density lipoprotein (LDL) levels, elevated cholesterol, and normal triglyceride levels; type IIA HLP exhibits elevated LDL and very low-density lipoprotein (VLDL) levels, and both cholesterol and triglycerides are high. In type III-HLP abnormal plasma lipoproteins (density below 1.006 g/ml) are present, and both cholesterol and triglycerides are high. Type IV HLP is characterized by elevated VLDL and triglycerides and normal or elevated cholesterol. Type V HLP exhibits elevations in VLDL and chylomicrons. Cholesterol levels may be normal or elevated but triglyceride levels are very high.

Gofman et al. (28) pointed out that the level of a particular fraction in which cholesterol is transported may be of greater importance to the etiology of coronary disease than the total cholesterol level. The particular fraction most closely related to the risk of developing ASHD is the low-density lipoprotein (LDL), Gofman's S_f 0-12 or the electrophoretic beta-lipoprotein. The LDL is the major transport vehicle for cholesterol. Generally, the levels of total plasma cholesterol reflect LDL levels, and recent studies suggest that elevated LDL may be an independent risk

Table 1 Diet and Coronary Heart Disease in Three Ongoing Studies

| | Population[a] | | | | | |
| | Puerto Rico | | Honolulu | | Framingham | |
	CHD	No CHD	CHD	No CHD	CHD	No CHD
Number	286	7932	264	7008	79	780
% MI or death	57	—	62	—	65	—
Total calories	2289[c]	2395	2210[b]	2319	2488	2622
Protein, g	85	86	95	95	99	101
Fat, g	94	95	86	87	112	114
P/S	0.49	0.45	0.57	0.54	0.41	0.39
Cholesterol, mg	419	417	549	555	534	529
Carbohydrate, g	262[b]	280	249[b]	264	248	252
Sugar, g	50	52	45	46	78	72
Starch, g	167[c]	180	155[c]	165	118	117
Other, g	45	48	48	52	52[b]	61
Alcohol, g	8	12	8[c]	14	12[c]	25

[a]CHD: coronary heart disease.
[b]$p \leqslant .05$.
Source: After Ref. 27. Values are age-adjusted means.

Table 2 Characteristics of Plasma Lipoproteins

	Class[a]			
	Chylomicrons	VLDL	LDL	HDL[b]
Physical characteristics				
Diameter	75-1000	30-80	19-25	4-10
MWT $\times 10^{-6}$ (dalton)	$>4 \times 10^2$	5-6	2.3	0.18-0.36
Density	<1.006	0.95-1.006	1.006-1.063	1.063-1.21
Electrophoresis	Origin	Pre-beta	Beta	Alpha
S_f (ultracentrifuge)	$400-10^6$	20-40	0-12	HDL
Chemical composition (%)				
Protein	2	10	23	55
Triglyceride	85	50	10	4
Free cholesterol	1	7	8	2
Ester cholesterol	3	12	37	15
Phospholipid	9	18	20	24
Major apoproteins	A-I B C	B C-I C-II C-III E	B	A-I A-II

[a] VLDL: very low-density lipoprotein; LDL: low-density lipoprotein; HDL: high-density lipoprotein.
[b] HDL_2: d = 1.063-1.12; HDL_3: = d 1.12-1.21.

factor for ASHD (29,30). Barr et al. (31) studied the lipoprotein
fractions in normal and atherosclerotic subjects by electrophoresis and
concluded that the ratio of alpha/beta-lipoprotein was more indicative
of risk than total plasma cholesterol or beta-lipoprotein. Their concept
lay dormant for more than a decade until it was revived and amplified
(32). The revival generated interest in the alpha- or high-density
lipoprotein (HDL) and in its possible protective role in ASHD. Several
epidemiological studies have suggested an inverse correlation between
ASHD and plasma HDL levels (33-35). In the late 1960s, Rothblat and
his colleagues (36) demonstrated that cholesterol was taken up by tissue culture cells from a medium whose composition resembled that of
LDL; when the medium resembled HDL, egress of cholesterol from cells
was observed. The work and hypotheses arising from it were summarized in a review (36).

The experience from the Framingham study gave rise to "the lipid
hypothesis," which proposes that a specific reduction of serum cholesterol could give a predictable reduction of risk. This hypothesis
has been the basis for a number of primary prevention trials. Oliver
(37) has reviewed critically a number of these trials and states, "In
most, it was not possible to relate the degree of reduction of serum
cholesterol level concentrations to the degree of non-fatal myocardial
infarction but, where this could be achieved (as in the WHO Primary
Prevention Trial), the two would appear to be significantly correlated."

In a recently completed trial (38,39), in which the drug cholestyramine was administered to hypercholesterolemic men (plasma cholesterol
292 mg/dl), the reductions of cholesterolemia and myocardial infarction, compared with a placebo-fed control, was taken as validation of
the lipid hypothesis. Whether or not these findings justify sweeping
general dietary recommendations, or if special advice should be limited
to the known susceptible subjects, is the subject of current debate
(40-42). It has been suggested that the desirable cholesterol level
should be below 220 mg/dl (43) possibly as low as 190 mg/dl (2).

III. THE DIET

The American diet has undergone a number of changes in the past 70
years. Least among these has been the change in total caloric availability.

Friend et al. (44) have discussed changes in American food consumption patterns between 1909 and 1976, and these are summarized in Table
3. Within this period the number of calories available fell by 5%; protein
availability was virtually unchanged, but the ratio of animal/vegetable
protein rose from 1.06 to 2.26; carbohydrate availability fell by 24%;
and fat availability rose by 20%. Rizek et al. (45) have published
newer data on the patterns of fat in the American diet through 1980.

Table 3 Nutrients Available in the United States (per capita per day)

Year	Calories	Protein, g	Carbohydrate, g	Fat, g	A/V[a]	Cholesterol, mg	P/S[b]
1909–1913	3480	102	492	125	4.86	495	0.18
1935–1939	3260	90	436	132	2.76	476	0.22
1947–1949	3230	95	403	140	2.94	563	0.25
1957–1959	3140	95	375	143	2.40	553	0.28
1965	3150	96	372	144	1.74	517	0.33
1970	3300	100	380	156	1.61	526	0.40
1975	3220	99	370	152	1.34	489	0.43
1976	3300	101	376	157	1.30	498	0.44

[a] Animal/vegetable.
[b] Polyunsaturated/saturated.
Source: After Ref. 44.

In 1909-1913 the percentage of energy available from protein, carbohydrate, and fat was 12, 56, and 32, respectively; in 1980 it was 12, 46, and 42, respectively. The ratio of animal/vegetable fat available in the American diet was 4.95 in 1909-1913 and 1.41 in 1980, obviously the big change was in availability of vegetable fat. Grams of saturated fat available rose by 14% between 1909-1913 and 1980; grams of oleic acid by 33%; and grams of linoleic acid by 179%. There was, overall, an increase of 57% in fat from fats and oils; 37% in fats from meat, poultry, and fish; and 7% in fat from dairy products. Total cholesterol available was 509.7 mg/day in 1909-1913 and 508.9 mg/day in 1980. Cholesterol availability peaked between 1947 and 1959 (about 570 mg/day). The amount of cholesterol available from dairy products was unchanged between 1909 and 1980; that available from eggs fell slightly (6%); and that from meat, poultry, and fish rose by 38%.

The mortality from heart disease in the United States rose from 163 per 100,000 in 1900 to a peak of about 308 per 100,000 in 1950. By 1970 mortality from heart disease had fallen to 254 per 100,000 and by 1980 to 202 per 100,000. Compared with 1900, consumption of eggs is unchanged, consumption of butter has fallen by about 40%, but beef consumption has risen (46). In the face of the dramatic drop in coronary mortality, the consumption of eggs, the major source of dietary cholesterol, has not changed.

Recently three leading epidemiologists were asked to comment on the reduction in coronary mortality as related to diet. Marmot et al. (47) attributed the reduction in coronary disease to changes in fat consumption, recognition and treatment of hypertension, and decline in cigarette smoking. They observed that factors other than these had also played a role. Feinleib et al. (48) attributed the changes to primary prevention and improved medical care. Stallones (49) felt the change was due to reduction in cigarette smoking but stated "...one may reasonably hypothesize that other, unknown mechanisms are the most important causes of the change."

IV. FAT

The hypercholesterolemic effect of saturated fat is well documented. Comparison of cholesteremic effects of various dietary fats led Keys et al. (50,51) and Hegsted et al. (18) to formulate equations for predicting changes in serum cholesterol based on changes in dietary fat. Both equations attributed cholesterol increasing coefficients to saturated fatty acids and decreasing ones to unsaturated fatty acids. Ahrens (52) demonstrated that serum cholesterol levels in human subjects fed formula diets generally rose as fat saturation increased, but there were exceptions. Thus lard, with an iodine value of 120 or 115, was hypercholesterolemic compared with corn oil with an iodine value of 126.

Cocoa butter (iodine value 35) was much less cholesterolemic than either butter (iodine value 40) or palm oil (iodine value 52). Cocoa butter is also less atherogenic than palm oil in rabbits fed cholesterol or cholesterol-free, atherogenic diets (53,54). The difference may be due to the fatty acid composition of cocoa butter which contains appreciably more stearic acid than most fats.

The most cholesterolemic fat was coconut oil (iodine value 10), which gave cholesterol levels 52 ± 13% above baseline values. The general public perception of fat is that animal is saturated, hence harmful, and vegetable fat is polyunsaturated, hence beneficial. It should be pointed out that coconut oil and palm oil are vegetable fats. The mechanism by which unsaturated fat lowers plasma cholesterol levels is still unclear (55,56). There is some evidence for increased cholesterol excretion, but there is also evidence for redistribution from the plasma compartment (57,58).

Diets containing saturated fat are more atherogenic for rabbits than those containing unsaturated fat (59). This observation is true whether, the diets contain cholesterol (60,61) or not (62,63). One fat that is anomalous in this respect is peanut oil. Peanut oil has no untoward effect on human cholesterol levels (52) but is inordinately atherogenic for rats (64,65), rabbits (66), and monkeys (67,68). Peanut oil may contain 6 to 7% of long-chain saturated fatty acids (arachidic, 20:0; behenic, 22:0; lignoceric, 24:0), and these were, at one time, considered to be responsible for its atherogenicity. However it has been shown that the structure of the component triglycerides of peanut oil may be the determinant of its atherogenic effect. Randomization (autointeresterification) of peanut oil provides a fat whose fatty acid spectrum and iodine value are identical with those of the starting oil, but whose atherogenicity is reduced significantly (68,69). Analysis of the component triglycerides of the various oils is being carried out in efforts to find a clue to peanut oil's atherogenicity (70,71). Lipolysis of native and randomized oils proceeds at the same rate (72), and there appears to be no difference between corn and peanut oils vis-a-vis cholesterol transport in rats (73). The atherosclerotic lesion observed in animals fed peanut oil is more fibrous than the fatty lesion seen when corn oil is fed. The long-chain fatty acids may have an effect on connective tissue metabolism.

Hegsted et al. (18) found that the effects of dietary fat on plasma cholesterol levels could be explained by four variables—polyunsaturated fat, myristic acid, palmitic acid, and cholesterol. They attempted to delineate the roles of the fatty acids by feeding diets containing special fats enriched with these fatty acids. The fats were obtained by interesterification of medium-chain triglycerides with specific long-chain fatty acids. Unfortunately, no clear-cut results were obtained (74), possibly because interesterification of fat changes its natural structure and its metabolic properties. A similar study was carried out in rabbits, and in this case, too, results were equivocal (75).

The double bonds in most naturally occurring unsaturated fatty acids
are in the cis configuration, but trans-unsaturated fatty acids can be
found in the milk of many animal species and in plants. Hydrogenation
of fat will, under some conditions, cause inversion of double bonds
yielding fats containing trans unsaturation. Kummerow (76) suggested
that the increases in coronary disease since 1920 could be due to increased availability of trans-unsaturated fatty acids (TFA). The hypothesis is that TFA are less readily metabolized than cis-unsaturated
fatty acids and might thus affect membrane transport and other metabolic
processes. There are no differences in absorption or oxidation of the
isomeric fatty acids (77,78). Lipolysis of triglycerides containing cis
or trans fatty acids proceeds at similar rates (79) and while cholesteryl
esters of TFA are synthesized at a slower rate (80,81), they are also
hydrolyzed more slowly (82).

Effects of TFA in atherosclerosis are not a recent concern. When rabbits were fed cholesterol and oleic or elaidic acid or fed olive oil and
elaidinized olive oil, it was observed that rabbits fed TFA exhibited
higher serum cholesterol, but the severity of their atheromata was not
significantly different from the controls (83,84). Rabbits were fed a
semipurified, cholesterol-free atherogenic diet containing either 3.2 or
6.0% TFA; those fed 6.0% TFA had higher serum cholesterol levels, but
the average atherosclerosis in the two TFA-fed groups and the controls
was the same (85). The activities of five hepatic enzymes (glucose-6-
phosphatase, fatty acid synthetase, malate dehydrogenase, beta-hydroxybutyrate dehydrogenase, and monoamine oxidase) were the same in all
of these groups. Tissue fatty acid spectra reflected the diet. Swine
cholesterol and triglyceride levels were elevated compared with those of
swine fed beef tallow, but their atherosclerosis was 42% less severe (86).

Vervet monkeys were fed 3.2 or 6.0% TFA for a year or fed the TFA-
containing diets for 6 months, then returned to a control diet. There
were no differences in atherosclerosis or arteriosclerosis. Serum and
tissue fatty acid spectra reflected the particular diets. When monkeys
fed TFA were returned to a normal diet, the TFA virtually disappeared
(87). A similar phenomenon is seen when TFA-fed rats are returned to
a control diet (88).

There have been two studies of tissue TFA in normal and atherosclerotic human subjects; in neither case were differences observed (89,90).
Houtsmuller (91) reviewed the biological effects of TFA and concluded
that their biochemical properties were somewhere between those of
saturated and monounsaturated fats.

V. PROTEIN

In 1957 Yerushalmy and Hilleboe (92) opined that animal protein intake
was better correlated with risk of atherosclerosis than animal fat. The

first purely nutritional studies of atherosclerosis were carried out by Ignatowski in the first decade of this century (93,94). Ignatowski believed that a toxic metabolite of animal protein was responsible for the atherosclerosis seen in rabbits fed horsemeat or beef. The discovery that pure cholesterol was atherogenic (10,11) turned attention to that component of the diet, and work on dietary protein was in eclipse for many years. In the 1920s Newburgh and his colleagues (95-98) showed that both casein and beef were atherogenic for rabbits. They also demonstrated that the level of cholesterol present in the beef-containing diets was not sufficient to be atherogenic by itself (99).

The first comparison of the atherogenic effects of animal-vegetable protein was carried out by Meeker and Kesten (100,101). They maintained rabbits on a basal diet (wheat flour, alfalfa-leaf meal, linseed meal, salt mix, and ground carrots) which provided (by weight) 15% protein, 55% carbohydrate, and 5% fat. They compared this diet with one high (38%) in animal protein (casein, wheat flour, alfalfa-leaf and linseed meal, salt mix, and carrots) and one high (39%) in vegetable protein. The vegetable protein contained soy flour, the other ingredients being the same as those present in the basal and high animal protein diets. Some of the animals received a daily dose of 60 or 250 mg of cholesterol in 1 ml of olive oil. The cholesterol supplement was fed for 3 months and a cholesterol-free diet for a further 3 months. Soy protein was significantly less atherogenic than casein. Few plasma cholesterol data were given but extrapolation from bar graphs suggests that plasma cholesterol levels (mg/dl) in rabbits fed the cholesterol-free diets were: basal, 53 ± 11; casein, 125 ± 13; and soy protein, 64 ± 9. Howard et al. (102) fed rabbits a semipurified diet containing 25% casein (24.5% of calories) and 20% beef fat (44.1% of calories). Casein was significantly more cholesterolemic and atherogenic than soy protein. Replacement of casein with soy flour, which may have contained less protein and more carbohydrate, virtually eliminated atherosclerosis (present in only one of 11 rabbits).

Interest in dietary protein was stimulated by the work of Carroll and Hamilton (103) who fed rabbits diets containing 30% defatted protein (animal or vegetable) and 1% corn oil for a month and found a wide range of cholesterolemic effects (Table 4). Kritchevsky et al. (104) fed rabbits semipurified diets containing corn protein, wheat gluten, or lactalbumin. The lactalbumin was 101% more cholesterolemic and 113% more atherogenic than the two vegetable proteins. Analysis of the serum lipoproteins showed the greatest difference to be increases in VLDL and intermediate-density lipoprotein (IDL) in the rabbits fed lactalbumin (105,106).

What are the chemical or metabolic differences between animal and vegetable protein that elicit the different responses? The intestinal flora appear to have no influence on protein effects (107). Huff et al. (108) fed rabbits mixtures of the L-amino acids which make up casein

Table 4   Influence of Protein on Serum
Cholesterol Levels (mg/dl) in Rabbits
(30% defatted protein, 1% corn oil)

Protein	Serum cholesterol
Animal	
Egg yolk	260
Skim milk	230
Turkey	217
Lactalbumin	210
Casein	200
Whole egg	183
Fish	167
Beef	153
Chicken	147
Pork	107
Egg white	100
Vegetable	
Wheat gluten	80
Peanut	80
Oat	77
Cottonseed	73
Sesame seed	70
Alfalfa	67
Soy isolate	67
Sunflower seed	53
Pea	40
Fava bean	30

Source: After Ref. 103.

or soy protein. They also fed partial enzymic hydrolysates of casein
and soy. The casein amino acid mixture had no effect on cholesterol
levels, and the soy mixture elevated serum cholesterol by 80%. Both
hydrolysates lowered cholesterol levels, casein by 16% and soy by 41%.
    We (109,110) hypothesized that the ratio of lysine/arginine (approx-
imately 0.9 in soy protein; 2.0 in casein) could be of importance. Lysine
is a metabolic antagonist of arginine in the rat (111) and chick (112).
Lysine inhibits arginase activity (113), and arginine deficiency leads to
fatty liver in chicks (114). It is possible that a lysine-arginine effect
could be observed in the rabbit. We fed rabbits diets containing casein,
soy protein, casein plus enough arginine to give the lysine/arginine ratio
of soy, and soy plus enough lysine to give the lysine/arginine ratio of
casein (105,106,115-117) (Table 5). Addition of arginine to casein did

Table 5 Influence of Lysine and Arginine on Atherosclerosis in
Rabbits Fed Casein or Soy Protein

	Group[a,b]			
	Casein	Soy	Cas-A	Soy-L
Number	20	25	20	25
Serum lipids, mg/dl				
Cholesterol	260	157	254	209
Triglycerides	105	63	123	74
Average atherosclerosis[c]				
Aortic arch	1.61	0.70	1.30	0.94
Thoracic aorta	1.07	0.40	0.94	0.70

[a]Diets contained: 40% sucrose, 25% protein, 14% coconut oil, 15% cellulose, 5% mineral mix, 1% vitamin mix.
[b]Lysine/arginine ratios: casein: 2.0; soy: 0.9; Cas-A: 0.9;
Soy-L: 2.0.
[c]Graded on a 0-4 scale.
Source: After Ref. 106.

not affect cholesterolemia but reduced atherosclerosis by 25%; addition of lysine to soy protein raised cholesterol levels by about 50% and increased the severity of atherosclerosis by 64%. The major changes in serum lipoproteins were the raised levels of VLDL and IDL in the rabbits fed casein or soy plus lysine. Similar results were seen when wheat gluten and lactalbumin were compared (105).

To further test the hypothesis, three proteins whose lysine content is similar, but whose lysine/arginine ratios are different, were fed to rabbits. The proteins (lysine/arginine) were fish protein (1.44), casein (1.89), and whole milk protein (2.44). The severity of atherosclerosis was correlated with increasing lysine/arginine ratio; when the severity was plotted against the lysine/arginine ratio a straight line (r = 0.9979, p < .05) was obtained (118).

Dilution of casein with soy protein decreases its cholesterolemic effect so that a 1:1 mixture of casein and soy is no more cholesterolemic than 100% soy (108). Diets containing beef protein or casein were significantly more cholesterolemic than a diet containing textured vegetable protein (TVP) (185 mg/dl vs. 37 mg/dl) and twice as atherogenic. A 1:1 mixture of beef protein and TVP was somewhat more cholesterolemic than TVP (61 mg/dl vs. 37 mg/dl) and no more atherogenic (119).

Rabbits fed casein exhibit higher cholesterol levels, significantly larger body pools of cholesterol, slower cholesterol turnover, and slower cholesterol oxidation compared with rabbits fed soy protein (120,121). Casein-fed rabbits also absorb more cholesterol and excrete less than rabbits fed soy protein (120,122). Addition of lysine to soy protein increases the rapidly turning over body pool of cholesterol by 67%, the slowly turning over pool by 59%, and turnover time by 83% (106).

The level of dietary protein also affects atherogenicity. Newburgh and Clarkson (98) found that rabbits fed 36% beef protein developed more severe atherosclerosis than ones fed 27% protein. Lofland et al. (123) fed White Carneau pigeons four different fats, 30 mg/kg cholesterol, either 8 or 30% of a vegetable protein (wheat gluten), or an animal protein (casein-lactalbumin, 85:15). In every dietary combination, prevalence of atherosclerosis was higher in birds fed 30% protein. Strong and McGill (124) fed baboons diets that were high or low in cholesterol and protein and contained either saturated or unsaturated fat. In all but one instance the high-protein diets were more sudanophilic. Terpstra et al. (125) fed rabbits 10, 20, or 40% casein; after 4 weeks serum cholesterol levels were 120, 380, and 920 mg/dl, respectively.

Hardinge and Stare (126) compared serum cholesterol levels in true vegetarians, lacto-ovo vegetarians, and nonvegetarians. The average level of serum cholesterol were : true vegetarians, 206 mg/dl; lacto-ovo vegetarians, 256 mg/dl; and nonvegetarians, 292 mg/dl. Hodges et al. (127) found that a diet containing mixed protein was more cholesterolemic than one containing only vegetable protein. Walker et al. (128) reported that young women fed 50 g/day of vegetable protein had lower levels of cholesterol than those eating animal protein. Sirtori et al. (129,130) studied subjects with type II hyperlipoproteinemia who were fed for 3 weeks on a diet containing 21% protein, of which only 7% was animal protein (1.5% of total protein), and 63% (13% of total protein) was textured soy protein. In type IIA patients total cholesterol and LDL cholesterol levels fell by 16 and 19%, respectively, while HDL-cholesterol and triglyceride levels were unchanged. In type IIB patients total cholesterol levels fell by 20%; LDL cholesterol by 22%; HDL cholesterol by 16%; and triglycerides by 10%. The diet was effective regardless of the level of saturation of the dietary fat. Soy protein-rich diets are most effective in hyperlipidemic subjects (131).

Diets containing animal protein are generally more cholesterolemic and atherogenic for rabbits than diets containing vegetable protein. The protein effect can be modified by other dietary components, such as fiber or type of fat (123,132). Diets containing animal and vegetable protein in equal proportions are no more cholesterolemic or atherogenic than those containing vegetable protein alone. Cholesterol turnover is slower, excretion diminished, and body pools increased in rabbits fed animal protein.

VI. CARBOHYDRATE

There are data which suggest that plasma triglycerides are a risk factor for atherosclerosis (133,134), but current opinion (2) questions the possibility that triglycerides are an independent risk factor. High triglyceride levels, however, are associated with those lipoprotein aberrations which are considered atherogenic, namely, low HDL (135), high LDL (136) and beta-VLDL (137). High-carbohydrate diets cause elevated triglycerides in man (138,139) and substitution of sucrose or fructose for starch leads to hypertriglyceridemia (140-142).

Fructose is triglyceridemic for rats (143,144). Sucrose is more cholesterolemic and atherogenic than glucose in cholesterol-fed chickens (145) or rabbits (146). Lactose is very atherogenic when fed to rabbits together with cholesterol (147). Kritchevsky and Tepper (148) have devised a cholesterol-free, semipurified diet that is hypercholesterolemic, hyperlipoproteinemic, and atherogenic for rabbits. The diet contains 40% carbohydrate, 25% protein, 14% coconut oil, 15% cellulose, 5% salt mix, and 1% vitamin mix. When fed as a part of this diet sucrose and fructose are more lipidemic and atherogenic than glucose or lactose (149,150). St. Clair et al. (151) maintained miniature swine on diets containing 41% carbohydrate and 0.5% cholesterol for 2 years. Although no significant differences were observed, sucrose seemed to be the most atherogenic sugar.

The influence of dietary carbohydrate on atherosclerosis in primates has been reviewed recently (152). Portman et al. (153) found no differences in cholesterolemia between cebus monkeys fed starch or sucrose. Corey et al. (154) fed spider monkeys a series of diets containing glucose, sucrose, or starch, 0.5% cholesterol, and either 10% safflower oil or coconut oil-safflower oil 9:1. At 7 weeks, sucrose was more cholesterolemic than glucose or starch regardless of the dietary fat. At 48 weeks, sucrose was significantly more cholesterolemic than glucose when the dietary fat was the coconut oil-safflower oil mixture, but when the fat was safflower oil, the two sugars gave identical cholesterol levels. Lang and Barthel (155) fed diets containing 66% sucrose or starch to three species of monkeys for 16 months. In ringtail monkeys sucrose was more triglyceridemic, but starch resulted in more coronary disease. In rhesus monkeys starch was more cholesterolemic and gave more coronary disease than sucrose. There were no differences between the two carbohydrates when they were fed to stump-tail macaques. Semipurified diets (148) containing fructose, sucrose, or glucose were fed to vervet monkeys (156) for 6 months. Cholesterol levels (mg/dl ± SEM) were: fructose, 205 ± 10 mg/dl, sucrose, 194 ± 19 mg/dl, and glucose 141 ± 5 mg/dl. Cholesterol levels in control monkeys were 117 ± 12 mg/dl. Average levels of aortic sudanophilia (percentage of surface ± SEM) were: fructose, 20.3 ± 11.5%, sucrose, 2.5 ± 0.8%, and glucose, 4.5 ± 2.0%. Several of the aortas of the monkeys fed fructose had raised atherosclerotic plaques.

Baboons were fed the same semipurified diet containing fructose, sucrose, starch, or glucose for 1 year. Cholesterol levels in the four groups were similar (155 ± 3 mg/dl), but triglycerides were (mg/dl ±

SEM): fructose, 129 ± 11 mg/dl; sucrose, 116 ± 8 mg/dl; starch 108 ±
5 mg/dl; and glucose, 105 ± 7 mg/dl. Aortic sudanophilia was low, the
percentage area ± SEM being: fructose, 11.2 ± 5.7%; sucrose, 6.7 ± 4.7%;
starch, 9.3 ± 4.3%; and glucose, 6.2 ± 4.8% (157). In a second experiment (158) baboons were fed the same diets for 17 months. Two changes
were made—a lactose-fed group was added, and all the diets contained
0.1% cholesterol. The serum lipid levels and aortic sudanophilia are detailed in Table 6. The extent of sudanophilia in the fructose and sucrose
groups was not changed by addition of cholesterol to the diet, but three
of six fructose-fed baboons and two of six sucrose-fed baboons exhibited
atherosclerotic lesions. Sudanophilia in the starch and glucose groups
rose by 129 and 177%, respectively, and one of six starch-fed baboons
had atherosclerotic lesions. In the lactose-fed group 66% of the aortic
area was sudanophilic and five of six animals had atherosclerotic plaques.
In rabbits, lactose is not atherogenic in the absence of cholesterol (150)
but becomes very atherogenic when cholesterol is added to the diet (147).
Comparison of baboons fed lactose or lactose plus 0.1% cholesterol showed
that, as in rabbits, the sugar was only weakly sudanophilic in the absence of cholesterol. Since serum cholesterol and lipoprotein levels are
not especially high in the lactose- and cholesterol-fed baboons, it has
been suggested that the sugar's effect may be mediated at the level of
aortic glycosaminoglycan metabolism. The studies in primates suggest
that carbohydrate effects are dependent upon the type of lipid present
in the diet.

VII. FIBER

The fiber hypothesis states that populations ingesting diets high in fiber
suffer fewer Western diseases, such as atherosclerosis, cancer, and
diabetes. Its principal proponents have been Burkitt et al. (159) and
Trowell (160). Walker and Arvidsson (161) proposed, in 1954, that the
absence of heart disease in African black populations could be attributed
to the high levels of fiber in their diets.

Dietary fiber is discussed extensively in Chapter 6 (162). Suffice it
to repeat here that fiber is a generic term covering a number of substances of unique chemical structure and specific physiological effects.
Bran is without effect on human cholesterolemia (163) and so is cellulose
(164). Pectin, on the other hand, exhibits hypocholesterolemic properties (164-167). Addition of fiber to a standard diabetic diet will reduce
cholesterol levels (168-171).

Hardinge and Stare (126) found true vegetarians had cholesterol levels
of 206 mg/dl; lacto-ovo vegetarians had cholesterol levels of 256 mg/dl,
and in omnivores the levels were 292 mg/dl. The average fiber intakes
of the three groups were 23.4 g/day, 14.5 g/day, and 9.6 g/day, respectively (172). Sacks et al. (173) found vegetarians to have

Table 6 Influence of Carbohydrate on Atherosclerosis in Baboons Fed Semipurified Diets Containing 0.1% Cholesterol[a] (Values ± SEM)[b]

	Carbohydrate				
	Fructose	Sucrose	Starch	Glucose	Lactose
Serum lipids, mg/dl					
Cholesterol	144 ± 5 a	143 ± 3 b	155 ± 5 c	155 ± 6	170 ± 5 abc
Triglycerides	122 ± 5 abcd	91 ± 4 a	96 ± 4 b	105 ± 5 c	98 ± 3 d
Liver lipids, mg/g					
Cholesterol	4.9 ± 0.4 ab	3.7 ± 0.3 a	4.4 ± 0.6	2.9 ± 0.5 bc	5.1 ± 0.7 c
Triglycerides	25.6 ± 2 a	41.3 ± 7 b	27.8 ± 6 c	13.3 ± 1 abc	28.0 ± 7
Aorta					
Sudanophilia (%)	11 ± 4 a	10 ± 5 b	21 ± 9 c	17 ± 10 d	66 ± 14 abcd
Plaques	3/6	2/6	1/6	0/6	5/6

[a]Six baboons per group fed 40% carbohydrate, 25% casein, 13.9% coconut oil, 15% cellulose, 5% mineral mix, 1% vitamin mix and 0.1% cholesterol for 17 months.
[b]Values in horizontal row bearing the same letter are significantly different (p ≤ .05).
Source: After Ref. 158.

significantly lower plasma cholesterol and triglyceride levels. Levels of LDL were 73 ± 2 mg/dl in vegetarians and 118 ± 3 mg/dl in controls; the HDL levels in the two groups were comparable (43 ± 1 vs. 49 ± 1 mg/dl). Oat bran lowers cholesterol levels by 13% and LDL cholesterol levels by 14% (174).

Pectin or vegetable gums will lower liver cholesterol levels in rats fed 1% cholesterol, but cellulose or agar will raise them (175-177). Pectin will also reduce cholesterol levels in rabbits fed diets rich in (178, 179), or free of (180) cholesterol. Pectin reduces severity of atherosclerotic lesions in chickens fed cholesterol (181). Wheat bran has no effect on cholesterol levels in rats (176) or cynomolgus monkeys (182). Cookson et al. (183) found that alfalfa inhibited cholesterolemia and atherosclerosis in cholesterol-fed rabbits. Howard et al. (184) reported that dilution of an atherogenic diet by an equal weight of stock diet also inhibited atherosclerosis in rabbits.

Over 20 years ago it was observed that atherosclerosis could be established in rabbits fed cholesterol-free diets containing saturated fat (185,186). A review of the literature (187) showed that saturated fat was atherogenic when fed as part of a semipurified diet but not when it was added to stock diet. The difference was thought to be due to the different fibers present in the two diets, and the suggestion was verified by experiment (148,188). Moore (189) fed rabbits semipurified diets containing butter and several types of fiber and found that the severity of atherosclerosis was lower in animals fed wheat straw than in those fed cellulose or cellophane. Rabbits fed alfalfa exhibited lower cholesterol levels and less severe atherosclerosis than those fed wheat straw or cellulose (132). Vervet monkeys fed semipurified diets containing cellulose exhibit significantly higher total serum cholesterol and LDL-cholesterol levels than monkeys fed either alfalfa or wheat straw (190).

The mode of action of fiber is unclear. In rabbits it increases cholesterol excretion and decreases absorption (191). Fiber affects bile acid turnover (192) and may increase bile acid (193) and neutral steroid excretion (194). Fiber binds bile acids (195-198) which may inhibit lipid absorption and thus reduce cholesterolemia.

VIII. TRACE MINERALS

Water hardness may have a protective role in atherosclerosis (199-203). Schroeder et al. (204) found that tissue chromium levels fell with increasing age and that this element was missing in atherosclerotic aortas. They suggested chromium deficiency could be a factor in the development of atherosclerosis. Calcium (205), selenium (206), and magnesium (207-209) deficiency have all been suggested as etiological factors in development of heart disease. Klevay (210) has proposed that a high dietary zinc/copper ratio may be a significant risk factor for coronary disease. Mertz (211) reviewed the literature on trace minerals and cardiovascular disease and pointed out that sodium, calcium, magnesium, zinc, copper, vanadium, chromium, iron, and iodine all can negatively affect metabolic processes involved in atherosclerosis. At present there are few available data bearing on this point. This particular area of investigation is in its infancy, and we may expect much experimental work to appear in the next few years. This effort may permit us to focus attention on specific trace elements and their metabolic effects.

IX. CONCLUSION

The influence of diet on vascular disease is complicated. In animals it is possible to obtain data based on feeding of specific nutrients administered under static conditions. The foregoing discussion has provided examples of interactions among nutrients that can affect the influence of single substances on lipid metabolism and atherosclerosis. In man dietary effects are modulated by the influences of other risk factors, such as smoking, obesity, or hypertension.

Currently, efforts at intervention are aimed at reduction of plasma lipid levels. Decreased cardiovascular mortality has been observed in clinical trials in which cholesterol lowering was marked (21) or minimal (213). The recently completed Lipid Research Clinics trial (38, 39) has been hailed as proof of the validity of the lipid hypothesis. It was carried out in a high-risk, hypercholesterolemic population, but it may offer clues to control of atherosclerosis. The major effects were achieved, using a pharmaceutical agent. Surgical intervention via partial ileal bypass leads to dramatic lowering of cholesterol levels (214) with arteriographic evidence of plaque regression in about 14% of the subjects (215). With development of noninvasive means of visualizing the actual state of coronary vessels and arteries, it should become possible to provide a true assessment of the effects of diet and drugs. For the present it would seem wise to be prudent (given current knowledge) but not hysterical (given the gaps in that knowledge).

ACKNOWLEDGMENT

Supported, in part, by a Research Career Award (HL-00734) from the National Institutes of Health and by funds from the Commonwealth of Pennsylvania.

REFERENCES

1. J. Davignon, The lipid hypothesis. Pathophysiological basis. *Arch. Surg. 113*:28 (1978).
2. AHA Committee Report, Rationale for the Diet-Heart Statement of the American Heart Association. *Circulation 65*:839A (1982).
3. T. Strasser, Atherosclerosis and coronary heart disease: The contribution of epidemiology. *WHO Chron. 26*:7 (1972).
4. P. N. Hopkins and R. R. Williams, A survey of 246 suggested coronary risk factors. *Atherosclerosis 40*:1 (1981).
5. A. Kagan, W. B. Kannel, T. R. Dawber, and N. Revotskie, The coronary profile. *Ann. N.Y. Acad. Sci. 97*:883 (1962).
6. L. A. Carlson and L. E. Bottiger, Ischaemic heart disease in relation to fasting values of plasma triglycerides and cholesterol: Stockholm Prospective Study. *Lancet 1*:865 (1972).
7. D. Kritchevsky, Diet and atherosclerosis. *Am. J. Pathol. 84*: 615 (1976).
8. R. W. Wissler and D. Vesselinovitch, Differences between human and animal atherosclerosis, in *Atherosclerosis III* (G. Schettler and A. Weizel, eds.). Springer Verlag, Berlin, 1974. pp. 319-325.
9. D. Kritchevsky, Animal models for atherosclerosis research, in *Hypolipidemic Agents* (D. Kritchevsky, ed.). Springer Verlag, Berlin, 1975, pp. 216-228.
10. N. Anitschkow and S. Chalatow, Ueber experimentelle Cholester-insteatose und ihre Bedeutung fuer die Entstehung einiger pathologischer Prozesse. *Zentralbl. Allg. Pathol. Anat. 24*:1 (1913).
11. L. Wacker and W. Hueck, Ueber experimentelle Atherosklerose und Cholesterinamie. *Muench. Med. Wochenschr. 60*:2097 (1913).
12. S. M. Grundy and A. L. Metzger, A physiological method for estimation of hepatic secretion of biliary lipids in man. *Gastroenterology 62*:1200 (1972).
13. J. A. Kaplan, G. E. Cox, and C. B. Taylor, Cholesterol metabolism in man. *Arch. Pathol. 76*:359 (1963).
14. S. M. Grundy and H. Y. I. Mok, Determination of cholesterol absorption in man by intestinal perfusion. *J. Lipid Res. 68*:263 (1977).
15. E. Quintao, S. M. Grundy, and E. H. Ahrens, Jr., Effects of dietary cholesterol on the regulation of total body cholesterol in man. *J. Lipid Res. 12*:233 (1971).
16. S. M. Grundy, E. H. Ahrens, Jr., and J. Davignon, The interaction of cholesterol absorption and cholesterol synthesis in man. *J. Lipid Res. 10*:304 (1969).
17. A. Keys, J. T. Anderson, and F. Grande, Serum cholesterol response to changes in the diet. II. The effect of cholesterol in the diet. *Metabolism. 14*:759 (1965).

18. D. M. Hegsted, R. B. McGandy, M. L. Myers, and F. J. Stare, Quantitative effects of dietary fat on serum cholesterol in man. *Am. J. Clin. Nutr.* *17*:281 (1965).

19. W. E. Connor, R. E. Hodges, and R. E. Bleiler, The serum lipids in men receiving high cholesterol and cholesterol-free diets. *J. Clin. Invest.* *40*:894 (1961).

20. F. H. Mattson, B. A. Erickson, and A. M. Kligman, Effect of dietary cholesterol on serum cholesterol in man. *Am. J. Clin. Nutr.* *25*:589 (1972).

21. M. M. Gertler, S. M. Garn, and P. D. White, Diet, serum cholesterol and coronary artery disease. *Circulation 2*:696 (1950).

22. G. Slater, J. Mead, G. Dhopeshwarkar, S. Robinson, and R. B. Alfin-Slater, Plasma cholesterol and triglycerides in men with added eggs in the diet. *Nutr. Rep. Int.* *14*:249 (1976).

23. M. W. Porter, W. Yamanaka, S. D. Carlson, and M. A. Flynn, Effect of dietary egg on serum cholesterol and triglyceride of human males. *Am. J. Clin. Nutr.* *30*:490 (1977).

24. T. R. Dawber, R. J. Nickerson, F. N. Brand, and J. Pool, Eggs,serum cholesterol and coronary heart disease. *Am. J. Clin. Nutr.* *36*:617 (1982).

25. W. B. Kannel and T. Gordon, eds., The Framingham Study. Section 24, in *The Framingham Diet Study: Diet and the Regulation of Serum Cholesterol.* United States Department of Health, Education and Welfare, Washington, D.C., 1970.

26. A. B. Nichols, C. Ravenscroft, D. E. Lamphiear, and L. D. Ostrander, Jr., Independence of serum lipid levels and dietary habits, the Tecumseh study, *J. Am. Med. Assoc.* *236*:1948 (1976).

27. T. Gordon, A. Kagan, M. Garcia-Palmieri, W. B. Kannel, W. J. Zukel, J. Tillotson, P. Sorlie, and M. Hjortland, Diet and its relation to coronary heart disease and death in three populations. *Circulation 63*:500 (1981).

28. J. W. Gofman, F. Lindgren, H. Elliott, W. Mantz, J. Hewitt, B. Strisower, V. Herring, and T. P. Lyon, The role of lipids and lipoproteins in atherosclerosis. *Science 111*:166 (1950).

29. T. Gordon, W. P. Castelli, M. C. Hjortland, W. B. Kannel, and T. R. Dawber, Predicting coronary heart disease in middle-aged and older persons. The Framingham Study. *J. Am. Med. Assoc.* *238*:497 (1977).

30. P. W. Wilson, R. J. Garrison, W. P. Castelli, M. Feinlieb, P. M. McNamara, and W. B. Kannel, Prevalence of coronary heart disease in the Framingham Offspring Study: Study of lipoprotein cholesterol. *Am. J. Cardiol.* *46*:649 (1980).

31. D. P. Barr, E. M. Russ, and H. A. Eder, Protein-lipid relationships in human plasma. II. Atherosclerosis and related conditions. *Am. J. Med.* *11*:480 (1951).

32. G. J. Miller and N. E. Miller, Plasma high density lipoprotein concentration and development of ischaemic heart disease. *Lancet* *1*:16 (1975).

33. G. G. Rhoads, C. L. Gulbrandsen, and A. Kagan, Serum lipoproteins and coronary heart disease in a population study of Hawaii Japanese men. *N. Engl. J. Med.* *294*:293 (1976).

34. T. Gordon, W. P. Castelli, M. C. Hjortland, W. B. Kannel, and T. R. Dawber, High density lipoprotein as a protective factor against coronary heart disease. The Framingham Study. *Am. J. Med.* *62*:707 (1977).

35. W. B. Kannel, High density lipoproteins: Epidemiologic profile and risks of coronary artery disease. *Am. J. Cardiol.* *52*:9B (1983).

36. G. H. Rothblat and D. Kritchevsky, The metabolism of free and esterified cholesterol in tissue culture cells: A review. *Exp. Mol. Pathol.* *8*:314 (1968).

37. M. F. Oliver, Coronary heart disease prevention. Trials using drugs to control hyperlipidemia, in *Liproproteins, Atherosclerosis and Coronary Heart Disease* (N. E. Miller and B. Lewis, eds.). Elsevier Biomedical Press, Amsterdam, 1981, pp. 165-195.

38. Lipid Research Clinics Program, The Lipid Research Clinics coronary primary prevention trial results. 1. Reduction in incidence of coronary heart disease. *J. Am. Med. Assoc. 251*: 351 (1984).

39. Lipid Research Clinics Program, The Lipid Research Clinics coronary primary prevention trial results. II. The relationship of reduction in incidence of coronary heart disease to cholesterol lowering. *J. Am. Med. Assoc. 251*:365 (1984).

40. E. H. Ahrens, Jr., The management of hyperlipdemia—whether rather than how. *Ann. Inter. Med. 85*:87 (1976).

41. E. H. Ahrens, Dietary fats and coronary heart disease. Unfinished business. *Lancet 2*:1345 (1974).

42. D. J. McNamara, Diet and hyperlipidemia. A justifiable debate. *Arch. Intern. Med. 142*:1121 (1982).

43. M. F. Oliver, Diet and coronary heart disease. *Br. Med. Bull.* *73*:49 (1981).

44. B. Friend, L. Page, and R. Martson, Food consumption patterns in the United States: 1909-1913 to 1976, in *Nutrition Lipids and Coronary Heart Disease* (R. I. Levy, B. M. Rifkind, B. H. Dennis, and N. D. Ernst, eds.). Raven Press, New York, 1979, pp. 489-522.

45. R. L. Rizek, S. O. Welsh, R. M. Marston, and E. M. Jackson, Levels and sources of fat in the U.S. food supply and in diets of individuals, in *Dietary Fats in Health* (E. G. Perkins and W. J. Visek, eds.). *Am. Oil. Chem. Soc.*, Champaign, Ill. 1983, pp. 13-43.

46. R. Olson, Is there an optimum diet for the prevention of coronary heart disease? in *Nutrition, Lipids and Coronary Heart Disease* (R. I. Levy, B. M. Rifkind, B. H. Dennis, and N. D. Ernst, eds.). Raven Press, New York, 1979, pp. 349-364.

47. M. G. Marmot, M. Booth, and V. Beral, International trends in heart disease mortality. *Atherosclerosis Rev. 9*:19 (1982).

48. M. Feinleib, T. Thom, and R. J. Havlik, Decline in coronary heart disease mortality in the United States. *Atherosclerosis Rev. 9*:29 (1982).

49. R. A. Stallones, Mortality due to ischemic heart disease: Observations and explanations. *Atherosclerosis Rev. 9*:43 (1982).

50. A. Keys, J. T. Anderson, and F. Grande, Prediction of serum cholesterol responses of man to changes in fats in the diet. *Lancet 2*:959 (1957).

51. A. Keys, J. T. Anderson, and F. Grande, Effect on serum cholesterol in man of mono-ene fatty acid (oleic acid) in the diet. *Proc. Soc. Exp. Biol. Med. 98*:387 (1958).

52. E. H. Ahrens, Jr., Nutritional factors and serum lipid levels. *Am. J. Med. 23*:928 (1957).

53. D. Kritchevsky and S. A. Tepper, Cholesterol vehicle in experimental atherosclerosis. 7. Influence of naturally occurring saturated fats. *Med. Pharmacol. Exp. 12*:315 (1965).

54. D. Kritchevsky, S. A. Tepper, G. Bises, and D. M. Klurfeld, Experimental atherosclerosis in rabbits fed cholesterol-free diets. 10. Cocoa butter and palm oil. *Atherosclerosis 41*:279 (1982).

55. S. M. Grundy and E. H. Ahrens, Jr., The effects of unsaturated dietary fats on absorption, excretion, synthesis and distribution of cholesterol in man. *J. Clin. Invest. 49*:1135 (1970).

56. S. M. Grundy, Dietary fats and sterols, in *Nutrition, Lipids and Coronary Heart Disease* (R. I. Levy, B. M. Rifkind, B. H. Dennis, and N. D. Ernst, eds.). Raven Press, New York, 1979, pp. 89-118.

57. I. D. Frantz, Jr. and J. B. Carey, Jr., Cholesterol content of human liver after feeding of corn oil and hydrogenated coconut oil. *Proc. Soc. Exp. Biol. Med. 106*:800 (1961).

58. T. Gerson, F. B. Shorland, and Y. Adams, The effects of corn oil on the amounts of cholesterol and excretion of sterol in the rat. *Biochem. J. 81*:584 (1961).

59. D. Kritchevsky, Role of cholesterol vehicle in experimental atherosclerosis. *Am. J. Clin. Nutr. 23*:1105 (1970).

60. D. Kritchevsky, A. W. Moyer, W. C. Tesar, J. B. Logan, R. A. Brown, M. C. Davies, and H. R. Cox, Effect of cholesterol vehicle in experimental atherosclerosis. *Am. J. Physiol. 178*:30 (1954).

61. D. Kritchevsky, A. W. Moyer, W. C. Tesar, R. F. J. McCand-
     less, J. B. Logan, R. A. Brown, and M. Englert, Cholesterol
     vehicle in experimental atherosclerosis. II. Effect of unsatura-
     tion. *Am. J. Physiol. 185*:279 (1956).

62. J. P. Funch, B. Krogh, and H. Dam, Effect of butter, some
     margarines, and arachis oil in purified diets on serum lipids
     and atherosclerosis in rabbits. *Br. J. Nutr. 14*:355 (1960).

63. D. Kritchevsky, S. A. Tepper, H. K. Kim, J. A. Story, D.
     Vesselinovitch, and R. W. Wissler, Experimental atherosclerosis
     in rabbits fed cholesterol-free diets. 5. Comparison of peanut,
     corn, butter and coconut oils. *Exp. Mol. Pathol. 24*:375 (1976).

64. G. A. Gresham and A. N. Howard, The independent production
     of atherosclerosis and thrombosis in the rat. *Br. J. Exp. Pathol.
     41*:395 (1960).

65. R. F. Scott, E. S. Morrison, W. A. Thomas, R. Jones, and
     S. C. Nam, Short term feeding of unsaturated versus saturated
     fat in the production of atherosclerosis in the rat. *Exp. Mol.
     Pathol. 3*:421 (1964).

66. D. Kritchevsky, S. A. Tepper, D. Vesselinovitch, and R. W.
     Wissler, Cholesterol vehicle in experimental atherosclerosis. 11.
     Peanut oil. *Atherosclerosis 14*:53 (1971).

67. D. Vesselinovitch, G. S. Getz, R. H. Hughes, and R. W. Wis-
     sler, Atherosclerosis in the rhesus monkey fed three food fats.
     Atherosclerosis 20:303 (1974).

68. D. Kritchevsky, L. M. Davidson, M. Weight, N. P. J. Kriek,
     and J. P. duPlessis, Influence of native and randomized peanut
     oil on lipid metabolism and aortic sudanophilia in the vervet
     monkey. *Atherosclerosis 42*:53 (1982).

69. D. Kritchevsky, S. A. Tepper, D. Vesselinovitch, and R. W.
     Wissler, Cholesterol vehicle in experimental atherosclerosis. 13.
     Randomized peanut oil. *Atherosclerosis 17*:225 (1973).

70. J. J. Myher, L. Marai, A. Kuksis, and D. Kritchevsky, Acyl-
     glycerol structure of peanut oils of different atherogenic poten-
     tial. *Lipids 12*:775 (1977).

71. F. Manganaro, J. J. Myher, A. Kuksis, and D. Kritchevsky,
     Acylglycerol structure of genetic varieties of peanut oils of
     varying atherogenic potential. *Lipids 16*:508 (1981).

72. H. K. Kim and D. Kritchevsky, Lipolysis of corn, peanut and
     randomized peanut oils. *Lipids 18*:842 (1983).

73. P. Tso, G. Pinkston, D. M. Klurfeld, and D. Kritchevsky,
     The absorption and transport of dietary cholesterol in the pres-
     ence of peanut oil and randomized peanut oil. *Lipids 19*:11
     (1984).

74. R. B. McGandy, D. M. Hegsted, and M. L. Meyers, Use of
     semi-synthetic fats in determining effects of specific dietary
     fatty acids on serum lipids in man. *Am. J. Clin. Nutr. 23*:1288
     (1970).

75. D. Kritchevsky and S. A. Tepper, Cholesterol vehicle in experimental atherosclerosis. X. Influence of specific saturated fatty acids. *Exp. Mol. Pathol.* 6:394 (1967).
76. F. A. Kummerow, Current studies on relation of fat to health. *J. Am. Oil Chem. Soc.* 51:255 (1974).
77. K. Ono and D. S. Fredrickson, The metabolism of ^{14}C-labeled cis and trans isomers of octadecanoic and octadecadienoic acids. *J. Biol. Chem.* 239:2482 (1964).
78. R. L. Anderson, Oxidation of the geometric isomers of 9,12-octadecadienoic acids by rat liver mitochondria. *Biochim. Biophys. Acta 152*:531 (1968).
79. R. J. Jensen, J. Sampugna, and R. L. Pereira, Pancreatic lipase: Lipolysis of synthetic triglycerides containing a trans fatty acid. *Biochim. Biophys. Acta 84*:481 (1964).
80. D. S. Sgoutas, Hydrolysis of synthetic cholesterol esters containing trans fatty acids. *Biochim. Biophys. Acta 164*:317 (1968).
81. D. S. Sgoutas, Effect of geometry and position of ethylenic bond upon acyl coenzyme-cholesterol-*O*-acyltransferase. *Biochemistry 9*:1826 (1970).
82. D. Kritchevsky and A. R. Baldino, Pancreatic cholesteryl ester synthetase: Effects of trans unsaturated and long chain saturated fatty acids. *Artery 4*:480 (1978).
83. B. I. Weigensberg, G. C. McMillan, and A. C. Ritchie, Elaidic acid: Effect on experimental atherosclerosis. *Arch. Pathol. 72*: 126 (1961).
84. G. C. McMillan, M. D. Silver, and B. I. Weigensberg, Elaidinized olive oil and cholesterol atherosclerosis. *Arch. Pathol. 76*:106 (1963).
85. H. Ruttenberg, L. M. Davidson, N. A. Little, D. M. Klurfeld, and D. Kritchevsky, Influence of trans unsaturated fats on experimental atherosclerosis in rabbits. *J. Nutr. 113*:835 (1983).
86. R. L. Jackson, J. D. Morrisett, H. Pownall, A. M. Gotto, Jr., A. Kamio, H. Imai, R. Tracy, and F. A. Kummerow, Influence of dietary trans fatty acids on swine lipoprotein composition and structure. *J. Lipid Res. 18*:182 (1977).
87. D. Kritchevsky, Trans fatty acid effects in experimental atherosclerosis. *Fed. Proc. 41*:2813 (1982).
88. C. E. Moore, R. B. Alfin-Slater, and L. Aftergood, Incorporation and disappearance of trans fatty acids in rat tissues. *Am. J. Clin. Nutr. 33*:2318 (1980).
89. H. Heckers, M. Korner, T. W. L. Tusihen, and F. W. Melcher, Occurence of individual trans isomeric fatty acids in human myocardium, jejunum and aorta in relation to different degrees of atherosclerosis. *Atherosclerosis 28*:389 (1977).
90. L. H. Thomas, P. R. Jones, J. A. Winter, and H. Smith, Hydrogenated oils and fats: The presence of chemically modified fatty acids in human adipose tissue. *Am. J. Clin. Nutr. 34*:877 (1981).

91. U. M. T. Houtsmuller, Biochemical aspects of fatty acids with trans double bonds. *Fette Seifen Anstrichm. 80*:162 (1978).

92. J. Yerushalmy and H. E. Hilleboe, Fat in the diet and mortality from heart disease: A methodologic note. *N.Y. State J. Med. 57*:2343 (1957).

93. A. Ignatowski, Influence de la nourriture animale sur l'organisme des lapins. *Arch. Med. Exp. Anat. Pathol. 20*:1 (1908).

94. A. Ignatowski, Uber die wirkung des Tierischen eiweisses auf die Aorta und die parenchymatosen Organe der Kaninchen. *Virchows Arch. Pathol. Anat. Physiol. Klin. Med. 198*:248 (1909).

95. L. H. Newburg, The production of Bright's disease by feeding high protein diets. *Arch. Inter. Med. 24*:359 (1919).

96. L. H. Newburgh and T. L. Squier, High protein diets and arteriosclerosis in rabbits. A preliminary report. *Arch. Intern. Med. 26*:38 (1920).

97. L. H. Newburgh and S. Clarkson, Production of arteriosclerosis in rabbits by diets rich in animal protein. *J. Am. Med. Assoc. 79*:1106 (1922).

98. L. H. Newburgh and S. Clarkson, The production of arteriosclerosis in rabbits by feeding diets rich in meat. *Arch. Intern. Med. 31*:653 (1923).

99. S. Clarkson and L. H. Newburgh, The relation between atherosclerosis and ingested cholesterol in the rabbit. *J. Exp. Med. 43*:595 (1926).

100. D. R. Meeker and H. D. Kesten, Experimental atherosclerosis and high protein diet. *Proc. Soc. Exp. Biol. Med. 45*:543 (1940).

101. D. R. Meeker and H. D. Kesten, Effect of high protein diet on experimental atherosclerosis in rabbits. *Arch. Pathol. 31*: 147 (1941).

102. A. N. Howard, G. A. Gresham, D. Jones, and I. W. Jennings, The prevention of rabbit atherosclerosis by soya bean meal. *J. Atheroscler. Res. 5*:330 (1965).

103. K. K. Carroll and R. M. G. Hamilton, Effects of dietary protein and carbohydrate on plasma cholesterol levels in relation to atherosclerosis. *Food Sci. 40*:18 (1975).

104. D. Kritchevsky, S. A. Tepper, S. K. Czarnecki, J. A. Story, and J. B. Marsh, Experimental atherosclerosis in rabbits fed cholesterol-free diets. 11. Corn protein, wheat gluten and lactalbumin. *Nutr. Rep. Int. 26*:931 (1982).

105. S. K. Czarnecki, *Effects of Dietary Proteins on Lipoprotein Metabolism and Atherosclerosis in Rabbits.* Ph.D. Dissertation, University of Pennyslvania, Philadelphia, 1982.

106. D. Kritchevsky, S. A. Tepper, S. K. Czarnecki, D. M. Klurfeld, and J. A. Story, Effects of animal and vegetable protein in experimental atherosclerosis, in *Animal and Vegetable Proteins*

in *Lipid Metabolism and Atherosclerosis* (M. J. Gibney and D. Kritchevsky, eds.). Alan R. Liss, Inc., New York, 1983, pp. 85-100.

107. D. Kritchevsky, R. R. Kolman, R. M. Guttmacher, and M. Forbes, Influence of dietary carbohydrate and protein on serum and liver cholesterol in germ-free chickens. *Arch. Biochem. Biophys. 85*:444 (1959).

108. M. W. Huff, R. M.G. Hamilton, and K. K. Carroll, Plasma cholesterol levels in rabbits fed low fat, cholesterol-free semi-purified diets: Effects of dietary proteins, protein hydrolysates and amino acid mixture. *Atherosclerosis 28*:187 (1977).

109. D. Kritchevsky, S. A. Tepper, and J. A. Story, Influence of soy protein and casein on atherosclerosis in rabbits. *Fed. Proc. 37*:747 (1978).

110. D. Kritchevsky, Vegetable protein and atherosclerosis. *J. Am. Oil Chem. Soc. 56*:135 (1979).

111. J. D. Jones, R. Wolters, and P. C. Burnett, Lysine-arginine electrolyte relationships in the rat. *J. Nutr. 89*:171 (1966).

112. J. D. Jones, Lysine-arginine antagonism in the chick. *J. Nutr. 84*:313 (1964).

113. A. Hunter and C. E. Downs, The inhibition of arginase by amino acids. *J. Biol. Chem. 157*:427 (1945).

114. J. A. Milner and A. S. Hassan, Species specificity of arginine deficiency induced hepatic steatosis. *J. Nutr. 111*:1067 (1981).

115. S. K. Czarnecki and D. Kritchevsky, The effect of dietary proteins on lipoprotein metabolism in atherosclerosis in rabbits. *Am. Oil Chem. Soc. 56*:388A (1979).

116. D. Kritchevsky, Dietary protein in atherosclerosis, in *Diet, Drugs and Atherosclerosis* (G. Noseda, B. Lewis, and R. Paoletti, eds.). Raven Press, New York, 1980, pp. 9-14.

117. S. K. Czarnecki and D. Kritchevsky, Effects of dietary protein on lipoprotein metabolism and atherosclerosis in rabbits. *Fed. Proc. 39*:342 (1980).

118. D. Kritchevsky, S. A. Tepper, S. K. Czarnecki, and D. M. Klurfeld, Atherogenicity of animal and vegetable protein: Influence of the lysine to arginine ratio. *Atherosclerosis 41*:429 (1982).

119. D. Kritchevsky, S. A. Tepper, S. K. Czarnecki, D. M. Klurfeld, and J. A. Story, Experimental atherosclerosis in rabbits fed cholesterol-free diets. 9. Beef protein and textured vegetable protein. *Atherosclerosis 39*:169 (1981).

120. M. W. Huff and K. K. Carroll, Effects of dietary protein on turnover, oxidation and absorption of cholesterol and on steroid excretion in rabbits. *J. Lipid Res. 21*:546 (1980).

121. R. J. J. Hermus, *Experimental Atherosclerosis in Rabbits on Diets with Milk Fat and Different Proteins.* Centre for Agriculture Publications and Documentation, Wageningen, The Netherlands, 1975.

122. R. Fumagalli, R. Paoletti, and A. N. Howard, Hypocholestero-laemic effect of soy. *Life Sci. 22*:947 (1978).

123. H. B. Lofland, T. B. Clarkson, L. Rhyne, and H. O. Good-man, Interrelated effects of dietary fats and proteins on atherosclerosis in the pigeon. *J. Atheroscler. Res. 6*:395 (1966).

124. J. P. Strong and H. C. McGill, Jr., Diet and experimental atherosclerosis in baboons. *Am. J. Pathol. 50*:669 (1967).

125. A. H. M. Terpstra, C. J. H. Woodward, C. E. West, and H. G. Van Boven, A longitudinal cross-over study of serum cholesterol and lipoproteins in rabbits fed on semi-purified diets containing either casein or soya bean protein. *Br. J. Nutr. 47*:213 (1982).

126. M. G. Hardinge and F. J. Stare, Nutritional studies of vege-tarians. 2. Dietary and serum levels of cholesterol. *Am. J. Clin. Nutr. 2*:83 (1954).

127. R. E. Hodges, W. A. Krehl, D. B. Stone, and A. Lopez, Dietary carbohydrates and low cholesterol diets: Effects on serum lipids of man. *Am. J. Clin. Nutr. 20*:198 (1967).

128. G. R. Walker, E. H. Morse, and V. A. Oversley, The effect of animal protein and vegetable protein diets having the same fat content on the serum lipid levels of young women. *J. Nutr. 72*:317 (1960).

129. C. R. Sirtori, E. Agradi, F. Conti, O. Mantero, and E. Gatti, Soybean protein in the treatment of type II hyperlipoprotein-aemia. *Lancet 1*:275 (1977).

130. C. R. Sirtori, E. Gatti, O. Mantero, F. Conti, E. Agradi, E. Tremoli, M. Sirtori, L. Fraterrigo, L. Tavazzi, and D. Kritchevsky, Clinical experience with the soybean protein diet in the treatment of hypercholesterolemia. *Am. J. Clin. Nutr. 32*:1645 (1979).

131. W. L. Holmes, G. B. Rubel, and S. S. Hood, Comparison of the effect of dietary meat versus dietary soybean protein on plasma lipids of hyperlipidemic individuals. *Atherosclerosis 36*: 379 (1980).

132. D. Kritchevsky, S. A. Tepper, D. E. Williams, and J. A. Story, Experimental atherosclerosis in rabbits fed cholesterol-free diets. 7. Interaction of animal or vegetable protein with fiber. *Athero-sclerosis 26*:397 (1977).

133. L. A. Carlson, Serum lipids in men with myocardial infarction. *Acta Med. Scand. 167*:397 (1960).

134. M. J. Albrink, J. W. Meigs, and E. B. Man, Serum lipids, hypertension and coronary artery disease. *Am. J. Med. 31*:4 (1961).

135. S. Eisenberg, Type III hyperlipoproteinaemia. *Clin. Endocrinol. Metab. 2*:111 (1973).

136. S. B. Hulley, R. H. Rosenman, R. D. Bawol, and R. J. Brand, Epidemiology as a guide to clinical decisions. The association between triglycerides and coronary heart disease. *N. Engl. J. Med. 302*:1383 (1980).

137. N. R. Phillips, R. J. Havel, and J. P. Kane, Levels and interrelationships of serum and lipoprotein cholesterol and triglycerides. Association with adiposity and the consumption of ethanol, tobacco and beverages containing caffeine. *Arteriosclerosis 1*:13 (1981).

138. J. L. Knittle and E. H. Ahrens, Carbohydrate metabolism in two forms of hyperglyceridemia. *J. Clin. Invest. 43*:485 (1964).

139. E. L. Bierman and D. Porte, jr., Carbohydrate intolerance and lipemia. *Ann. Intern. Med. 68*:926 (1968).

140. I. MacDonald and D. M. Braithwaite, The influence of dietary carbohydrates on the lipid pattern in serum and in adipose tissue. *Clin. Sci. 27*:23 (1964).

141. P. T. Kuo, Dietary sugar in the production of hypertriglyceridemia in patients with hyperlipemia and atherosclerosis. *Trans. Assoc. Am. Physicians 78*:97 (1965).

142. E. A. Nikkila and R. Pelkonen, Enhancement of alimentary hypertriglyceridemia by fructose and glycerol in man. *Proc. Soc. Exp. Biol. Med. 123*:91 (1966).

143. E. A. Nikkila and K. Ojala, Induction of hyperglyceridemia by fructose in the rat. *Life Sci. 4*:937 (1965).

144. H. Bar-On and Y. Stein, Effect of glucose and fructose administration on lipid metabolism in the rat. *J. Nutr. 94*:95 (1968).

145. D. Kritchevsky, W. C. Grant, M. J. Fahrenbach, B. A. Riccardi, and R. F. J. McCandless, Effect of dietary carbohydrate on the metabolism of cholesterol-4-C^{14} in chickens. *Arch. Biochem. Biophys. 75*:142 (1958).

146. W. C. Grant and M. J. Fahrenbach, Effect of dietary sucrose and glucose on plasma cholesterol in chicks and rabbits. *Proc. Soc. Exp. Biol. Med. 100*:250 (1959).

147. W. W. Wells and S. C. Anderson, The increased severity of atherosclerosis in rabbits on a lactose-containing diet. *J. Nutr. 68*:541 (1959).

148. D. Kritchevsky and S. A. Tepper, Factors affecting atherosclerosis in rabbits fed cholesterol-free diets. *Life Sci. 4*:1467 (1965).

149. D. Kritchevsky, P. Sallata, and S. A. Tepper, Experimental atherosclerosis in rabbits fed cholesterol-free diets. 2. Influence of various carbohydrates. *J. Atheroscler. Res. 8*:697 (1968).

150. D. Kritchevsky, S. A. Tepper, and M. Kitagawa, Experimental atherosclerosis in rabbits fed cholesterol-free diets. 3. Comparison of fructose with other carbohydrates. *Nutr. Rep. Int. 7*:193 (1973).

151. R. W. St. Clair, N. C. Bullock, N. D. M. Lehner, T. B. Clarkson, and H. B. Lofland, Long term effects of dietary sucrose and starch on serum lipids and atherosclerosis in minature swine. *Exp. Mol. Pathol. 15*:21 (1971).

152. D. Kritchevsky, L. M. Davidson, and J. J. van der Watt, The influence of carbohydrates on atherosclerosis in primates, in *Metabolic Effects of Utilizable Dietary Carbohydrate* (S. Reiser, ed.). Marcel Dekker, Inc., New York, 1982, pp. 141-174.

153. O. W. Portman, D. M. Hegsted, F.J. Stare, R. Murphy, and L. Sinisterra, Effect of level and type of dietary fat on the metabolism of cholesterol and beta-lipoproteins in the cebus monkey. *J. Exp. Med. 104*:817 (1956).

154. J. E. Corey, K. C. Hayes, B. Dorr, and D. M. Hegsted, Comparative lipid response of four primate species to dietary changes in fat and carbohydrate. *Atherosclerosis 19*:119 (1974).

155. C. M. Lang and C. H. Barthel, Effects of simple and complex carbohydrates on serum lipids and atherosclerosis in non-human primates. *Am. J. Clin. Nutr. 25*:470 (1972).

156. D. Kritchevsky, L. M. Davidson, H. K. Kim, D. A. Krendel, S. Malhotra, J. J. van der Watt, J. P. duPlessis, P. A. D. Winter, T. Ipp, D. Mendelsohn, and I. Bersohn, Influence of semipurified diets on atherosclerosis in African green monkeys. *Exp. Mol. Pathol. 26*:28 (1977).

157. D. Kritchevsky, L. M. Davidson, I. L. Shapiro, H. K. Kim, M. Kitagawa, S. Malhotra, P. P. Nair, T. B. Clarkson, I. Bersohn, and P. A. D. Winter, Lipid metabolism and experimental atherosclerosis in baboons: Influence of cholesterol-free, semi-synthetic diets. *Am. J. Clin. Nutr. 27*:29 (1974).

158. D. Kritchevsky, L. M. Davidson, H. K. Kim, D. A. Krendel, S. Malhotra, D. Mendelsohn, J. J. van der Watt, J. P. duPlessis, and P. A. D. Winter, Influence of type of carbohydrate on atherosclerosis in baboons fed semipurified diets plus 0.1% cholesterol. *Am. J. Clin. Nutr. 33*:1869 (1980).

159. D. P. Burkitt, A. R. P. Walker, and N. S. Painter, Dietary fiber and disease. *J. Am. Med. Assoc. 229*:1068 (1974).

160. H. Trowell, Ischemic heart disease and dietary fiber. *Am. J. Clin. Nutr. 25*:926 (1972).

161. A. R. P. Walker and U. B. Arvidsson, Fat intake, serum cholesterol concentration and atherosclerosis in the South African Bantu. I. Low fat intake and age trend of serum cholesterol concentration in the South African Bantu. *J. Clin. Invest. 33*:1358 (1954).

162. G. V. Vahouny, Dietary fiber—aspects of nutrition, pharmacology and pathology, in *Nutritional Pathology* (H. Sidransky, ed.) Marcel Dekker, New York, 1985, Chapter 6.

163. R. M. Kay and A. S. Truswell, Dietary fiber: Effects on plasma and biliary lipids in man, in *Medical Aspects of Dietary Fiber* (G. A. Spiller and R. M. Kay, eds.), Plenum Medical Book Co., New York, 1980, pp. 153-173.

164. A. Keys, F. Grande, and J. T. Anderson, Fiber and pectin in the diet and serum cholesterol concentration in man. *Proc. Soc. Exp. Biol. Med.* 106:555 (1961).

165. D. J. A. Jenkins, A. R. Leeds, C. Newton, and J. H. Cummings, Effect of pectin, guar gum and wheat fibre on serum cholesterol. *Lancet 1*:1116 (1975).

166. R. M. Kay and A. S. Truswell, Effect of citrus pectin on blood lipids and fecal steroid excretion in man. *Am. J. Clin. Nutr.* 30:171 (1977).

167. T. A. Miettinen and S. Tarpila, Effect of pectin on serum cholesterol, fecal bile acids and bilary lipids in normolipidemic and hyperlipidemic individuals. *Clin. Chim. Acta* 79:471 (1977).

168. D. B. Stone and W. E. Connor, The prolonged effects of a low cholesterol, high carbohydrate diet upon the serum lipids in diabetic patients. *Diabetes 12*:127 (1963).

169. T. G. Kiehm, J. W. Anderson, and K. Ward, Beneficial effects of a high carbohydrate, high fiber diet on hyperglycemic, diabetic man. *Am. J. Clin. Nutr.* 29:895 (1976).

170. P. A. Miranda and D. L. Horwitz, High fiber diets in the treatment of diabetes mellitus. *Ann. Intern. Med.* 88:482 (1978).

171. J. W. Anderson, Dietary fiber and diabetes, in *Medical Aspects of Dietary Fiber* (G. A. Spiller and R. M. Kay, eds.). Plenum Medical Book Co., New York, 1980, pp. 193-221.

172. M. G. Hardinge, A. C. Chambers, H. Crooks, and F. J. Stare, Nutritional studies of vegetarians. III. Dietary level of fiber. *Am. J. Clin. Nutr.* 6:523 (1958).

173. F. M. Sacks, W. P. Castelli, A. Donner, and E. H. Kass, Plasma lipids and lipoproteins in vegetarians and controls. *N. Engl. J. Med.* 292:1148 (1975).

174. R. W. Kirby, J. W. Anderson, B. Sieling, E. D. Rees, W. J. Chen. R. E. Miller, and R. M. Kay, Oat bran intake selectively lowers serum low density lipoprotein cholesterol concentration of hypercholesterolemic men. *Am. J. Clin. Nutr.* 34:824 (1981).

175. A. F. Wells and B. H. Ershoff, Beneficial effects of pectin in prevention of hypercholesterolemia and increase in liver cholesterol in cholesterol-fed rats. *J. Nutr.* 74:87 (1961).

176. J. A. Story, A. Baldino, S. K. Czarnecki, and D. Kritchevsky, Modification of liver cholesterol accumulation by dietary fiber in rats. *Nutr. Rep. Int.* 24:1213 (1981).

177. A. C. Tsai, J. Elias, J. J. Kelley, R. S. C. Lin, and J. R. K. Robson, Influence of certain dietary fibers on serum and tissue cholesterol levels in rats. *J. Nutr.* 106:118 (1976).

178. B. H. Ershoff, Effects of pectin NF and other complex carbo-
     hydrates on hypercholesterolemia and atherosclerosis. *Exp.
     Med. Surg.* 21:108 (1963).
179. L. M. Berenson, R. R. Bhandaru, B. Radhakrishnamurthy,
     S. B. Srinivasan, and G. S. Berenson, The effect of dietary
     pectin on serum lipoprotein cholesterol in rabbits. *Life Sci.* 16:
     1533 (1975).
180. R. M. G. Hamilton and K. K. Carroll, Plasma cholesterol levels
     in rabbits fed low fat, low cholesterol diets. Effects of dietary
     proteins, carbohydrates and fibre from different sources.
     Atherosclerosis 24:47 (1976).
181. H. Fisher, W. G. Soller, and P. Griminger, The retardation
     by pectin of cholesterol-induced atherosclerosis in the fowl.
     J. Atheroscler. Res. 6:292 (1966).
182. M. R. Malinow, P. McLaughlin, L. Papworth, H. K. Naito,
     and L. A. Lewis, Effect of bran and cholestyramine on plasma
     lipids in monkeys. *Am. J. Clin. Nutr.* 29:905 (1976).
183. F. B. Cookson, R. Altschul, and S. Fedoroff, The effects of
     alfalfa on serum cholesterol and in modifying or preventing
     cholesterol-induced atherosclerosis in rabbits. *J. Atheroscler.
     Res.* 7:69 (1967).
184. A. N. Howard, G. A. Gresham, I. W. Jennings, and D. Jones,
     The effect of drugs on hypercholesterolaemia and atherosclerosis
     induced by semisynthetic, low cholesterol diets. *Prog. Biochem.
     Pharmacol.* 2:117 (1967).
185. G. F. Lambert, J. P. Miller, R. T. Olsen, and D. V. Frost,
     Hypercholesteremia and atherosclerosis induced in rabbits by
     purified high fat rations devoid of cholesterol. *Proc. Soc. Exp.
     Biol. Med.* 97:544 (1958).
186. H. Malmros and G. Wigand, Atherosclerosis and deficiency of
     essential fatty acids. *Lancet* 2:749 (1959).
187. D. Kritchevsky, Experimental atherosclerosis in rabbits fed
     cholesterol-free diets. *J. Atheroscler. Res.* 4:103 (1964).
188. D. Kritchevsky and S. A. Tepper, Experimental atherosclerosis
     in rabbits fed cholesterol-free diets: Influence of chow compo-
     nents. *J. Atheroscler. Res.* 8:357 (1968).
189. J. H. Moore, The effect of the type of roughage in the diet
     on plasma cholesterol levels and aortic atherosis in rabbit. *Br.
     J. Nutr.* 21:207 (1967).
190. D. Kritchevsky, L. M. Davidson, D. A. Krendel, J. J. van
     der Watt, D. Russell, S. Friedland, and D. Mendelsohn, Influ-
     ence of dietary fiber of aortic sudanophilia in vervet monkeys.
     Ann. Nutr. Metab. 25:125 (1981).
191. D. Kritchevsky, S. A. Tepper, H. K.Kim. D. E. Moses, and
     J. A. Story, Experimental atherosclerosis in rabbits fed cho-
     lesterol-free diets. 4. Investigation into the source of choles-
     teremia. *Exp. Mol. Pathol.* 22:11 (1977).

192. D. Portman and P. Murphy, Excretion of bile acids and hydroxysterols by rats. *Arch. Biochem. Biophys.* 76:367 (1958).
193. G. A. Leveille and H. E. Sauberlich, Mechanism of the cholesterol-depressing effect of pectin in the cholesterol-fed rat. *J. Nutr.* 88:209 (1966).
194. D. Kritchevsky, S. A. Tepper, and J. A. Story, Isocaloric, isogravic diets in rats. III. Effect of non-nutritive fiber (alfalfa or cellulose) on cholesterol metabolism. *Nutr. Rep. Int.* 9:301 (1974).
195. M. A. Eastwood and D. Hamilton, Studies on the adsorption of bile salts to nonabsorbed components of diet. *Biochim. Biophys. Acta* 152:165 (1968).
196. D. Kritchevsky and J. A. Story, Binding of bile salts in vitro by nonnutritive fiber. *J. Nutr.* 104:458 (1974).
197. H. J. Birkner and F. Kern, Jr., In vitro adsorption of bile salts to food residues, salicylazosulfapyridine and hemicellulose. *Gastroenterology* 67:237 (1974).
198. J. A. Story and D. Kritchevsky, Comparison of the binding of various bile acids and bile salts in vitro by several types of fiber. *J. Nutr.* 106:1292 (1976).
199. H. A. Schroeder, Relation between mortality from cardiovascular disease and treated water supplies: Variations in states and 163 largest municipalities of the United States. *J. Am. Med. Assoc.* 172:1902 (1960).
200. J. N. Morris, M. D. Crawford and J. A. Heady, Hardness of local water supplies and mortality from cardiovascular disease in county boroughs of England and Wales. *Lancet* 1:860 (1961).
201. D. R. Peterson, D. J. Thompson, and J. M. Nam, Water hardness, arteriosclerotic heart disease and sudden death. *Am. J. Epidemiol.* 92:90 (1970).
202. R. Masironi, Cardiovascular mortality in relation to radioactivity and hardness of local water supplies in the USA. *Bull. WHO* 43:687 (1970).
203. R. Masironi, A. T. Miesch, M. D. Crawford, and E. I. Hamilton, Geochemical environments, trace elements and cardiovascular diseases. *Bull. WHO* 47:139 (1972).
204. H. A. Schroeder, A. P. Nason, and I. H. Tipton, Chromium deficiency as a factor in atherosclerosis. *J. Chronic Dis.* 23:123 (1970).
205. M. Schroll, B. Peterson, and C. Christiansen, Is hypocalcaemia protective against hyperlipidaemia? *Br. Med. J.* 3:226 (1975).
206. K. Schwarz, Silicon, fibre and atherosclerosis. *Lancet* 1:454 (1977).
207. I. Szelenzi, Magnesium and its significance in cardiovascular and gastrointestinal disorders. *World Rev. Nutr. Diet.* 17:189 (1973).

208. M. S. Seelig and H. A. Heggtveit, Magnesium interrelationships in ischemic heart disease. A review. *Am. J. Clin. Nutr. 27*: 59 (1974).
209. L. C. Neri and H. L. Johansen, Water hardness and cardiovascular mortality. *Ann. N.Y. Acad. Sci. 304*:203 (1978).
210. L. M. Klevay, Coronary heart disease: The zinc/copper hypothesis. *Am. J. Clin. Nutr. 28*:764 (1975).
211. W. Mertz, Trace minerals and atherosclerosis. *Fed. Proc. 41*: 2807 (1982).
212. O. Turpeinen, Effect of cholesterol-lowering diet on mortality from cardiovascular disease and other causes. *Circulation 59*:1 (1979).
213. Committee of Principal Investigators, A cooperative trial in the primary prevention of ischaemic heart disease using clofibrate. *Br. Heart J. 40*:1069 (1978).
214. H. Buchwald and R.L. Varco, Partial ileal bypass for hypercholesterolemia and atherosclerosis. *Surg. Gynecol. Obstet. 124*: 1231 (1967).
215. L. Knight, R. Scherbel, K. Amplatz, R. L. Varco, and H. Buchwald, A radiographic appraisal of the Minnesota partial ileal bypass study. *Surg. Forum 23*:141 (1972).

5

Pathological Effects of Malnutrition on the Central Nervous System

Brian L. G. Morgan and Myron Winick

Institute of Human Nutrition
Columbia University College of Physicians & Surgeons
New York, New York

I. INTRODUCTION

Human brain growth follows a characteristic sigmoidal pattern (1). The short period of rapid growth is known as the brain growth spurt, which is the time when the brain is most vulnerable to noxious stimuli capable of impairing growth. There are also critical periods of physiological, biochemical, and psychological development when these variables are maturing at their most rapid rates. During each of these types of growth there are different types of growth ongoing, each with its own critical period. A good example is that anatomical growth encompasses a period when cell division is most rapid, which does not coincide with the period of most rapid myelination or with that of maximal dendritic arborization. The timing of each of these critical periods also differs from region to region in the brain. For instance, the activation of a given enzyme may occur later in the cerebellum than in the cerebrum. All of these critical phases of growth take place during intrauterine life or in the first 2 years of life. Once the chronological time has passed for a given type of growth, neither nutritional supplementation nor environmental or any other kind of stimulation can restart it. It is true that all facets of the growth process are time-dependent.

II. CELLULAR GROWTH

Tissue growth including brain growth is composed of cell division and
cell enlargement. The diploid nucleus from the cell of any given species
contains a constant amount of DNA, which is different from that of any
other species (2,3). The nuclei in man contain 6.0 pg of DNA. There
are a few tetraploid Purkinje cells in the cerebral and cerebellar cortices,
but most human brain cells are diploid (4). Hence a good approximation
of brain cell number may be obtained by dividing the DNA content by
6.0. This measurement may be applied to total brain cell number or to
regional cell numbers. Brain RNA content increases directly in proportion to the increasing DNA content throughout the growing period (5).

Cell size can be estimated by biochemical analysis and is usually expressed as (protein/DNA), (lipid/DNA), or (weight/DNA). An increase
in any of these ratios may be an indication of cell enlargement.

Total DNA content gives an accurate figure for brain cell number but
does not give any indication of the numbers of the different types of
brain cells. There are various categories of cells within the central
nervous system, characterized by different dimensions, protein, and
RNA contents. However they can be broadly categorized into just two
groups, namely neurons and glial cells (6).

Brain weight increases until 6 years of age (7). In utero DNA synthesis is linear. After birth it continues until at least 6 years of age
but at a much slower rate. There are two peaks of DNA synthesis during human brain development (Figure 1). The first can be seen to occur at around 18 weeks of gestation. This represents the period of most
rapid neuronal division. The second peak, representing the time of
most rapid glial cell proliferation, can be seen to occur approximately
at the time of birth (7).

As shown in Figure 2 brain stem and forebrain DNA levels reach 70%
of mature levels by age 2 years. The remainder is slowly acquired
by age 6 years. Accretion of DNA proceeds more rapidly in the cerebellum and reaches adult levels by 2 years of age. Although the cerebellum constitutes only 10% of total brain weight, it contains 30% of
total brain DNA. In the cerebellum, and possibly in other brain areas
as well, the microneurons divide during the first 20 months of age (8).

Despite criticism from Dobbing's group (9) most workers accept
Winick's (10) theory of growth. As shown in Figure 3 it proposes that
growth of any organ can be divided into three distinct phases. In the
first phase there is a rapid division of cells with cell size remaining constant. This is followed by a second where DNA synthesis continues at a
slower rate than in the first phase, but with protein synthesis continuing
at the same rate. This phase is characterized by a smaller but significant
increase in cell numbers and a general increase in cell size. The following third phase includes an increase in cell size but no increase in cell
numbers. Growth finally stops at the end of phase 3 when protein synthesis equals protein degradation. This is defined as maturity. Winick

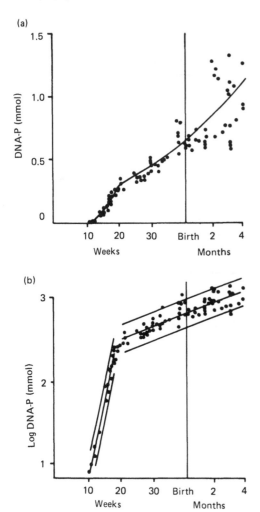

Figure 1 (a) Total DNA phosphate, equivalent to total cell number, in the forebrain of human fetuses and infants; (b) A semilogarithmic plot of the same data as shown in (a). In (b) regression lines with 95% confidence limits are added. (From Ref. 7.)

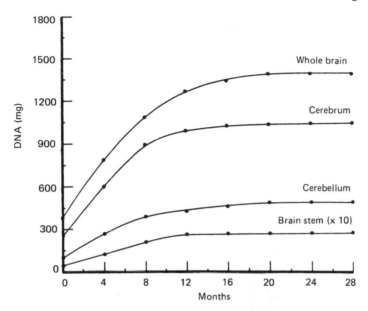

Figure 2 Total DNA content in the different regions of the human brain during development. (From Ref. 6.)

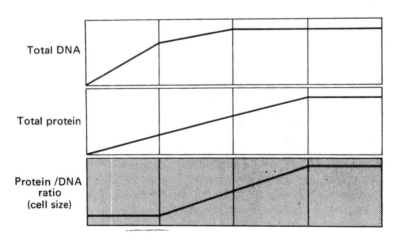

Figure 3 Changes in brain DNA, protein, and protein/DNA ratios during the periods of cellular growth. Going from left to right the first square represents the period of cell division alone, the second cell division and cell enlargement, the third cell enlargement alone, and the fourth maturity. (From Ref. 6.)

emphasizes that there is a gradual change from one phase to another. Hence at any one instant some cells will have reached stage 3 when others will still be in stage 2.

Sands et al. (9) refute the theory. They conclude from their rat data on kidney, liver, and heart that mean cell size increases at an earlier stage than proposed by Winick and quickly reaches a constant level. They also maintain that cell propagation continues unabated throughout tissue growth until maturity.

What controls the rate of DNA synthesis, or the timing of its synthesis, is under question. One clue to the question may be the finding that the activity of DNA polymerase correlates with the rate of cell division in the whole rat brain and its regions (11). The activity of the enzyme is highest when cellular proliferation is in its most rapid phase. Other phenomena have been shown to be associated with the growth process. Ribonucleic acid turnover increases in the growing brain. This is accomplished by the presence of an increased activity of alkaline RNase, which is an enzyme involved in RNA degradation (12). Lewis and Winick (13) made another interesting observation when they showed that there are two distinct species of cytoplasmic RNA that are present in the brain when there is ongoing cell division, but disappear at about the time it stops. They hypothesize that these 40S and 34S RNAs are involved in control of the timing of active cellular proliferation in the rat and possibly in other species.

The polyamines including spermidine, spermine, and their precursor putrescine, facilitate nucleic acid and protein synthesis by acting as organic cations to stabilize the macromolecules and cell organelles involved. In the human brain there is a rise in the level of spermidine in the different regions of the brain, occurring at the time of their characteristic periods of cell proliferation. It occurs before birth in the cerebellum and brain stem and perinatally in the forebrain (14). The concentration of putrescine also reaches its peak perinatally at about the same time as the first phase of the rapid increase in DNA content of the human brain, corresponding to the period of neuroblast replication. Russell (15) postulates that this increase in putrescine concentration is due to a rise in ornithine decarboxylase activity (ODC) which seems to accompany rapid cell proliferation in all tissues.

III. BRAIN LIPIDS

The brain has a high content of lipid. Many different types of lipid are present such as cholesterol, and complex lipids including phospholipids, glycolipids, and other esters containing a variety of fatty acids. The lipid content of human brain remains fairly constant until about 7 months of gestation. At this time lipid deposition increases in grey matter to achieve adult levels (16-18).

Myelin is synthesized in the membranes of the oligodendroglial cells once these cells have completed division (19). Glial cells surround the nerve axons in a spiral fashion and lipid is gradually deposited in the developing sheath (20). As a result, the oligodendroglial cell membranes transform into the adult myelin sheath with its characteristic lamellated structure. Eighty percent of the dry weight of adult myelin is lipid; 50% of the total lipid content of the adult brain is myelin (21). Myelination of the human brain occurs in cycles, with the process starting and proceeding to completion in different areas at different ages from late in gestation until about 30 years of age (22). Postnatally the rate of myelin synthesis becomes greater than that of DNA synthesis, which results in increased cholesterol/DNA and phospholipid/DNA ratios (23).

Various markers have been used to plot the development of myelin. Several substances found in high concentrations in myelin have been used. The best marker seems to be cerebroside sulfatide. It increases by 300 to 400% between 34 weeks of gestation and birth, with the most rapid rate of increase occurring between 12 and 24 weeks of postnatal life, by which time the brain contains 50% of its adult level. By 4 years of age a full complement of myelin has developed. Anything that interferes with the oligodendroglial cell proliferation or to the accretion of myelin at this early stage of development will lead to a permanent deficit in brain myelin (24-26).

Cholesterol and phospholipids are important constituents of all cell membranes, which makes them less than good indicators of myelin development (27). However cholesterol is a good indicator of brain maturity. As the brain develops the amount that is in the esterified form drops from about 5% in early fetal life in both the forebrain and cerebellum, to 1% by 200 days of gestation. Postnatally it rises slightly and then falls to disappear altogether in the mature brain. The brain stem seems very low in esterified cholesterol throughout life (28).

Choline phosphoglycerides are the major brain phospholipids found before birth in the human forebrain, cerebellum, and brain stem (29). After birth there is a change to ethanolamine phosphoglyceride (EPG) dominance in the brain stem, probably resulting from the preponderance of white matter in this region. Ethanolamine phosphoglyceride accounts for the major proportion of phospholipid in white matter (30). In human myelin the ratio of the plasmalogen/diocyl forms of EPG increases with age (31), with the most of the increase occurring in the first year of life. The increase in EPG seen in the brain stem is possibly a reflection of the increase of plasmalogens in myelin.

The myelin lipid is oriented around myelin protein to form a complex proteolipid. In fetal and young-infant brains this is richer in aspartic acid, proline, and leucine and poorer in phenylalanine and tyrosine than in brains from patients older than 18 months of age (32). Grey matter protein retains the amino acid spectrum found in white matter at all ages.

In addition to cerebrosides, the other major class of glycolipids found in the brain is the gangliosides. Gangliosides consist of *N*-acetylneuraminic acid (NeuNac), sphingosine, and three molecules of either glucose or galactose. Four major types of ganglioside are found in the brain as shown in Figure 4. These four types constitute over 95% of total human brain gangliosides. Although the gangliosides contain 65% of central nervous system NeuNAC, glycoproteins contribute an additional 32%, with the remainder present in the free form (33). Five to ten percent of the total lipid content of some nerve tissue cell membranes is in the form of gangliosides (34). This means that the brain has the highest concentration of gangliosides of any tissue in the body. Grey matter has as much as 15 times the level found in the liver (35).

Name	Symbol	Proposed structure
Monosialoganglioside	G_{MI}	Gal(1 → 3)GalNAc(1 → 4)Gal(1 → 4)Glu(1 → 1)Cer 3 ↑ 2 NeuNAC
Disialoganglioside	G_{DIa}	Gal(1 → 3)GalNAc(1 → 4)Gal(1 → 4)Glu(1 → 1)Cer 3 3 ↑ ↑ 2 2 NeuNAC NeuNAC
Disialoganglioside	G_{DIb}	Gal(1 → 3)GalNAc(1 → 4)Gal(1 → 4)Glu(1 → 1)Cer 3 ↑ 2 NeuNAC(8 ← 2)NeuNAC
Trisialoganglioside	G_{TI}	Gal(1 → 3)GalNAc(1 → 4)Gal(1 → 4)Glu(1 → 1)Cer 3 3 ↑ ↑ 2 2 NeuNAC NeuNAC(8 ← 2)NeuNAC

Key:
Gal = galactose
GalNAc = *N*-acetylgalactosamine
Glu = glucose
Cer = ceramide
NeuNAC = *N*-acetylneuraminic acid

Figure 4 Major gangliosides of normal human brain (Svennerholm's nomenclature). (From Ref. 105.)

There are about 10^{11} neurons in the human brain. These contain
the major portion of the brain's glycolipids (34,36) and sialoglycopro-
teins (37). Although small fractions of the grey matter gangliosides
are contained in the cell bodies (38,39), by far the largest portion are
located in the dendritic and axonal plexuses (40) at the synaptic plasma
membranes (41,37,42). The microsomal disialoganglioside G_{Dla} seems
to be a good marker for dendritic arborization (43).

It might be said that within the body gangliosides are lipids seeking
for a role. One function suggested is a permissive one in behavior
(44-47). Theoretical models have been described using sialocompounds
as functional units within the neuronal membrane that exert an ion-
binding and ion-releasing action. We have recently shown that they
are responsible for the uptake of neurotransmitters from the synaptic
cleft into the synaptic neuronal terminal (48).

Synaptogenesis includes the growth of the presynaptic axon; contact with, and "recognition" of, the appropriate postsynaptic neuron;
replacement of growth cone organelles; and assembly at the active zone
of pre- and postsynaptic dense material, complete with ion channels
and transmitter receptor molecules. There is some evidence to implicate
surface glycoproteins in the synaptic membrane in the recognition
process (49).

IV. BRAIN METABOLISM

Energy metabolism of the brain is very low at birth, peaks at six years,
and thereafter falls to adult levels (50,51). This is reflected in
the cerebral blood flow and accompanying oxygen consumption. Oxygen consumption reaches a maximum at 6 years of age when it is 60 ml/
min per whole brain and 5.2 ml/100 g/min, which is equivalent to half
the total body basal oxygen consumption. This is much higher than in
the adult and may account for the extra energy needed for brain growth
and development (52,53). A number of enzymes of oxidative metabolism
in the brain show a similar rise in activity level (54).

Observation of blood flow in different brain regions shows that it
reaches its peak at different times but in accordance with the speed
of maturation of each region (54). Those regions with large amounts
of white matter have their highest blood flow when myelination is at
its peak. Thereafter, blood flow gradually declines, in tandem with
metabolic rate, to its mature value.

Brain fuels also change with age. After parturition a baby is hypoglycemic and its blood ketone levels are low (55). As suckling is begun
the infant becomes ketogenic because milk is high in fat. Ketones are
readily used by the brain at this time and are an important fuel (56).
Enzymes involved in ketone metabolism are most active in the suckling
period and dramatically decrease in activity after weaning when ketosis

subsides. After weaning, glucose takes over as the major cerebral substrate for energy metabolism (55,57). A low-fat intake in the suckling period appears to impair cholesterol synthesis, leading to decreased myelination (57).

Glycolysis dominates glucose metabolism during the early stages of brain development, enabling the newborn brain to withstand hypoxia far longer than the adult brain (6). However respiration gradually becomes more and more important from birth and eventually assumes the dominant role in glucose metabolism in a caudocephalic direction from the spinal cord to the cerebral cortex (58). In the adult brain 0.3-1.0 μmol/kg/min of glucose is metabolized. Ninety percent goes through the glycolytic pathway, and the pyruvate formed is further metabolized in the citric acid cycle to carbon dioxide and water. Only a small amount is converted to lactate (6).

As shown in Figure 5 carbon skeletons produced from glucose metabolism are used in the synthesis of glutamate and aspartate. Aspartate

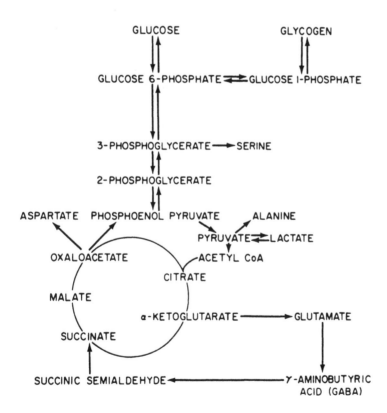

Figure 5 Some important pathways of glucose metabolism in the human brain.

is synthesized from pyruvate by transamination, glutamine from glutamate by amidation, and aminobutyrate from glutamate by decarboxylation. These amino acids are in dynamic equilibrium with their intermediates in the citric acid cycle, and thus they are able to reenter the cycle and be oxidized to carbon dioxide and water.

Several of the brain's metabolic pathways bypass one or more steps in the Krebs cycle. One example is the gamma-aminobutyrate (GABA) shunt, as indicated in the figure, which bypasses the succinyl-CoA step in the Krebs cycle (59,60). This pathway is believed to carry a significant amount of the flux from alpha-ketoglutarate to succinate in the citric acid cycle. The timing of its development is not known.

There is a good deal of glycogen in both neurons and glial cells, which has a high rate of turnover, but its specific function in development, if any other than as a supplier of energy, has yet to be defined (58).

The unique position of glucose in brain metabolism in the adult brain is partly due to the blood-brain barrier, which prevents many potentially useful energy substrates, such as citrate, from entering the brain. Little is known of the development of the blood-brain barrier. The timing of permeability development for different substrates is not known.

During the neonatal period amino acids readily pass into the brain (61) possibly because of the absence of a blood-brain barrier (62). As the brain develops, its level of glutamine and the acidic amino acids gradually increase and the basic and neutral amino acids decrease (63, 64). This indicates the presence of a differential effect on transportation of amino acids with maturation, possibly a result of the development of amino acid transport systems within the blood-brain barrier, which are distinguished from one another by an ability to transport only a specific group of amino acids. The decreased entry of amino acids into the brain with development has a similar timing to the decreased protein synthesis, which suggests an important role of amino acid supply in controlling protein synthesis and growth (10).

The adult brain has three separate, saturable, stereospecific carrier-mediated transport systems for neutral, basic, and acidic amino acids, respectively (65). Within each group there is competition such that the transport of one member of a given group is competitively inhibited by the other members of the same group. Such a system has inherent advantages, as the brain is able to use it to concentrate amino acids against a tissue gradient. As these systems are active they are very susceptible to oxygen and metabolic inhibitors, which reduce the flow of amino acids (66-68).

The brain takes up all amino acids except taurine, but the branched-chain amino acids are transported more rapidly than the others (69). L-Phenylalanine, L-tyrosine, L-leucine, L-methionine, and L-threonine are transported the most efficiently and are the brain's essential amino acids (70,71). However the transport systems have a limited capacity to transport amino acids. They have low-transport capacities and are saturated at low concentrations. The low levels of arterial amino acid concentrations also limit the availability of amino acids for brain metabolic processes (72).

Amino acids reaching the brain are utilized for amine neurotransmitter synthesis, for protein synthesis, and to a minor extent, for energy production. Nonessential amino acids produced as a by-product of glucose metabolism or taken up from the blood can be oxidized as can the essential amino acids (72).

The accretion of protein in the human brain is linear from the sixth month of gestation until the second year of life, after which it decreases. However, because of the complexity of brain proteins, it is unlikely that there is a critical period of protein development in the brain, in the true sense. Proteins, as well as being an integral part of the structure of brain tissue, are closely involved in metabolism as enzymes and as substrate-carriers in the blood-brain barrier. In association with lipids, proteins regulate the functions of membranes, including the interchange of ions and molecules both within cells and between cellular and extracellular fluid. Many proteins are made throughout life and rapid changes in their various levels can occur within the diurnal variation of a single day. Winick (6) has estimated that the half-life of human brain proteins is 14 days, which means that 4 g are produced each day. One theory of memory involves changes in protein synthesis (73).

V. EFFECTS OF MALNUTRITION ON BRAIN GROWTH AND DEVELOPMENT

Although, in general, the growth of the brain is affected less than the rest of the body by undernutrition, during the period of maximal cellular proliferation even moderate nutritional deprivation will give rise to growth retardation (74).

Over 30 different factors are required for normal brain growth (75). Even a less-than-optimal level of a single nutrient during the period of the brain growth spurt can cause significant delays in mitosis. One good example of this effect is the cessation of protein synthesis and the prevention of cell growth, which is caused by a drop in the level of a single amino acid in the intracellular pool below a given critical level (usually 10-40 pmol). If the missing substance is provided at a later date during the normal period of brain growth, although cell proliferation may be reinstituted, the impact of the missing amino acid on normal brain growth may not be reversible (75).

Growth of any kind, including that of the brain, is time-dependent, and any nutritional insufficiency during the period of growth will end in a permanent deficit in brain size. Nutritional rehabilitation after that time will not lead to catch-up growth. If the nutritional insult is of limited duration, and the infant is rehabilitated before the end of the growth period, then only those phases of growth ongoing at the time of the insult will be the ones permanently affected. Thus undernutrition from the third to fifth fetal month will lead to a reduction in neuronal division, whereas malnutrition during the fifth fetal month to the third postnatal month will have an effect primarily on the glia (76).

Because of the complex programming of brain growth and development, interference with one part of the growth cycle may interfere with all subsequent steps that are dependent upon the first one. For instance, malnutrition during the first 8 to 9 months after birth will inhibit lipid synthesis and the growth of oligodendroglia, which in turn, will result in retarded myelination (6).

VI. BRAIN SIZE

Body weight and brain weight are lower in malnourished children than in their well-fed counterparts (77). Conversely, brain/body weight ratios are often higher in the malnourished children. This is probably a result of a brain-sparing effect and also partly owing to the presence of edema (78).

Malnutrition during gestation often results in small-for-date babies with small brain weights and head circumferences as shown in Figure 6

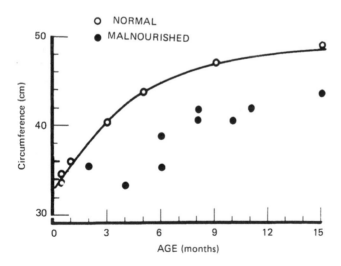

Figure 6   Head circumference vs. postnatal age in normal and malnourished children. (From Ref. 6.)

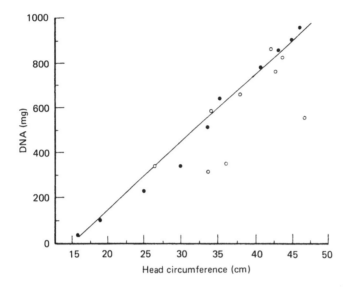

Figure 7 Brain DNA content vs. head circumference in normal and malnourished children. ● = normal; ○ = malnourished. (*Source*: Ref. 6.)

(79). Although when these variables are expressed as ratios with body weight they are usually within the normal range (80). Children with head circumferences less than the 10th percentile at birth often have poor growth, later microcephaly, and neurological deficit (81).

Head circumference measurements have long been used by pediatricians as the indicator of brain growth. Winick (6) showed that head circumference measurements could be correlated with brain cell numbers as measured by DNA content (Figure 7). Nutritionally deprived children often have a reduced head circumference (82,83). Such children, contrary to expectations, often have normal IQs (6). This has been explained by the beneficial effects imposed by environmental stimulation (84,47). Using an animal model Morgan and Winick (46) have shown that the only significant change in brain chemistry resulting from environmental stimulation accompanying the characteristic improvement in learning is an elevated N-acetylneuraminic acid (NeuNAC) content. As a result they hypothesized that brain NeuNAC levels are a better measure of brain function than brain size (85). They have further shown that serum NeuNAC levels correlate with brain NeuNAC levels (Figures 8-10) and have suggested that serum NeuNAC levels are a better in vivo measure of functional brain development than the head circumference measurements routinely used at the present time.

Morgan and Winick

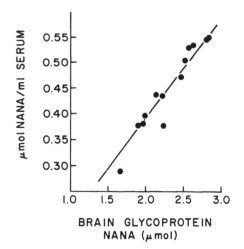

Figure 8 Serum *N*-acetylneuraminic acid (NANA; µmol/ml) vs. total brain glycoprotein NANA content. Each point represents the NANA levels from serum and brain tissue pooled in litters. The equation of the line is y = 0.39 + 4.23 x. (From Ref. 85.)

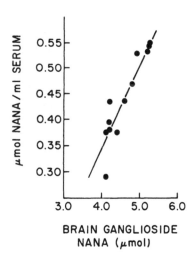

Figure 9 Serum *N*-acetylneuraminic acid (NANA; µmol/ml) vs. whole brain ganglioside NANA content. Each point represents the NANA levels from serum and brain tissue pooled in litters. The equation of the line is y = 2.38 + 5.06. (From Ref. 85.)

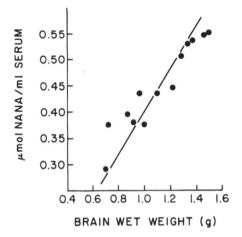

Figure 10 Serum *N*-acetylneuraminic acid (NANA; μmol/ml) vs. brain weight (g). Each point represents the mean weight of brains from a single litter vs. the NANA level of 1 ml serum taken from the pooled serum from the same litter. Equation of the line is y = 3.08 × −0.29. (From Ref. 85.)

VII. BRAIN COMPOSITION

A. Cellular Growth

1. *Effects of Intrauterine Undernutrition*

Intrauterine undernutrition usually results from placental insufficiency as a direct effect of the decrease in the flow of nutrients from the mother to the fetus. This can further be exacerbated by maternal undernutrition. The most common nutritional deficiencies suffered by the human fetus are low intake of energy and/or protein. Such deficiencies are fairly common and usually arise from depleted maternal nutrient stores because of poor dietary intake before and during pregnancy, serious illness, or frequent close pregnancies. Mothers who are underweight before pregnancy and those who gain insufficient weight during pregnancy are in the high-risk category (86).

If undernutrition is imposed early in gestation the number of neurons will be reduced. If it is imposed late in gestation, after 26 weeks, when neuronal division is completed, only glial cell numbers will be significantly affected.

Reduced glial cell numbers have a much smaller effect on intelligence than a reduced neuronal complement (21,87,88). However a loss of glia will be associated with a marked reduction in the degree of myelination. A loss of neurons from the cerebrum will be more detrimental to intelligence than a similar loss from the cerebellum. This was well demonstrated by Dow (89) who found that the loss of the cerebellum resulted in impaired movement but no impairment to intellect, perception, or sensory function.

2. *Effects of Postnatal Malnutrition*

Little is known of the effects of protein-energy malnutrition (PEM) on cell division in the human brain. Severely malnourished infants develop into children with smaller heads than expected for their age, which are often larger than normal for their corresponding body weight (Figure 11). Brain DNA content is also much lower in malnourished children (Figure 12), but DNA per gram of brain is often normal (Figure 13).

In young children severe malnutrition can be clinically classified into two diseases, marasmus and kwashiorkor. In PEM they overlap and the patient suffers from a mixed syndrome. Marasmus typically occurs during the first year of life in the infant who is starved as a result of early weaning or breast-fed by a mother with inadequate milk because of being semistarved herself. It also occurs as a result of prolonged diarrhea. The condition is more threatening to brain development than kwashiorkor, as it occurs at an earlier critical age (90). Kwashiorkor occurs in older children of 2 to 3 years of age, who have been weaned onto low-protein diets, providing more than enough energy as starch (91,92).

B. Myelination and Myelin Lipids

1. *Effects of Intrauterine Undernutrition*

Small-for-gestational-age babies have reduced brain myelin lipids such as cerebroside sulfatide (25) and galactose cerebroside (93). They also have lower than normal activity of galactolipid sulfatransferase, which is the rate-limiting enzyme in sulfatide synthesis (92).

2. *Effects of Postnatal Undernutrition*

Using cerebroside-sulfatide (25) or cholesterol (23) as a marker, it would seem evident that chronically malnourished children have low brain myelin levels for their chronological age (Figure 14). However cholesterol concentration remains the same (43). Since DNA content per gram of brain is the same in normal and malnourished babies (Figure 13), this means that the degree of myelination of each individual neuron is probably the same (Figure 15.)

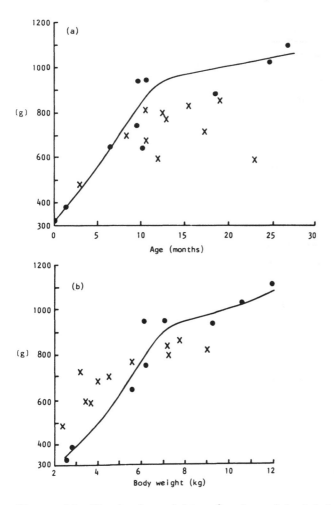

Figure 11 The brain weights of malnourished (x) and control (●) Jamaican children plotted (a) against age, and (b) against body weight. (From Ref. 103.)

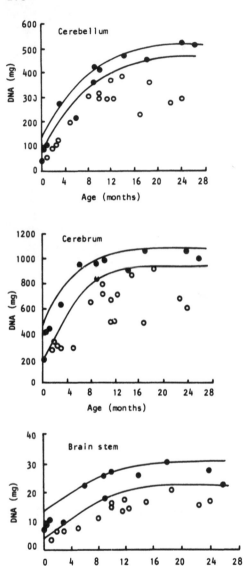

Figure 12 The effect of protein-energy malnutrition on the DNA content of the human cerebellum, cerebrum, and brain stem. Values for malnourished children shown (○) and those for normal children shown (●). (From Ref. 90.)

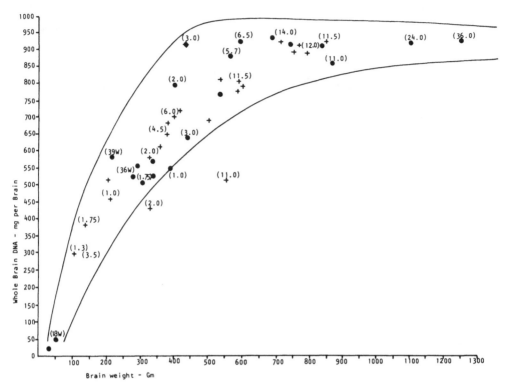

Figure 13 The DNA content of the human brain plotted against the brain weight. Values for malnourished children shown +, and those for "controls" shown •. Age of fetuses in weeks (w), others in months. (From Ref. 105.)

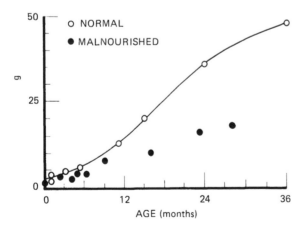

Figure 14 Total human brain cholesterol used as an indicator of myelination vs. age, showing decreased levels in malnourished children. (From Ref. 6.)

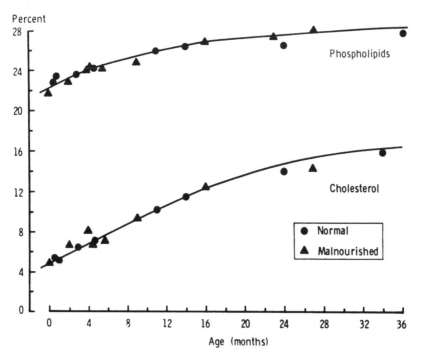

Figure 15 Percentage cholesterol and phospholipids in the brains of normal and malnourished children, showing that there is no difference in degree of myelination per brain cell. (From Ref. 6.)

VIII. MALNUTRITION AND SYNAPTIC DEVELOPMENT

During the period of most rapid neuronal proliferation (12 to 26 weeks of pregnancy) neurons are being produced at 250,000/min (76,94). At birth the brain contains its full complement of 100 billion different neurons (95). Most of these neurons form a complex network to store information and to transmit messages (96). However there are about 3 million neurons with a motor sensory function in the brain (97).

Each neuron is formed of a round soma or cell body surrounded by a dense network of fine tube-like processes, the dendrites. These constitute the main surface area for contact with other neurons at synaptic junctions where the neuron receives incoming messages. The axon is a thinner and larger extension of the soma. Coordinated brain activity is produced by the passage of electrical impulses from the cell body and down the axon to other parts of the brain. At the axon terminal the synaptic gap separates the axon from the cell membrane of the adjoining neuron. A given neuron has the capacity to make

1000 to 10,000 synaptic connections by which it may receive information from up to 1000 adjoining neurons. Typically synapses are formed between the axon of one cell and the dendrite of another, but a number are also made between axon and axon and between axon and cell body (98).

On arrival at the terminal of an axon a nerve impulse causes calcium ions to release either an excitatory or an inhibitory transmitter. The transmitter crosses the synaptic gap and binds to receptor sites on the cell membrane of the adjacent neuron. In this way either an excitatory or an inhibitory message is transmitted. Most of the neurotransmitter bound to the cell membrane than dissociates and is taken up again by its axon of origin. The remainder is broken down by enzymes outside the axon terminal in the synaptic cleft. For instance, acetylcholinesterase can degrade 25,000 molecules of acetylcholine per second within the cleft (99).

Each neuron synthesizes its characteristic transmitter from precursor molecules within its own axon. The transmitter is stored in vesicles within the synaptosome. Each terminal contains thousands of vesicles containing between 10,000 and 100,000 molecules of transmitter per vesicle (100). Approximately 20 substances have been identified as transmitters including serotonin, acetylcholine, catecholamines, histamine, gamma-aminobutyric acid, glycine, glutamate, and aspartate. Certain brain areas are enriched in specific neurotransmitters, e.g., the corpus striatum is enriched in dopaminergic terminals originating from cell bodies in the substantia nigra (101). However it is not clear whether each nerve cell releases only one type of neurotransmitter or whether some cells release a mixture.

For normal brain function the transmission of information from one cell to another, and from one part of the brain to another, must go on at the optimal rate. This means that there must be an optimal number of neurons with sufficient synaptic connections to cater to the needs of an individual. Postnatal undernutrition impairs the development of the dendrites, which occurs mainly in the postnatal period (102). Gangliosides are the best marker of dendritic arborization and are present in high concentration in grey matter. Studies on Jamaican children dying from malnutrition revealed that there was a deficit in total gangliosides in the forebrain (Figure 16). This deficit was found to be solely in the disialoganglioside fraction known as the G_{Dla} fraction (103,104). As the G_{Dla} fraction is mainly located in the dendrities (105) this observation indicated that these children had fewer dendrites in their forebrain, which would limit the transmission and storage of information.

As mentioned earlier, gangliosides may be the link between biochemistry and behavior in other ways than simply as structural components of membranes. Evidence to support this view has come from experiments

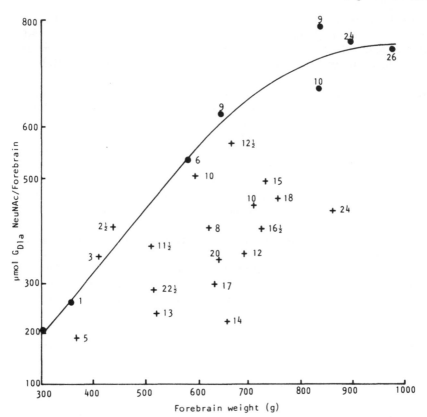

Figure 16 The effect of protein-energy malnutrition on the amount of the disialoganglioside, G_{Dla} (expressed as $\mu mol\ G_{Dla}$ NeuNAC, that is G_{Dla} *N*-acetylneuraminic acid), in the human forebrain. Values for malnourished children shown +, and those for 'controls' shown ●. Age of children given in months. (From Ref. 105.)

in which undernourished rat pups were subjected to early environmental stimulation (47). Environmental stimulation was achieved by assigning a trained "aunt" (maiden female) to spend 8 hr a day with rat dams and their litters from day 3 to day 21 of lactation. Stimulation of pups during the first 21 days of life reduced the change in open-field behavior caused by undernutrition. The open-field used was an 80 × 80-cm area enclosed within wooden walls 60-cm high. The floor was divided into 8-cm squares. A central area (32 cm × 32 cm) was designated by squares in the center of the open field. A ball was suspended within 2 cm of a corner square designated as the novel square. The following behaviors were recorded by an observer during each open-field test: (a) latency—the number of seconds before a rat moved its paw out of the four squares that defined the starting corner; (b) the number of squares traversed by both front and rear paws; (c) the number of vertical extensions of the head, body, and forelimbs, either free-standing or against the wall, excluding those vertical extensions associated with grooming (rears); (d) the number of times the forelimbs entered the "central area" (centers); (e) the presence or absence of defecation; and (f) the total time the rat passed with its nose in the square over which the object was suspended. This change was associated with a significantly higher ganglioside and glycoprotein NeuNAC content in the brain (Figure 17).

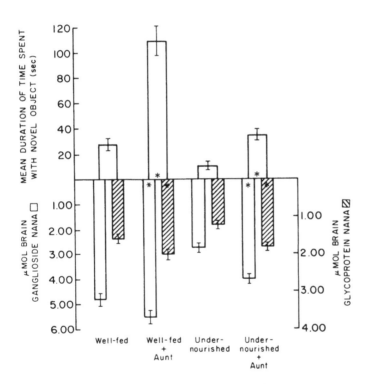

Figure 17 Open-field behavior in well-fed and undernourished 21-day-old rats that had been subject to environmental stimulation during the first 21 days of life (well-fed + aunt and undernourished + aunt), compared with control animals (well-fed and undernourished) and the corresponding brain content of ganglioside and glycoprotein NANA (Neu-NAC). Bars represent $\bar{x} \pm$ SE of values obtained from analysis of 16 separate brains. Where indicated (*), the differences between stimulated and control groups were statistically significant (P < .001). Differences between well-fed and their corresponding undernourished groups were all statistically significant, P < .001. (From Ref. 47.)

This was a permanent effect as at age 6 months, after the rats had been nutritionally rehabilitated, the persistent effects of early stimulation were shown in an improved ability to learn a Y maze (Figure 18). The Y maze used in these studies was electrified with a metal-grid floor. An opaque barrier divided the stem of the Y into a start-box and an alley. To avoid shock, the rat had to move from the start-box, down the alley into the correct arm of the maze within 5 sec. Shock was given if after 5 sec the rat had not entered the correct arm of the maze. For half of the rats, the right arm was correct, and for the remainder the left arm was correct. Each animal was trained at its own

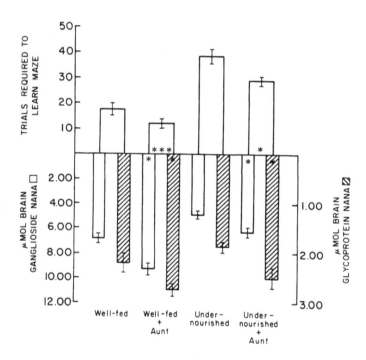

Figure 18 Y-maze performance in adult animals that had been subject to environmental stimulation during the first 21 days of life, compared with performance of control animals and the corresponding brain content of ganglioside and glycoprotein NANA (NeuNAC). Bars represent mean ± SE of values obtained from analysis of 16 separate brains. Where indicated (*), the differences between stimulated and control groups were statistically significant (P < .001). Differences between well-fed and their corresponding undernourished groups were all statistically significant, P < .001. (From Ref. 47.)

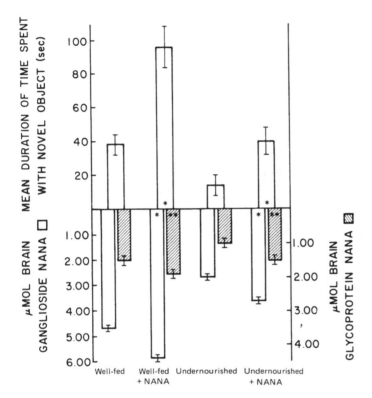

Figure 19 Open-field behavior in experimental animals (given daily intraperitoneal injections of NANA during postnatal days 7-14) of 21 days of age and the corresponding brain content of ganglioside and glycoprotein NANA. Bars represent ($\bar{x} \pm$ SE) of values obtained from analysis of 16 separate brains. Where indicated the differences between control and experimental groups were statistically significant (*P < .001; **P < .01). Differences between control groups and their corresponding undernourished groups were all statistically significant (P < .001). (From Ref. 46.)

shock-threshold level. Training continued until each rat had performed nine consecutive correct responses.

In another experiment by Morgan and Winick (46), NeuNAC given intraperitoneally was shown to cause a rise in brain ganglioside and glycoprotein NeuNAC. Administration of NeuNAC daily from 14 to 21 days of life brought about these changes without altering brain weight, cell size, or cell numbers. Changes in behavior identical to those resulting from early stimulation were seen (Figures 19,20).

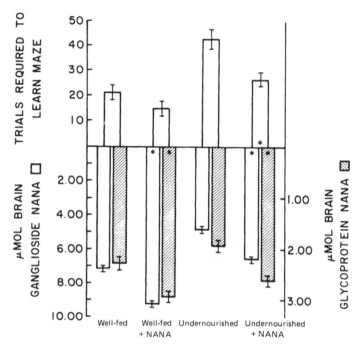

Figure 20   Y-maze performance in adult animals (which had been given
daily injections of NeuNAC during postnatal days 7-14) and the corresponding brain content of ganglioside and glycoprotein NANA (Neu-
NAC).   Bars represent ($\bar{x} \pm$ SE) of values obtained from analysis of 16
separate brains.   Where indicated the differences between control
groups and their corresponding underfed groups were all statistically
significant, P < .001.   (From Ref. 46.)

    Because both environmental stimulation and administration of NeuNAC
caused the same changes in brain NeuNAC and behavior, it suggests
that brain NeuNAC is a determining factor in the expression of behavior.
Morgan and Sinai (48) have further shown that this might be explained
by a role of NeuNAC in neurotransmission.   Ganglioside NeuNAC is involved in the binding of neurotransmitter molecules to synaptic membranes.   Undernourished animals have a lower level of ganglioside in
their synaptic membranes, which leads to inefficient binding and hence
inefficient transmission.   Morgan and Naismith (106) have shown that
undernutrition delays the onset of development of enzymes involved in
ganglioside and glycoprotein synthesis.   In well-fed control rats the
UDP-glucosyltransferase enzyme activity increased gradually between
days 6 and 12, then decreased rapidly to level out at day 24 (Figure
21).   Undernutrition delayed the attainment of peak activity and the

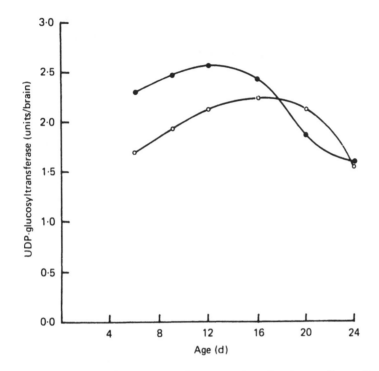

Figure 21 Activity (units/brain) of UDP-glucosyltransferase (EC 2.4.1.52) in the brains of well-fed (●----●) and undernourished (○----○) rat pups. Mean values are for six litters of eight pups. One unit of activity is defined as 1 nmol cerebroside formed per hour. (From Ref. 106.)

results suggested that the mature level of activity would be below that of control rats. Similar results were obtained for sialidase, UDP-galactosyl transferase, and CMP-N-acetylneuraminic acid synthetase. The marked displacement of peak enzyme activity attainment time shown by these data was a most striking finding. Previously only the extent, not the timing, of the brain growth spurt was thought to be influenced by undernutrition (107). This may well be true for the multiplication of nerve cells, as measured by the rate of increment of DNA in the brain (108). These observations suggest, however, that it may not be true for the growth of the cellular processes.

IX. MALNUTRITION AND ENERGY METABOLISM

Undernutrition during pregnancy, which leads to intrauterine growth retardation, often involves vascular insufficiency and consequently

anoxia. Any period of anoxia during the prenatal period or at birth
can reduce the rate of cell division in the brain (6) and hence increases
the risk for impaired intellectual performance.

X. FATTY ACIDS AND BRAIN DEVELOPMENT

The high fat content of the brain is mainly due to phospholipids, glycolipids, and sphingolipids present in the brain cell membranes. There
is little triglyceride in the brain. The brain contains the essential
fatty acid linoleic acid, which must be supplied by the diet, and many
other polyunsaturated fatty acids, which can theoretically be made
by the brain by chain elongation and desaturation (109,110). However these long-chain, polyunsaturated fatty acids are not readily
made from linoleic acid during the period of rapid development (111),
and since they are not present in vegetable foods Crawford et al.
(112) have hypothesized that animal fat is an essential dietary constituent for normal development of the human brain. The spectrum of
polyunsaturated fatty acids in the human brain changes with age. As
the grey matter increases so does the long-chain polyunsaturated fatty
acid content. This mainly occurs in utero but does continue at a significant rate during the early postnatal period. Hence the composition
of the fat contained in infant formulas could be important in this respect. Formulas containing only vegetable fat may provide inadequate
nutrition. The brain also contains a large amount of cholesterol, and
in rats both a reduced cholesterol and/or saturated fat intake has been
shown to lead to impaired myelination (57,1).

XI. MALNUTRITION AND BRAIN PROTEIN

Little is known about how undernutrition affects human brain protein.
One interesting question that needs to be resolved is the correct
balance of amino acids required by an infant being fed parenterally.
Should a premature infant be fed a mixture of amino acids that causes
its serum protein levels to mimic those found in a baby born at term,
or should the infant be fed a mixture that leads to a spectrum of
plasma amino acids similar to that found in utero? Because the carrier
mechanisms transporting amino acids across the blood-brain barrier
and the barrier itself are very immature at this stage of development,
the constituents of the blood are readily accessible to the brain tissue.
Hence plasma amino acid levels could be critical to normal brain development.

Phenylketonuria is a phenomenon that increases the level of phenylalanine in the blood and that available to the developing brain. This
inborn error of metabolism is characterized by a deficiency of the

enzyme phenylalanine hydroxylase, leading to an impairment in phenyl-alanine degradation.

If the disease is not diagnosed early in life and the child is given a low-phenylalanine diet, the growth and development of the brain is impaired because of the accumulation in the blood and brain of high levels of phenylalanine. This is a good example of how excesses of one amino acid can decrease the levels of other amino acids and protein synthesis in the brain.

XII. AMINO ACIDS AND NEUROTRANSMITTERS

The intracellular free amino acids in the brain provide the substrate for protein and neurotransmitter synthesis. Such amino acids are either obtained from the plasma or produced locally from precursors. In the adult human brain four amino acids are present at concentrations in excess of 1 μmol/g. These are taurine (1.25 μmol/g), glutamic acid (7.19 μmol/g), aspartic acid (1.10 μmol/g), and glutamine (5.55 μmol/g) (113). A reduction in the supply of protein would be expected to limit both neurotransmitter and protein synthesis. Such measurements are difficult to make in children. However a number of experiments on the effects of malnutrition on brain amino acids have been done in other species including monkeys (114,115). The results of these studies show that in young animals the plasma levels of certain amino acids, especially the branched-chain amino acids and tryptophan, are reduced by protein-energy malnutrition. These changes are reflected in the amino acid pools in the brain. Plasma amino acids are also reduced in kwashiorkor presumably leading to reduced brain levels. In older animals subjected to protein-energy malnutrition the changes in the plasma and brain amino acids are less severe (116).

XIII. IRON DEFICIENCY

Brain iron is present as part of the prosthetic groups (heme or flavin) for a number of enzymes as well as in a storage form that is bound to a protein similar to ferritin. With increasing age all areas of the brain except the medulla show an increase in nonheme iron (117). There is a great deal of regional variation, but in general, levels of nonheme iron are at about 10% of adult values at birth, 50% by 10 years of age, and at maximal level between 20 and 50 years (4-5 mg/100 g wet weight of tissue). Iron-containing brain cytochromes rise rapidly to maximal levels at the same rate as the development of myelin (118).

Iron plays an important role in many metabolic processes including its role as a cofactor for a number of enzymes critical to oxidative metabolism and neurotransmitter metabolism. Recent studies have shown

that changes in activity of these enzymes occur during the early
stages of iron deficiency (119). These enzyme effects may account
for the behavioral changes that have been associated with even mild
iron deficiency (120). In children mild iron deficiency affects atten-
tional processes in that they are less attentive to those environmental
clues that aid problem solving. In iron-deficient infants, remedial
deficits in attention-related behaviors have been described (121).
These iron-related changes in cognitive function are quite small and
may not carry clinical relevance (120).

XIV. CATCH-UP GROWTH

This review clearly shows that malnutrition in early life has a substantial effect on the chemical composition of the brain. Several investigators have reported that brain catch-up growth is possible in
children. For instance, Engsner (78) reported that children with
kwashiorkor exhibited some catch up in head circumference on rehabilitation. He also showed that a smaller degree of catch-up was
possible in children with marasmus. Catch-up growth has also been
shown in children exposed to intrauterine undernutrition (122). However in all of these studies catch-up was measured in terms of head
circumference, which gives no indication of the location of the catch-
up growth or of the type of growth. In fact, one of the big unanswered questions is: Does this growth represent neurons or glia?
Furthermore does it occur in all brain regions equally?

XV. BRAIN FUNCTION—BEHAVIOR AND INTELLIGENCE

Another area of controversy is whether or not undernutrition in early
life has long-term effects on intellectual performance. The controversy
arising from the contradictory results available can possibly be explained
by other environmental factors that go along with poverty (123,124).
Only one study has been carried out that controlled for these confounding variables. This study examined the effects of the famine in Holland
during World War II (125). It showed that infants exposed to undernutrition in utero had normal mental functions in later life. However
this study was not very representative of typical prenatal undernutrition as the infants had normal head circumferences despite a 10% reduction
in body weight.

Studies of small-for-date infants, showing the same level of growth
retardation as these infants affected by prenatal malnutrition, have
indicated that many such children have low IQ scores, poor school
performance, behavioral problems, and other signs of minimal brain
damage (126). Observations of twins, which controls for environmental
factors, also has sometimes confirmed these observations (127-129).

Children nutritionally cured of marasmus or kwashiorkor show definite signs of sensory-motor retardation (129). There may be evidence of developmental deficits in speaking and hearing, eye-hand coordination, interpersonal relations, and the ability to solve problems (130). It appears that nutritional therapy must be begun during the phase of increased brain-cell growth and continued throughout the period of myelination and dendritic arborization to completely overcome the adverse effects of marasmus and kwashiorkor.

Several studies have shown that the behavioral effects of postnatal undernutrition can be partially overcome by environmental stimulation. Children suffering from diseases such as cystic fibrosis and pyloric stenosis in affluent societies are malnourished but tend not to suffer from permanent behavioral deficits (131). The older the child, and hence the more complete the brain growth, the less likely that there will be behavioral deficits.

Winick and his colleagues (84,132) selected two groups of Korean female children from a population of children that had been taken into the Holt adoption agency either before or after age 2 years. Each group was then subdivided into an additional three groups on the basis of growth on admission to the agency. Those below the 3rd percentile for height and weight were categorized as severely malnourished. Those lying on the 3rd through the 24th percentile were considered moderately malnourished, and the remainder, which were above the 25th percentile, were assigned to the well-fed group. All children were rehabilitated at the agency before being brought to the United States and fostered to middle-class American families. To be included in the study the first group of children had to have been adopted by Americans before the end of their brain growth period, namely between 18 months and 3 years; and the second group of children was somewhere between the ages of 3 and 5, which is after most brain growth has been completed. When given IQ and achievement tests after 6 years in the United States, and hence in a stimulating environment, the early-adopted children had higher scores than average American children of the same age. But the severely malnourished group did significantly less well than the other two groups (Figure 22).

The results for the children adopted at an older age were slightly different. Here it was found that the well-nourished and the moderately malnourished girls reached and surpassed the American average, but the severely malnourished group fell short of it. By comparing the results in Figure 23 for the early- and late-adopted children we find that in every case the early-adopted group scored better than the late-adopted children.

These studies support the view that malnourished children who have been nutritionally rehabilitated after the age of rapid brain growth do not perform as well as those rehabilitated before the growth spurt. The weakness of the study is that the children's environment before

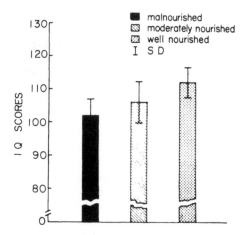

Figure 22   Relationship of IQ of female orphans adopted before age
3 and nutritional status.   There were 42 children in the malnourished
group, 52 in the moderately nourished group and 47 in the well-nourished group.   (From Ref. 84.)

adoption and rehabilitation could have had a more profound effect on
their intellectual development than their plane of nutrition—the most
undernourished being proportionally more environmentally deprived
than the moderately or well-nourished children.   This could have been
a complicating factor, but there is no doubt whatsoever that given a
stimulating environment all children performed beyond expectations.
Thus environmental stimulation seems to partially overcome the effect
of nutritional deprivation and is, in fact, the only therapy we have
for children who have been malnourished during the period of their
brain growth spurt.

The prevalence of severe malnutrition in the United States is very
small but a substantial number of people are at risk for developing
nutritional problems (133).   Mild undernutrition during the perinatal
period could conceivably lead to mild subnormal mental faculties.   To
prove cause and effect is extremely difficult, because many people with
such deficits have no diagnosable pathology (134).

There is no way of determining retrospectively that mild brain damage
at a cellular level occurred during the perinatal stage or that early
environmental deficits occurred.   However marginal undernutrition
during the first 2 years of life does seem to be associated with impaired
later intellectual and psychological function in such areas as problem
solving, object awareness, quickness of perception, and of the thought
process itself (135).

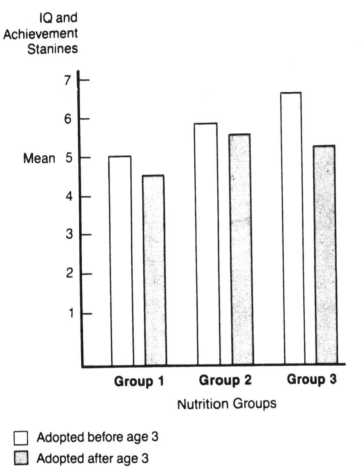

IQ and
Achievement
Stanines

□ Adopted before age 3
▨ Adopted after age 3

Figure 23 Relationship of IQ and achievement of female orphans and timing of adoption (by nutritional status). Group 1: severely malnourished (below the 3rd percentile for height and weight); group 2; moderately malnourished (3rd through 24th percentile); group 3: well-nourished (control) at or above the 25th percentile. There were 42, 52, and 47 in the early-adopted groups 1, 2, and 3, respectively and 57, 109, and 74 in the late-adopted groups, respectively. (Data adapted from Refs. 84, 132.)

Development tests given to children under 2 years old bear no correlation with IQ tests given to older children (136). Hence such tests,
even if given routinely to populations suspected of being at risk for
nutritional deficiencies, would not permit the identification of children
with learning problems. As a result we have no way of diagnosing
such children at a time when their brain is still rapidly growing.

XV. CONCLUSIONS

Early growth of the human brain is a chronologically programmed series
of biochemical, physiological, and psychological events. The anatomical
growth of the brain passes through a phase characterized by a rapid
increase in cell number, it then goes through a transitional phase,
where both cell number and cell size increase, and finally through a
phase of cell enlargement, where cell size increases but cell number
remains constant (6).

Winick (6) demonstrated that the long-term impact of short-term
nutritional insults on brain development was dependent upon the type
of cell growth occurring at the time of the insult: the period of rapid
cell division was the most vulnerable. Further research demonstrated
that the more rapid the rate of cell division, the earlier and more profound the adverse effects of undernutrition. If such undernutrition
occurred during peak periods of dendritic arborization and myelination,
the effects were not completely reversible.

Animal studies have shown that either prenatal or early postnatal malnutrition in rats leads to reduction in brain cell number and size. When
the malnutrition exists over both the prenatal and postnatal periods,
the effects are even more pronounced. Both prenatal and postnatal
malnutrition lead to a 60% reduction in brain size (6).

Since different parts of the brain develop at different rates, the
cell growth retardation induced by undernutrition may occur in localized areas, rather than diffusely, throughout the brain. Such local
effects would lead to a disproportional growth failure in which one
brain region would be more profoundly affected than another. There
is evidence, for example, that after 8 days of malnutrition beginning
at birth (in the rat) the cerebellum has a reduced cell number, whereas the cerebrum is not affected until 14 days of undernutrition. Thus
malnutrition lasting only 8 days after birth would result in a brain
with a much proportionally smaller cerebellum. This disproportionate
growth may even occur *within* defined brain areas (6). For example,
one histological study has shown that a local region of a cerebral cortex, so altered by undernutrition, may have up to 40% fewer synapses
per neuron (134). Reduced rates of cell division attributable to undernutrition are found locally in the cerebrum, as well as in the cerebellum
and the brain stem (6).

Attempts have been made to correlate changes in cellularity of various areas of the brain with the changes in behavior displayed by animals undernourished during the period of their major brain growth. However, using available biochemical techniques, one can do little more than make gross estimates of cerebral, cerebellar, and brain stem cellularity. Histological studies, on the other hand, are time consuming and can only provide us with a series of two-dimensional images. Hence, to be able to make precise correlations between brain structure and function, a technique is needed that has the capacity to measure and portray in a three-dimensional image structural changes in discrete anatomical areas. Reduced levels of any given nutrient can delay mitosis for many days (75). If such a nutrient is reinstated in the diet, cell proliferation is similarly reinstituted, although frequently not in the normal pattern. Because of the complex programming involved in brain growth and development, interference with one step of the process may not be reversible if its "time" is passed. This, in turn, can interfere with any and all subsequent steps dependent upon the first one.

The responses of the developing human brain to undernutrition appear to be similar to those found in laboratory animals (137). It is known that undernutrition in the first 8 or 9 months of postnatal life reduces the rate of cell division and inhibits lipid synthesis and the growth of oligodendroglia, which in turn results in retarded myelination (6). Neuronal growth processes, dendritic arborization, and synaptic formation may also be retarded in the first year or two of life (6).

The most common nutritional problem affecting the human fetus is inadequate maternal intake of energy and protein. It is evidenced in the mother by inadequate weight gain, which is magnified if the mother is underweight before pregnancy (138). Women who fail to gain enough weight during pregnancy may deliver infants who suffer from various degrees of fetal growth retardation, apart from the brain growth retardation, evidenced by a reduced number of cells and smaller head size (79); more specific effects of prenatal undernutrition on the human brain are not known. Similarly, little is known about the long-term consequences, but a high proportion of small-for-gestational-age infants shows low IQ scores, problems with academic achievement, behavioral disturbances, and various other signs of minimal brain damage.

Children suffering from marasmus, a result of protein-energy undernutrition, have a measurable reduction in head circumference. In behavioral terms, this translates to a high frequency of irritability and/or apathy (139). Even when a child is cured of marasmus or kwashiorkor through nutritional means, signs of sensory-motor retardation often remain. Impaired development may be seen in the processes of speaking and hearing, eye-hand coordination, interpersonal relations, and problem-solving (139).

Even after laboratory animals are nutritionally rehabilitated, they show alterations in normal behavioral patterns, such as increased excitability, decreased exploratory behavior, decreased ability to perform in certain maze tests, and a slow extinction of a conditioned response (140).

Several studies have shown that the behavioral effects of postnatal undernutrition can be partially overcome by environmental stimulation. Children in affluent societies who are suffering from diseases such as cystic fibrosis and pyloric stenosis are certainly malnourished, but tend not to suffer from permanent behavioral disturbances (141).

Korean orphans who were severely malnourished during the first 6 months of life were compared with adequately nourished Korean orphans of the same age after both groups of children had been adopted by middle-class American families. If the malnourished child was adopted before 3 years of age, the average IQ and school performance were the same as United States norms. Even though the IQs and school performance of previously well-nourished children were somewhat higher, the severe behavioral effects, which have been described as the stigmata of early malnutrition, in children raised in a deprived environment were not seen. Thus providing an enriched environment appears to prevent, or at least minimize, the expected behavioral outcomes of early nutrition (84,132).

When a child suffers from gross undernutrition the changes in brain growth may be measured by standard head circumference measurements. However subtle changes in growth caused by marginal undernutrition are not sufficient to cause reduced head circumference, although they may have permanent detrimental effects on behavior. At present, we are developing a radiation-free technique, involving the use of nuclear magnetic resonance in conjuction with finite analysis, that will enable us for the first time to follow brain growth in children during the first 2 years of life. This will provide the physician with a tool to pick up brain growth retardation in at-risk children, such as small-for-gestational-age and premature babies, at a time when nutritional intervention or environmental stimulation (or other appropriate intervention) could be of value in making up the deficit.

REFERENCES

1. A. N. Davison and J. Dobbing, Myelination as a vulnerable period in brain development. *Br. Med. Bull.* 72:40-44 (1966).
2. A. Boivin, R. Vendrely, and C. Vendrely, L'acide désoxyribonucleique du noyou cellulaire, dépositaire des caractères héréditaires: Arguments d'ordre analytique. *C. R. Acad. Sci.* 226: 1061-1063 (1948).

3. M. Enesco and C. P. Leblond, Increase in cell number as a factor in the growth of the organs of the young male rat. *J. Embryol. Exp. Morphol. 10*:530-562 (1962).

4. L. W. Lapham, Tetraploid DNA content of Purkinje neurons of human cerebellar cortex. *Science 159*:310-312 (1968).

5. M. Winick and A. Noble, Quantitative changes in DNA, RNA, and protein during prenatal and postnatal growth. *Devel. Biol. 12*:451-466 (1965).

6. M. Winick, Nutrition and cellular growth of the brain, in *Malnutrition and Brain Growth*. Oxford University Press, New York, 1976, pp. 63-97.

7. J. Dobbing and J. Sands, The quantitative growth and development of the human brain. *Arch. Dis. Child. 48*:757-767 (1973).

8. J. Raaf and J. W. Kernshan, A study of the external granular layer in the cerebellum: The disappearance of the external granular and the growth of the molecular and internal granular layers in the cerebellum. *Am. J. Anat. 75*:151-172 (1944).

9. J. Sands, J. Dobbing, and C. A. Gratrix, Cell number and cell size: Organ growth and development and the control of catch-up growth in rats. *Lancet 2*:503-505 (1979).

10. M. Winick, Malnutrition and mental development in *Nutrition Pre- and Postnatal Development* (M. Winick, ed.). Plenum Publishing Corp., New York, 1979, pp. 41-60.

11. J. A. Brasel, R. A. Ehrenkranz, and M. Winick, DNA polymerase activity in rat brain during ontogeny. *Dev. Biol. 23*:424-432 (1970).

12. P. Rosso and M. Winick, Effects of early undernutrition and subsequent refeeding on alkaline ribonuclease activity of rat cerebrum and liver. *J. Nutr. 105*:1104-1110 (1975).

13. C. G. Lewis and M. Winick, Pattern of cytoplasmic RNA in brain and liver of immature rats. *Proc. Soc. Exp. Biol. Med. 156*:158-161 (1977).

14. P. A. McAnulty, H. K. M. Yusuf, J. W. T. Dickerson, E. N. Hey, and J. C. Waterlow, Polyamines of the human brain during normal fetal and postnatal growth and during postnatal malnutrition. *J. Neurochem. 28*:1305-1310 (1977).

15. D. H. Russell, Discussion: Putrescine and spermidine biosynthesis in growth and development. *Ann. N.Y. Acad. Sci. 171*:772-782 (1970).

16. G. Brante, Studies on lipids in nervous system with special reference to quantitative chemical determination and typical distribution. *Acta Physiol. Scand. 18*(Suppl. 63):1-189 (1949).

17. A. H. Tingey, Human brain lipids at various ages in relation to myelination. *J. Ment. Sci. 102*:429-432 (1956).

18. J. N. Cummings, H. Goodwin, E. M. Woodward, and G. Curzan, Lipids in the brains of infants and children. *J. Neurochem. 2*:289-297 (1958).

19. J. Altman, DNA metabolism and cell proliferation, in *Handbook of Neurochemistry*, Vol. 2 (A. Lajtha, ed.). Plenum Press, New York, 1969, pp. 31–77.

20. R. P. Bunge, Glial cells and the central myelin sheath. *Physiol. Rev.* 48:197-251 (1968).

21. H. P. Chase, Undernutrition and growth and development of the human brain, in *Malnutrition and Intellectual Development* (J. D. Lloyd-Still, ed.). MTP, Lancaster, 1976, PB-38.

22. P. A. Yakovlev and A. R. Lecours, The myelogenetic cycles of regional maturation of the brain, in *Regional Development of the Brain in Early Life* (A. Minkowski, ed.), Blackwell, Oxford, 1967, pp. 3-65.

23. P. Rosso, J. Hormazabal, and M. Winick, Changes in brain weight, cholesterol, phospholipid and DNA content in marasmic children. *Am. J. Clin. Nutr.* 23:1275-1279 (1970).

24. J. Dobbing, The influence of early nutrition on the development and myelination of the brain. *Proc. R. Soc. B 159*:503-509 (1964).

25. H. P. Chase, J. Dorsey, and G. McKhan, The effect of malnutrition on the synthesis of a myelin lipid. *Pediatrics 40*:551-559 (1967).

26. N. E. Ghittoni and F. Raveglia, Effects of malnutrition on the lipid composition of cerebral cortex and cerebellum in the rat. *J. Neurochem.* 21:983-987 (1973).

27. D. Kritchevsky and W. L. Holmes, Occurrence of desmosterol in developing rat brain. *Biochem. Biophys. Res. Commun.* 7: 128-131 (1962).

28. H. K. M. Yusuf, *Studies of Brain Development in Man and the Rat*. Ph.D. Thesis, University of Surrey, England. (1976).

29. H. K. M. Yusuf and J. W. T. Dickerson, The effect of growth and development on the phospholipids of the human brain. *J. Neurochem. 28*:783-788 (1977).

30. L. Svennerholm and M. T. Vanier, The distribution of lipids in the human nervous system. II. Lipid composition of human fetal and infant brain. *Brain Res.* 47:457-468 (1972).

31. M. A. Fishman, H. C. Agrawal, A. Alexander, J. Golterman, R. E. Martenson, and R. F. Mitchell, Biochemical maturation of human central nervous system myelin. *J. Neurochem.* 24:689-694 (1975).

32. A. L. Prensky and H. W. Moser, Changes in the amino acid composition of proteolipids of white matter during maturation of the human nervous system. *J. Neurochem. 14*:117-121 (1967).

33. E. G. Brunngraber, L. A. Witting, C. Haberland, and B. Brown, Glycoproteins in Tay-Sachs disease: Isolation and carbohydrate composition of glycopeptides. *Brain Res. 38*:151-162 (1972).

34. R. W. Ledeen, Ganglioside structure and distribution: Are they localized at the nerve endings? *J. Supramol. Struct.* 8: 1-17 (1978).

35. L. Svennerholm, Deamination of nucleotides and the role of their deamino form in ammonia formation from amino acids, in *Handbook of Neurochemistry*, Vol. 3 (A. Lajtha, ed.). Plenum Press, New York, 1970, pp. 425-439.

36. H. Suzuki, Formation and turnover of the major brain gangliosides during development. *J. Neurochem.* 14:917-925 (1967).

37. H. Dekirmenjian, E. G. Brunngraber, N. Lemkey-Johnson, and L. M. H. Larramendi, Distribution of gangliosides, glycoprotein-NANA and acetylcholinesterase in axonal and synaptosomal fractions of cat cerebellum. *Exp. Brain Res.* 8:97-104 (1969).

38. A. Hamberger and L. Svennerholm, Composition of gangliosides and phospholipids of neuronal and glial cell enriched fractions. *J. Neurochem.* 18:1821-1829 (1971).

39. W. T. Norton and J. F. Poduslo, Neuronal perikarya and astroglia of rat brain: Chemical composition during myelination. *J. Lipid Res.* 12:84-90 (1971).

40. H. H. Hess, N. H. Bass, C. Thalheimer, and R. Devarakonda, Gangliosides and the architecture of human frontal and rat somatosensory isocortex. *J. Neurochem.* 26:1115-1121 (1976).

41. E. G. Lapetina, E. F. Soto, and E. DeRobertis, Lipids and proteolipids in isolated subcellular membranes of rat brain cortex. *J. Neurochem.* 15:437-445 (1968).

42. B. L. G. Morgan and M. Winick, The subcellular localization of administered *N*-acetylneuraminic acid (NANA) in the brains of well-fed and undernourished rats. *Br. J. Nutr.* 46:231-238 (1981).

43. J. W. T. Dickerson, Effect of growth and undernutrition on the chemical composition of the brain, in *Nutrition, Proc. 9th Int. Congr. Nutrition, Mexico* 2 (A. Chavez, H. Bourges, and S. Basta eds.). Karger, Basel, 1975, pp. 132-138.

44. L. N. Irwin and F. E. Samson, Content and turnover of gangliosides in rat brain following behavioral stimulation. *J. Neurochem.* 18:203-211 (1971).

45. J. A. Dunn and E. L. Hogan, Brain gangliosides: Increased incorporation of 1-^{3}H-glucosamine during training. *Pharmacol. Biochem. Behav.* 3:605-612.

46. B. L. G. Morgan and M. Winick, Effects of administration of *N*-acetylneuraminic acid (NANA) on brain NANA content and behavior. *J. Nutr.* 110:416-424 (1980).

47. B. L. G. Morgan and M. Winick, Effects of environmental stimulation on brain *N*-acetylneuraminic acid content and behavior. *J. Nutr.* 110:425-432 (1980).

48. B. L. G. Morgan and J. Sinai, The role of gangliosides in the uptake of serotonin and the synaptosomal membrane. (In press, 1984.)

49. R. P. Rees, The morphology of interneuronal synaptogenesis: A review. *Fed. Proc.* *37*:2000-2009 (1978).

50. H. E. Himwich and J. F. Fazekas, Comparative studies of the metabolism of the brain of infant and adult dogs. *Am. J. Physiol.* *132*:454-461 (1941).

51. C. Kennedy, G. D. Grave, J. W. Jehle, and L. Sokoloff, Changes in blood flow in the component structures of the dog brain during postnatal maturation. *J. Neurochem.* *19*:2423-2433 (1972).

52. C. Kennedy and L. Sokoloff, An adaptation of the nitrous oxide method to the study of cerebral circulation in children; normal values for cerebral blood flow and cerebral metabolic rate in childhood. *J. Clin. Invest.* *36*:1130-1133 (1957).

53. L. Sokoloff, Cerebral circulatory and metabolic changes associated with aging. *Res. Publ. Assoc. Nerv. Ment. Dis.* *41*:237-239 (1966).

54. L. Sokoloff, Changes in enzyme activities in neural tissues with maturation and development of the nervous system, in *The Neurosciences, Third Study Program* (F. O. Schmitt and F. G. Worden, eds.). M. I.T. Press, Cambridge, 1974, pp. 885-896.

55. H. A. Krebs, D. H. Williamson, M. W. Bates, M. A. Page, and R. A. Hawkins, The role of ketone bodies in caloric homeostasis. *Adv. Enzyme Regul.* *9*:387-409 (1971).

56. B. Persson, G. Settergren, and G. Dahlquist, Cerebral arteriovenous differences of acetoacetate and D-β-hydroxybutyrate in children. *Acta Pediatr. Scand.* *61*:273-278 (1972).

57. S. Carney-Crane and B. L. G. Morgan, The effect of alterations in ketone body availability on the utilization of β-hydroxybutyrate by developing rat brain. *J. Nutr.* *113*:1063-1072 (1983).

58. G. Porcellati, Biochemical processes in brain and nervous tissue. *Bibl. Nutr. Diet a* *17*:16-35 (1972).

59. A. Lajtha, S. Berl, and H. Wallsch, Amino acid and protein metabolism of the brain. IV. The metabolism of glutamic acid. *J. Neurochem.* *3*:322-328 (1959).

60. G. M. McKhann, R. W. Albers, L. Sokoloff, O. Mickelson, and D. B. Tower, The quantitative significance of the gamma-aminobutyric acid pathway in cerebral oxidative metabolism, in *Inhibition in the Nervous System and Aminobutyric Acid* (E. Roberts, ed.). Pergamon Press, Oxford, 1960, pp. 169-186.

61. K. Seta, H. Sershen, and A. Lajtha, Cerebral amino acid uptake in vivo in newborn mice. *Brain Res.* *47*:415-425 (1972).

62. G. Sessa and M. M. Perez, Biochemical changes in rat brain associated with the development of the blood-brain barrier. *J. Neurochem.* *25*:779-782 (1975).

63. Y. Machiyama, R. Balazs, and T. Julian, Oxidation of glucose through the γ-aminobutyrate pathway in brain. *Biochem. J. 96:* 68P (1965).
64. H. C. Agrawal, J. M. Davis, and W. A. Himwich, Postnatal changes in free amino acid pools of rat brain. *J. Neurochem. 13:*607-615 (1966).
65. W. H. Oldendorf and J. Szabo, Amino acid assignment to one of the three blood-brain barrier amino acid carriers. *Am. J. Physiol. 230:*94-98 (1976).
66. G. Furaff, W. King, and S. Undenfriend, The uptake of tyrosine by rat brain in vitro. *J. Biol. Chem. 236:*1773-1777 (1961).
67. Y. Tsukada, Y. Nagata, S. Hirano, and T. Matsutani, Active transport of amino acids into cerebral cortex slices. *J. Neurochem. 10:*241-256 (1963).
68. D. K. Neame, Uptake of histidine, histamine and other imidazole derivatives by brain slices. *J. Neurochem. 11:*655-662 (1964).
69. P. Felig, J. Wahren, and G. Ahlborg, Uptake of individual amino acids by the human brain. *Proc. Soc. Exp. Biol. Med. 142:* 230-231 (1973).
70. W. H. Oldendorf, Brain uptake of radio labelled amino acids, amines, and hexoses after arterial injection. *Am. J. Physiol. 221:* 1629-1639 (1971).
71. G. Baños, P. M. Daniel, S. R. Moorhouse, and O. E. Pratt, The influx of amino acids into the brain of the rat in vivo: The essential compared with non-essential amino acids. *Proc. R. Soc. Lond. B. 183:*59-70 (1973).
72. L. Sokoloff, G. G. Fitzgerald, and E. E. Kaufman, Cerebral nutrition and energy metabolism, in *Nutrition and the Brain*, Vol. 1 (R. J. Wurtman and J. J. Wurtman, eds.). Raven Press, New York, 1977, pp. 87-140.
73. H. Hyden and E. Egyhazi, Changes in RNA content and base composition in cortical neurons of rats in a learning experiment involving transfer of handedness. *Proc. Nat. Acad. Sci. 52:* 1030-1035 (1964).
74. J. Cravioto and E. R. Delicardie, Nutrition, mental development and learning, in *Human Growth* (F. Falkner and J. M. Tanner, eds.). Plenum, New York, 1979, pp. 481-511.
75. R. Balazs, P. D. Lewis, and A. J. Patel, Nutritional deficiencies and brain development, in *Human Growth, Vol. 3: Neurobiology and Nutrition* (F. Falkner and J. M. Tanner, eds.). Plenum, New York, 1979, pp. 512-562.
76. J. Dobbing, The later development of the brain and its vulnerability, in *Scientific Foundations of Pediatrics* (J. A. Davis and J. Dobbing, eds.). Heinemann, London, 1974, pp. 565-577.
77. R. E. Brown, Organ weight in malnutrition with special reference to brain weight. *Dev. Med. Child. Neurol. 8:*512-522 (1966).

78. G. Engsner, Brain growth and motor nerve conduction velocity in children with protein-calorie malnutrition. *Acta Univ. Ups. 180*:1-60 (1974).

79. P. Gruenwald, Chronic fetal distress and placental insufficiency. *Biol. Neonat. 5*:215-268 (1963).

80. F. McLean and R. Usher, Measurement of liveborn fetal malnutrition infants compared with similar gestation and similar birth weight controls. *Biol. Neonat. 16*: 215-221 (1970).

81. R. L. Gross, P. M. Newberne, and J. V. O. Reiv, Adverse effects on infant development associated with maternal folic acid deficiency. *Nutr. Rep. Int. 10*:241-243 (1974).

82. M. B. Stoch and P. M. Smythe, Does undernutrition during infancy inhibit brain growth and subsequent intellectual development? *Arch. Dis. Child 38*:540-552 (1963).

83. M. Winick and P. Rosso, Head circumference and cellular growth of the brain in normal and marasmic children. *J. Pediatr. 74*: 774-778 (1969).

84. M. Winick, K. K. Meyer, and R. C. Harris, Malnutrition and environmental enrichment by early adoption. *Science 190*:1173-1175 (1975).

85. B. L. G. Morgan, G. L. Boris, and M. Winick, A useful correlation between blood and brain *N*-acetylneuraminic acid. *Biol. Neonat. 42*:299-303 (1983).

86. P. Rosso, Nutrition and maternal-fetal exchange. *Am. J. Clin. Nutr. 34*:744-755 (1981).

87. R. Fancourt, S. Campbell, D. Harvey, and A. P. Norman, Follow up study for small-for-date babies. *Br. Med. J. 1*:1435-1437 (1976).

88. J. A. Low, R. S. Galbraith, D. Muir, H. Killen, J. Karchmur, and D. Campbell, Intrauterine growth retardation: A preliminary report of long-term morbidity. *Am. J. Obstet. Gynecol. 130*: 534-545 (1978).

89. R. S. Dow, Historical review of cerebellar investigations, in *The Cerebellum in Health and Disease* (W. S. Fields and W. D. Willis, eds.). Warren H. Green, Inc., St. Louis, 1970, pp. 5-9.

90. M. Winick, P. Rosso, and J. Waterlow, Cellular growth of cerebrum, cerebellum, and brain stem in normal and marasmic children. *Exp. Neurol. 26*:393-400 (1970).

91. H. P. Chase, C. A. Canosa, C. S.Dabiere, N. N. Welch, and D. O'Brien, Postnatal undernutrition and human brain development. *J. Ment. Defic. Res. 18*:355-366 (1974).

92. H. P. Chase, N. N. Welch, C. S. Dabiere, N. S. Vasan, and L. J. Butterfield, Alterations in human brain biochemistry following intrauterine growth retardation. *Pediatrics 50*:403-411 (1972).

93. M. K. J. Sarma and K. S. Rao, Biochemical composition of different regions in brains of small-for-date infants. *J. Neurochem. 22*: 671-677 (1974).

94. N. M. Cowan, The development of the brain. *Sci. Am. 241*: 112-133 (1979).
95. D. H. Hubel, The brain. *Sci. Am. 241*:45-53 (1979).
96. P. J. Morgane, M. Miller, T. Kempner, W. Stern, W. Forbes, R. Hall, J. Branzino, J. Kissane, E. Hawrylewicz, and O. Resnick, The effects of protein malnutrition on the developing central nervous system in the rat. *Neurosci. Biobehav. Rev.* 3:139-230 (1978).
97. W. J. H. Nauta and M. Feirtag, The organization of the brain. *Sci. Am. 241*:88-111 (1979).
98. C. E. Stevens, The neuron, in *The Brain*. Scientific American Books, W. H. Freeman, Oxford, 1979, pp. 15-25.
99. I. H. Ulus, The effect of choline on cholinergic function, in *Cholinergic Mechanisms and Psychopharmacology* (D. J. Jender, ed.). Plenum Press, New York, 1977, pp. 525-538.
100. L. L. Iversen, The chemistry of the Brain. *Sci. Am. 241*:134-149 (1979).
101. R. Y. Moore and F. E. Bloom, Central catecholamine neuron systems: Anatomy and physiology of the dopamine system. *Annu. Rev. Neurosci. 1*:129-169 (1978).
102. J. Dobbing, Nutritional growth restriction and the nervous system, in *The Molecular Basis of Neuropathology* (R. H. S. Thompson and A. W. Davison, eds.). Arnold, London, 1981, pp. 89-106.
103. A. Merat, *Effects of Protein-Calorie Malnutrition on Brain Gangliosides*. Ph.D. Thesis, University of Surrey, England, 1971.
104. J. W. T. Dickerson, Protein deficiency and the brain, in *Food and Health: Science and Technology* (G. G. Birch and K. J. Parker, eds.). Applied Science Publishers, London, 1980, pp. 487-500.
105. J. W. T. Dickerson, A. Merat, and H. K. M. Yusuf, Effects of malnutrition on brain growth and development, in *Brain and Behavioral Development* (J. W. T. Dickerson and H. McGurk, eds.). Surrey Univ. Press, Guildford, 1982, pp. 73-108.
106. B. L. G. Morgan and D. L. Naismith, Effect of early postnatal undernutrition on the growth and development of the rat brain. *Br. J. Nutr. 48*:15-23 (1982).
107. J. Dobbing, Undernutrition and the developing brain. *Am. J. Dis. Child. 120*:411-415 (1970).
108. J. Dobbing, The later development of the brain and its vulnerability, in *Scientific Foundations of Pediatrics* (J. A. Davis and J. Dobbing, eds.). Heinemann, London, 1974, pp. 565-577.
109. F. D. Collins, A. J. Sinclair, J. P. Royle, D. A. Coats, A. T. Maynard, and R. F. Leonard, Plasma lipids in human linoleic acid deficiency. *Nutr. Metab. 13*:150-167 (1971).

110. R. T. Holmann, Essential fatty acid deficiency in humans, in *Dietary Lipids and Postnatal Development* (C. Galli, G. Jacini, and A. Pecile, eds.). Raven Press, New York, 1973, pp. 127-142.

111. C. B. Cowey, J. M. Owen, J. W. Aldren, and C. Middleton, Studies on nutrition of flat-fish. The effect of different dietary fatty acids on the growth and fatty acid composition of turbot. (*Scaptholamus maximus*). *Br. J. Nutr. 36*:479-486 (1976).

112. M. A. Crawford, A. J. Sinclair, P. M. Msuya, and A. Murhambo, Structural lipids and their polyenoic constituents in human milk, in *Dietary Lipids and Postnatal Development* (C. Galli, G. Jacini, and A. Pecile, eds.). Raven Press, New York, 1973, pp. 41-56.

113. J. W. T. Dickerson, A. Merat, H. K. M. Yusuf,

   in, *Brain and Behavioral Development* (J. W. T. Dickerson and H. McGurk, eds.). Surrey University Press, Guildford, England, 1982, p. 97.

114. C. O. Enwonwu and B. S. Worthington, Accumulation of histidine, 3-methyl-histidine and homocarnocine in the brains of protein-calorie deficient monkeys. *J. Neurochem. 21*:799-807 (1973).

115. C. O. Enwonwu and B. S. Worthington, Regional distribution of homocarnocine and other ninhydrin-positive substances in brains of malnourished monkeys. *J. Neurochem. 22*:1045-1052 (1974).

116. T. S. Nowak and H. N. Munro, Effects of protein-calorie malnutrition on biochemical aspects of brain development, in *Nutrition and the Brain*, Vol. 2 (R. J. Wurtman and J. J. Wurtman, eds.). Raven Press, New York, 1977, pp. 194-260.

117. R. L. Liebel, D. B. Greenfield, and E. Pollitt, Iron deficiency: Behavior and brain biochemistry, in *Nutrition Pre- and Postnatal Development* (M. Winick, ed.). Plenum Press, New York. 1979, pp. 383-440.

118. P. R. Dallman, Tissue effects of iron deficiency, in *Iron in Biochemistry and Medicine* (A. Jacobs and M. Worwood, eds.). Academic Press, inc., New York, 1974, pp. 437-475.

119. P. R. Dallman, C. Refino, and M. J. Yland, Sequence of development of iron deficiency in the rat. *Am. J. Clin. Nutr. 35*: 671-677 (1982).

120. R. L. Liebel, D. B. Greenfield, and E. Pollitt, Iron deficiency: Behavior and brain biochemistry, in *Nutrition Pre- and Postnatal Development* (M. Winick, ed.). Plenum Press, New York, 1979, p. 400.

121. F. A. Oski and A. M. Honig, The effects of therapy on the developmental scores of iron deficient infants. *J. Pediatr. 92*: 21-25 (1977).

122. D. P. Davies, Some aspects of 'catch-up' growth in "light-for-date babies," in *Topics in Pediatrics. Vol. 2. Nutrition in Childhood* (B. Wharton, ed.). Pitman Medical Tunbridge Wells, 1980, pp. 72-80.

123. E. Pollitt and C. Thomson, Protein-calorie malnutrition and behavior: A review from psychology, in *Nutrition and the Brain*, Vol. 2 (R. J. Wurtman and J. J. Wurtman, eds.). Raven Press, New York, 1977, pp. 261-306.

124. J. Cravioto, E. R. Delicardie, and H. G. Birch, Nutrition growth and neurointegrative development: An experimental ecological study. *Pediatrics* 38:319-372 (1966).

125. Z. Stein, M. Susser, G. Saenger, and F. Marolla, *Famine and Human Development: The Dutch Hunger Winter of 1944/45.* Oxford University Press, London, 1975.

126. P. M. Fitzhardinge and E. M. Stevens, The small-for-date infant. II. Neurological and intellectual sequelae. *Pediatrics* 50:50-57 (1972).

127. T. Fujikura and L. A. Froehlich, Mental and motor development in monozygotic twins with dissimilar birth weights. *Pediatrics* 53:884-889 (1974).

128. G. S. Babson and D. S. Phillips, Growth and development of twins of dissimilar size at birth. *N. Engl. J. Med.* 289:937-940 (1973).

129. S. M. Pereira, S. Sundaraj, and A. Begum, Physical growth and neurointegrative performance of survivors of protein-energy malnutrition. *Br. J. Nutr.* 42:165-171 (1979).

130. J. Cravito and E. R. Delicardie, Nutrition, mental development and learning, in *Human Growth* (F. Falkner and J. M. Tanner, eds.). Plenum Press, New York, 1979, pp. 481-511.

131. J. D. Lloyd-Still, Clinical studies on the effects of malnutrition during infancy on subsequent physical and intellectual development, in *Malnutrition and Intellectual Development* (J. D. Lloyd-Still, ed.). M. T. P., Lancaster, 1976, pp. 103-159.

132. N. M. Lien, K. K. Meyer, and M. Winick, Early malnutrition and "late" adoption: A study of their effects on the development of Korean orphans adopted into American families. *Am. J. Clin. Nutr.* 30:1734-1739 (1977).

133. *Ten State Nutritional Survey 1968-1970, Highlights.* U.S. Department of Health Education and Welfare, 1972.

134. J. Dobbing, Human brain development and its vulnerability, in *Biologic and Clinical Aspects of Brain Development.* Mead Johnson Symposium on Perinatal and Developmental Medicine No. 6, 1974.

135. C. Trevarthen, Neuroembryology and the development of perception, in *Human Growth*, Vol. 3. *Neurobiology and Nutrition* (F. Falkner and F. Tanner, eds.). Plenum Press, New York, 1979, pp. 271-299.

136. L. E. Hicks, R. A. Langhorn, and J. Takenaka, Interpreta-
      tion of behavioral findings in studies of nutritional supplementa-
      tion. *Am. J. Public Health 73*:695-697 (1983).
137. E. M. Widdowson and R. A. McCance, Some effects of accelerat-
      ing growth: 1. General somatic effects. *Proc. R. Soc. B. 152*:
      188-206 (1960).
138. P. Rosso, Prenatal nutrition and fetal growth and development.
      Pediatr. Ann. 10:21-26 (1981).
139. N. S. Scrimshaw, Kwashiorkor, marasmus, and intermediate
      forms of protein-calorie malnutrition, in *Cecil Textbook of
      Medicine*, 16th ed. (J. B. Wyngaarden and L. H. Smith, Jr.,
      eds.). W. B. Saunders Co., Philadelphia, 1982, p. 536.
140. J. A. Brasel, Cellular changes in intrauterine malnutrition, in
      Nutrition and Fetal Development (M. Winick, ed.). John Wiley
      & Son, New York, 1974, pp. 13-26.
141. J. D. Lloyd-Still, Clinical studies on the effects of malnutrition
      during infancy on the subsequent physical and intellectual de-
      velopment, in *Malnutrition and Intellectual Development* (J. D.
      Lloyd-Still, ed.). M. T. P., Lancaster, 1976, pp. 103-159.

6

Dietary Fibers
Aspects of Nutrition, Pharmacology, and Pathology

George V. Vahouny

George Washington University Medical Center
Washington, D.C.

I. INTRODUCTION

In 1956, Surgeon Captain T. L. Cleave propounded much of the basic
theory of the importance of unrefined carbohydrates in his paper, *The
Neglect of Natural Principles in Current Medical Practice* (1). Cleave's
original "saccharine disease" hypothesis (2,3) was confirmed, refined,
and expanded by others, including Burkitt and Trowell (4) and Walker
(5). This concept was reexpressed as the dietary fiber hypothesis (4,6)
and has two primary statements: (a) A diet rich in foods containing
plant cell walls is protective against a wide range of diseases common
in Western cultures, including constipation, diverticular disease, colon
cancer, heart disease, diabetes, obesity, and gallstones; (b) In some
cases, a diet low in these dietary materials may be either causative in
certain diseases or may provide conditions in which other factors are
more active. Since the early 1970s, dietary fiber has become a topic
that has burgeoned from an esoteric interest of a few research laboratories to a subject that has aroused international interest. This growth
has been helped by an intense public interest in the potential benefit
to be gained by adding fiber to the diet—and may have been helped
by the perception that, for once, medicine was saying "do" instead of
"don't" (7). This welcomed expansion in the concern with nutritional
and medical aspects of dietary fibers has clearly indicated that these
materials cannot be considered inert components of the diet. Although
the mechanisms of action are not yet completely understood, dietary

fibers or specific fiber components have already found extensive therapeutic utility in the treatment of several medical disorders or diseases.
Among these are simple constipation, symptomology of diverticular
disease of the colon, and diabetes mellitus. Whether or not the lack of
fiber per se may be responsible, either directly or indirectly, for the
variety of diseases expressed in the hypothesis remains to be determined.

There are, as in any rapidly expanding area of medicine, a number
of problems associated with this field of nutrition and medicine. Some
of these are undoubtedly a result of a misunderstanding concerning
the definition of dietary fiber. In the original hypothesis, the term
was applied to diets rich in foods containing plant cell walls. In the
current lay and scientific communities, the term dietary fiber is casu-
ally applied not only to foods rich in plant cell walls but also to a
variety of manufactured products including brans, pectins, gums, and
even highly characterized polysaccharides, such as cellulose. The
physical and chemical properties of these materials are, in many cases,
unrelated to their properties when present in the original complex
matrix of the plant cell wall.

The complex nature of dietary fiber has led to extensive studies on
the properties and applications of these isolated products and polysaccharides. The result has been a major shift in emphasis from the
prophylactic or therapeutic value of dietary fiber per se, to what
should be appropriately considered as the pharmacological action of
plant polysaccharides. Although these types of studies have produced
valuable information, the use of these isolated polysaccharides as a
high-fiber supplement to natural foods may introduce changes in gastrointestinal responses, many of which were not anticipated or desirable.

The purpose of this chapter is to review the current status of dietary
fiber and plant cell wall components in nutrition and medicine, and to
provide some understanding of the potential utility, as well as some of
the undesirable consequences of increasing the intake of certain plant
cell wall polysaccharides. This approach is in no way meant to denigrate
the importance of dietary fiber in human nutrition as derived from the
extensive epidemiological and experimental evidence.

A more extensive assessment of the chemistry and analysis of dietary
fibers and the physiological and metabolic responses to fibers and their
components has been provided in several recent reviews (8-14).

II. DIETARY FIBERS AND THEIR COMPONENTS

A. Definitions and Terminology

The definition of dietary fiber has been a source of considerable confusion and debate, perhaps more so than in any other field of nutrition.

The original hypothesis (4,6) related to types of diets and the amounts of plant cell wall material they contained. However this definition does not account for plant storage polysaccharides, such as pectic substances, gums, and mucilages, which are often water-soluble and have important nutritional and pharmacological implications. Trowell et al. (15) therefore proposed a definition restricting dietary fiber to *the sum of lignin and plant polysaccharides that are undigested by endogenous secretions of the human digestive tract.* This definition also is not without shortcomings since it focuses attention on the indigestibility of dietary fiber and does not include a wide variety of materials associated with plant cell walls. In the first case, several of the "indigestible" polysaccharides of plant cells can indeed be metabolized by the colonic bacteria (16), and this action certainly contributes to the physiological and metabolic effects of dietary fibers. In the second case, associated substances such as phytic acid, silica, cuticular substances, and cell-wall protein may also contribute to the overall responses to dietary fiber and should not be overlooked.

In an attempt to clarify the confusion in terminology in the field of dietary fiber research, Southgate and White (17) have proposed the definitions shown in Table 1. The analytical definition includes the

Table 1 Terminology in Dietary Fiber Studies

Hypothesis	Mixture in diet	Terminology in metabolic studies on diet supplements	Analytical definition
Dietary fiber	Protein Lipids Inorganic constituents Lignin Cellulose Noncellulose Polysaccharides	Cellulose Pectic substances	Dietary fiber
	Polysaccharide food additives	Pectins Gums Mucilages Algal polysaccharides Modified cellulose	

Source: Ref. 18.

sum of lignins and both structural and storage plant polysaccharides.
Studies on isolated polysaccharides or their complexes should be referred to by their trivial or systematic names (e.g., pectin, cellulose,
etc.). Thus according to Heaton (13), "...dietary fibre cannot be
defined, only described, in chemical terms."

A glossary of common and chemical terms used in this field has been
prepared (18), and some of these terms relating directly to the present
discussion are listed at the end of this chapter (Section VIII).

B. Chemical and Physical Composition

The qualitative composition of dietary fibers is summarized in Table 2.
The major components include: cellulose; noncellulose polysaccharides;
a noncarbohydrate component, lignin; and associated substances. These
components are largely, but not entirely, involved in plant cell wall
structure. Cellulose (Table 2) is a linear polymer of glucose residues
in a beta-(1,4)-linkage, and represents the major carbohydrate in nature. The ability of adjacent polymers to associate by hydrogen bond-
ing results in a crystalline fibrillar structure, typical in plant cell
walls. Depolymerization by selective acid hydrolysis results in microcrystalline cellulose, which forms stable suspensions in water and is
employed as an emulsification agent. Derivatization to carboxymethyl
and methylcelluloses forms polymers with gelling properties, which are
employed to vary physical properties of foods.

Hemicelluloses, in contrast to cellulose, are soluble in aqueous alkali.
Primarily composed of beta-(1,4)-linked pyranosides containing a high
concentration of galactose and mannose, they are termed galactomannans.
The acid and neutral hemicelluloses are distinguished by their relative
content of uronic acids. These plant cell wall components are generally
in lower concentrations than cellulose, but they can form the matrix
for cellulose organization in the cell wall and are the storage form of
polysaccharides in many seeds.

The pectic substances include protopectin, pectic and pectinic acids,
and pectin. Galacturonic acid [alpha-(1,4)-D-galacturonans] is, the
principal constituent of the primary chain with rhamnose insertions,
which confer a kink in the primary chain. The carbonyl groups of the
galacturonic acid chains are variably methoxylated and have important
effects on the physical property of the polymer. Highly methoxylated
pectins are viscous and are used commercially in jams, while the poorly
methoxylated forms have little or no gelling ability.

The plant gums are complex polysaccharides (Table 2) which result
from injury. Typical examples include gum arabic, which contains a
primary chain of galactose, and guar gum, which is a galactomannan.

Lignin, the major nonpolysaccharide of plant cell walls, is a complex
series of aromatic polymers, derived from condensation of phenolic
alcohols. It is essentially inert to chemical and microbial degradation.

The primary plant cell wall consists mainly of cellulose fibers in a mixture of pectic substances and hemicelluloses. The secondary cell wall contains several layers of cellulose fibers in a parallel arrangement and in a matrix of hemicelluloses. Lignification occurs with maturation of the cell wall and provides additional rigidity in the plant structure. The plant gums and mucilages are products of specialized secretory cells and are not structurally involved in the plant cell wall. Although most plant cell walls are composed primarily of carbohydrate, varying amounts of protein and lipid are also present (Table 3) and can impart important properties to the specific fiber derivative under investigation.

C. Properties of Dietary Fibers

Generally, the physical properties of dietary fibers have been studied in vitro and have included properties of intact dietary fibers as well as those of individual commercial food supplements and the major fiber components. These properties, however, may not be identical to those expressed in vivo since other factors have not been considered. For example, these include osmolality effects of other food components, adsorption, and bacterial action. Thus the in vitro studies are only a reflection of possible physical properties of dietary fibers under specific experimental conditions. Nevertheless, they have provided important clues regarding the potential physiological responses to dietary fibers and fiber supplements.

1. *Water Holding Capacity*

This property reflects the extent of hydratability of the fiber. Hydration can occur either by adsorption to the fiber surface or by entrapment within the fiber matrix. This response will be determined by both the structure of the fiber components and their physical relationship within the cell wall, by particle size, and by the pH and electrolyte concentration of the solvent (21). In general, water-holding is increased when there are higher levels of free polar sugar residues exposed in the fiber material and is decreased with increasing intramolecular linking.

The practical application of knowledge on the water-holding capacity of fiber is of limited utility. Viscous fibers, such as guar gum and pectins may partition nutrients into their gel matrix owing to their marked affinity for water. This could therefore account, in part, for their effects on limiting nutrient availability and their ability to reduce the rate of sugar absorption (22). However application of information on water-holding capacity to other known physiological responses is not at all clear. First of all, water-holding capacity may vary with the maturity of the plant source (23) and with the extent of food preparation (24). Thus it has been demonstrated that the particle size of wheat bran can markedly alter water-holding capacity, its digestibility by colonic bacteria, and can modify the physiological response (stool output) to the fiber supplement (24,25).

Table 2 Composition of Dietary Fibers

Cell wall component	Main chemical components		Description
	Primary chains	Secondary chains	
Polysaccharides			
Cellulose	Glucose	—	Main structural component forming a crystalline matrix, extensively digested by colonic bacteria
Hemicelluloses	Mannose, galactose, glucose, xylose	Arabinose, galactose, glucuronic acid	15-30% of plant cell wall; under a variety of polysaccharide polymers with branched configuration; from the matrix of the plant cell wall (with pectins), which enmesh cellulose fibers; extensively digested by colonic bacteria
Pectins	Galacturonic acid	Rhamnose, arabinose, xylose, fucose	1-4% of plant polysaccharides; components of primary cell wall and middle lamellae varying in methylation; water-soluble and gelatinous; extensively digested by colonic bacteria

Mucilages	Galactose-mannose, glucose-mannose, arabinose-xylose, galacturonic acid	Galactose	Produced by plant secretory cells; associated with stored polysaccharides of plant seeds; emulsification and stabilization properties
Gums	Galactose, glucuronic-mannose, galacturonic acid	Xylose, fucose, galactose	Produced by secretory cells at sites of injury
Algal poly-saccharides	Mannose, xylose, glucuronic acid	Galactose	Vary in uronic acid and sulfate content; derived from algae and seaweed
Nonpolysaccharides			
Lignin	Sinapyl alcohol conferyl alcohol -coumaryl alcohol	—	Complex, cross-linked phenylpropane polymer; hydrophobic; increased content with plant age; resistant to bacterial degradation
Associated substances	Saponins, phytates, silica, proteins, lipids, ions	—	—

Source: Refs. 19 and 20.

Table 3 Major Constituents of the
Primary Cell Wall

Constituent	% of Fresh weight
Water	60
Cellulose	10-15
Hemicellulose	5-15
Pectic substances	2-8
Lipids	0.5-3
Protein	1-2

Source: Ref. 10.

Although this property of water-holding by fiber generally has been
assumed to be directly related to effects on fecal volume, the current
evidence suggests that there is no direct relationship between this property and the stool-bulking ability of the fiber. In fact, the data of
Stephen and Cummings (26) suggest that fiber preparations that hold
the least water as determined in vitro (e.g., bran and bagasse) produced the greatest changes in stool weight. Conversely, viscous pectin, which is extensively hydrated, has no effect on stool output.

2. *Cation-Exchange Properties*

Although no anion-exchange properties of dietary fibers have been
found, cation-exchange capacity has been extensively documented and
is largely related to the carboxyl groups of the acidic sugars of polysaccharides. This exchange capacity (up to 2 mEq/L) has been demonstrated for monovalent cations (27), and for calcium (28), zinc, and
iron (29). The possible implications of this property on mineral balance
is discussed in Section V.B.

3. *Adsorption of Bile Acids and Other Organic Molecules*

Eastwood and Boyd (30) reported that appreciable quantities of bile
acids were associated with the insoluble fraction of the small intestinal
contents of rats. Since then, there have been repeated demonstrations
of the binding of bile acids to various grains, food fiber sources, and
isolated fiber components (31-36). This binding appears to be primarily
an adsorption phenomenon, and is influenced by pH and osmolarity, bile
acid structure, and the physical and chemical forms of the fiber (31).
In general, binding of bile acids is greater at lower pH, where acidic
groups are un-ionized, is probably hydrophobic (31), and appears to

be reversible (37). Certain fiber sources appear to have preference
for binding the unconjugated bile acid, rather than the taurine or
glycine conjugates, and may show a decree of specificity for di- or tri-
hydroxylated bile acid analogues. However this is not a uniform phenomenon among the various fibers tested in vitro.

Among the fibers and their individual components, the viscous or gelling fibers, and the lignins appear to have the greatest bile acid-sequestering ability (38,39). With the viscous fibers, such as pectin, this
ability to affect the normal partitioning in the intestine has been demonstrated indirectly by the ability of dietary fiber to increase fecal bile
acids in vivo (38). With the insoluble fibers, the majority of studies
have been conducted in vitro and generally have considered only individual or mixed bile acids at micellar concentrations. Other studies
(36,39) however suggest that this sequestration phenomenon may be
more complex than originally proposed.

The sequestration ability of commercial bile acid-binding resins and
certain types of fibers does not appear to be confined to bile acids.
The binding of cholesterol from bile acid micelles by cholestyramine,
cereal fibers, and alfalfa has also been demonstrated, and at levels
roughly proportional to bile acid-binding (34,36,37). In more recent
studies (36,39), the ability of various fiber preparations to sequester
individual amphipaths and amphiphiles, which normally constitute mixed
micelles in the intestinal lumen, was tested. With this complex mixture
of bile salts, phospholipids, fatty acids, monoglycerides, and cholesterol,
the apparent binding of all components of mixed micelles was roughly
proportional (Table 4). As expected, the exchange resin, cholestyramine,
completely sequestered the available micellar bile acid and also removed
all other micellar components. Among the fiber material tested, the
viscous guar gum and insoluble lignin were effective micelle sequestrants
while cellulose was essentially inert in binding micelles. The importance
of this property of dietary fibers in altering lipid absorption and metabolism is discussed more fully in Section III.C.

Additional influences on bile acid sequestration have received only
limited attention. The ability of free carboxyl moieties of polyuronic
acids to form complexes with metal cations also may influence subsequent
"adsorption" properties of the fiber (40,41). In this connection, coordination complexes between pectin and either aluminum or ferric ions
have been reported to markedly influence binding properties of the
fiber in vitro. Undoubtedly, these and other interactions of fibers,
both ionic and hydrophobic, can have major effects on both in vitro and
in vivo properties of certain fiber components.

In general, the in vitro studies on the binding of bile acids by various
food fibers and individual fiber components are compatible with the ability of these same materials to influence fecal bile acid levels and to modify enterohepatic circulation of these cholesterol metabolites (42,43)
(see Section III.C).

Table 4 Binding of Bile Acids and Micellar Components

Dietary fiber source or component	% Binding of micellar components [a]				
	Taurocholate	Lecithin	Cholesterol	Monolein	Oleic acid
Cholestyramine control	82	99	95	96	92
White wheat bran	4	6	0	10	11
Alfalfa meal	7	3	1	15	19
Guar gum	36	22	23	31	23
Cellulose	1	0.5	8	6	4
Lignin	20	27	18	20	13

[a]Micellar mixtures contained 5 µM taurocholate, 625 µM lecithin, 250 µM monoolein, 50 µM oleic acid, and 250 µM cholesterol. Incubations were carried out with 40 mg of each test fiber.
Source: Refs. 36 and 39.

4. *Digestibility by Colonic Bacteria*

Although dietary fiber, by definition, is not digested by secretion of the stomach and small intestine, there may be extensive degradation in the colon by bacterial fermentation. This is, in part, determined by the nature of the fiber and the bacterial flora and can influence the overall response of the host. There are over 100 species of bacteria in the colon, essentially all of which are anaerobes, for which dietary fiber is a major nutrient and energy source. These contain the enzymes (e.g., cellulases, hemicellulases, pectinesterases) required to digest the plant polysaccharides, depending upon the surrounding milieu and on the physical and chemical state of the plant fiber. Subsequent metabolism of the di- and monosaccharides results in the growth of the microorganisms and the production of the major end-products of fermentation, which include volatile short-chain fatty acids (acetate, propionate, and butyrate) and gases (CO_2, H_2, and methane).

It appears that the addition or removal of fiber from the diet does not substantially influence the overall composition of the bacterial flora (44). However bacteria account for half of the fecal dry weight (45), and during fiber intake a large percentage of the increase in stool weight is explained by increased bacterial content.

The various fibers and their polysaccharide components are fermented to different degrees. Wheat bran, for example, may be degraded only to the extent of 30 to 35% (27,28), while cabbage fiber is almost completely metabolized. Among the individual fiber polysaccharides, hemicelluloses and pectins are essentially completely degraded, while the fermentation of cellulose is less extensive when present in bran or fed in purified form. In contrast, the cellulose of fruits and vegetables appears more susceptible to fermentation (46). The nonpolysaccharide product lignin appears to be completely resistant to fermentation in the colon, and in addition, does not influence overall fermentation of the fiber polysaccharides.

III. PHYSIOLOGICAL AND PHARMACOLOGICAL RESPONSES

A. Gastric Filling and Emptying

The physicochemical properties of dietary fibers and their components described earlier are, to some extent, predictive of the in vivo responses to these food components. These responses ultimately involve the entire alimentary tract (Table 5).

Diets containing higher levels of plant fibers are bulkier and require more time for ingestion (47). This will stimulate chewing, thereby increasing the flow of saliva and dilution of oral contents. The solid nature of the food fiber and the dilution effects are likely to have a positive influence on dental caries and other bacterial-dependent effects in the mouth (10). Chewing may also promote a sense of satiation by a

Table 5 Physicochemical Properties of Dietary Fibers and Predicted Responses

Property	Fiber characteristic	Predicted response
Bulk	Particle size; ionic properties	Increased chewing; slower intake; satiation; altered stool bulk and laxation; altered transit time
Hydratability (water-holding capacity)	Gel formation	Delayed gastric emptying; satiation; delayed nutrient availability; un-altered or increased transit time; altered bile acid metabolism
Cation-exchange capacity	Polyuronic acids; associated phytate levels	Modified cation availability; trace element imbalance
Bile acid binding	Pectins, gums, lignins	Laxation; altered enterohepatic circulation; modified lipid digestion and absorption
Bacterial digestibility	Polysaccharides	Altered water-holding capacity; altered microbial growth and chemical environment; osmotic catharsis

Source: Ref. 20.

direct effect on the satiety center in the hypothalamus (48) and by increasing gastrin secretion and the flow of gastric juices.

In the stomach dietary fiber, in general, can delay gastric emptying of nutrients because of the more solid nature of the ingested food (49). The increased volume of gastric juices and the greater distention of the stomach caused by the fiber bulk also can provide a sense of satiation (50). Although objective and scientifically acceptable methods for measurement of satiation have yet to be developed, the subjective evaluation of dietary fiber effects on satiety has been reported (51) and clearly demonstrates the overall effect of fruit fibers on satiety using energy-equivalent meals (Figure 1).

Among the specific fiber types, the viscous polysaccharides, such as pectin and guar, have been demonstrated to markedly delay gastric emptying (53-56). That this effect is largely, if not entirely, a function of fiber viscosity has also been documented (22,56). As shown in Figure 2, the retention of glucose in the stomach of animals given one of three types of guar gum with varying viscosities, clearly shows the direct relationship between delayed gastric emptying and fiber viscosity.

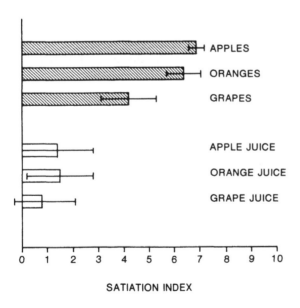

Figure 1 Fruit fibers and satiety. Subjective evaluation of satiation on a scale of 0-10 was made by healthy nonobese subjects comparing the effects of intact fruits and juices made from the same fruits consumed at energy-equivalent levels. (Modified from Ref. 52.)

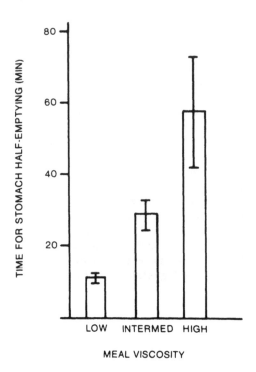

Figure 2 Effect of dietary intakes of low-, intermediate-, and high-viscosity guar on gastric emptying of glucose in the rat. (Modified from Ref. 56.)

Thus the presence of bulkier dietary fiber in the overall food consumption delays both the filling and emptying of the stomach. This can affect the rate of ingestion and the sense of satiation and has therefore found application in weight reduction and maintenance. Furthermore, delayed gastric emptying of nutrients will affect the rate of nutrient absorption and its sequelae, as discussed in Section III.D.

B. Total Gastrointestinal Transit Times and Fecal Bulk

The overall effect of dietary fibers or specific fiber components on gastrointestinal physiology might be expected to be dependent, in part, upon the overall rate of passage of materials from mouth to anus. The alteration in transit time has been assumed as a characteristic response to the intake of cereal fiber (57). This rate, however, varies greatly, not only between individuals but even in the same person (58). Such an effect might be expected since overall transit rates are dependent upon rates of gastric filling and emptying, on rate of passage in the small intestine, and on colonic metabolism and transit. For these reasons, and because of the available methods for measurement of overall transit times (59), there is no precise information of the effects of dietary fibers on transintestinal transit times. Furthermore, it is likely that total transit, as determined by appearance of markers in the stools or by breath hydrogen measurements, may be a result rather than a cause of colonic metabolism (16).

Studies by Cummings et al. (60) also emphasized the individual variation in colonic responses as well as demonstrated effects of dietary fiber on transit times. For example, bran intake reduced colonic transit from 73 ± 24 hr to 43 ± 7 hr, whereas the effects of cabbage and apples were less pronounced, and carrots were without effect. This, and similar studies, have served to emphasize the problems of assessing the role of gastrointestinal transit times on nutrient availability, intestinal adaptations, and other aspects of intestinal and colonic functions. Despite these shortcomings, the relationship between transit time and either mean stool weight or change in fecal bulk has been repeatedly emphasized (59,60) as at least one important aspect of dietary fiber influence on overall intestinal and colonic function.

The effects of dietary fibers on fecal bulk is somewhat less ambiguous, and these have been reviewed extensively elsewhere (59,60). These studies suggest that insoluble fiber types, such as wheat bran, have the greatest effect on stool weight, while viscous materials, such as pectins and guar, have minimal effects. As pointed out earlier, it appears that this effect on stool bulk is inversely related to the hydration capacity of the fiber source. Other factors influencing stool-bulking ability include particle size (21), bacterial metabolism of the fiber, cathartic responses to the short-chain fatty acid metabolites, and chemical composition of the fiber (60). In this latter regard, it has been proposed (61) that changes in fecal weight induced by dietary fibers most closely relates to the pentose content of the noncellulose components of the fiber.

C. Fecal Bile Acid and Neutral Steroid Output

The information derived from in vitro studies on the ability of dietary fibers to sequester bile acids (Section II.C-3) carries implications regarding altered enterohepatic circulation of these cholesterol metabolites, altered biliary bile composition, and modified digestion and absorption of lipids, including cholesterol. Thus the effects of specific types of dietary fibers on plasma cholesterol levels have been attributed, at least in part, to modified bile acid metabolism and to increases in fecal bile acid excretion. The extensive bibliography of positive and negative effects of dietary fibers and fiber products on these parameters of cholesterol metabolism is summarized elsewhere (11,43,62,63).

Pectin is the most studied fiber component in relation to bile acid metabolism. Leveille and Sauberlich (38) initially demonstrated that

the hypocholesteremic response of pectin-feeding in rats was accompanied by increased fecal bile acid excretion. Other studies also showed
that pectin directly interfered with cholesterol absorption, either when
administered acutely to rats (64) or following a short- or long-term
feeding regimen (65,66).

In most subsequent studies with other dietary fibers or fiber components in various species, including man, decreases in plasma cholesterol levels or increases in fecal acidic steroids (bile acids) and/or
neutral steroids (cholesterol and its bacterial metabolites) have been
used as indicators of effects of fibers on lipid absorption and metabolism. These approaches, albeit indirect, suggest that the viscous fibers
show the most consistent effects, while the effects of the particulate
fibers are more variable. Thus increased fecal output of acidic steroids
has been reported for fiber derivatives such as pectins (38,67-70),
guar gum (67), and psyllium seed colloid (Metamucil) (71,72). In humans, for example, Kay and Truswell (68) reported that daily administration of as little as 15 g of pectin for 3 weeks resulted in a 13% reduction of plasma cholesterol, 44% increase in fecal fat, increased fecal
neutral sterols (17%), and a 33% increase in fecal bile acids. Other
studies with pectin (30 to 36 g daily for 2 weeks) have reported increases
in fecal bile acids ranging from 34% (67) to 75% (70). In similar studies
(69), guar increased fecal acidic steroids by 84%.

Results with particulate fibers such as wheat bran, which does not
sequester bile acids in vitro (Table 4), generally suggest that there is
little if any effect of reasonable intakes of this fiber on fecal bile acid or
neutral steroid output or on serum cholesterol levels (19,20,42,43). In
contrast, oat bran, which is more mucilaginous due to its beta-glycan
content, has been reported to increase bile acid excretion in man (73).
Lignins, which can sequester bile acids by hydrophobic interactions,
also may increase fecal bile acid excretion (74).

In general therefore, there is a reasonable relationship between the
ability of certain fiber derivatives to sequester bile acids and micellar
components in vitro and their overall effect on fecal bile acid excretion.
It also seems reasonably clear that this represents at least one mechanism, but by no means the only one, by which these fibers can interfere
with lipid absorption in the intestine (75). There is now considerable
evidence for adaptive changes in intestinal metabolism in response to
dietary fiber intake (56), and these certainly play a role in various
aspects of intestinal lipid (and other nutrient) absorption, transport,
and metabolism (75). Many of these "adaptive" responses have important implications in the overall effects of fiber supplements.

As discussed at considerable length in Section VI, the ability of certain types of fibers to enhance fecal bile acid excretion and affect circulating lipid levels has a paradox, which is of major concern. These
fibers alter the balance of bile acids by removing them from a rapid
enterohepatic circulation to a slower cecohepatic cycle (76). This, in

effect, raises colonic bile acid concentration and therefore may be pre-disposing to enhanced carcinoma in this organ.

D. Delayed Nutrient Absorption

There is a variety of mechanisms by which dietary fiber can influence the digestion and absorption of major nutrients (Figure 3). The primary responses appear to be a delay in the major processes involved in the overall efficiency of nutrient bioavailability. Definitive studies concerning the effect of dietary fiber preparations on the rate and completeness

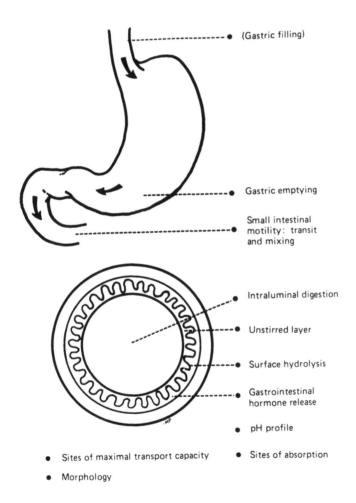

Figure 3 Mechanisms of sites at which dietary fiber may influence intestinal absorption of nutrients (56). (Courtesy of A. K. Leeds.)

of sugar absorption have been reported by Jenkins et al. (22). Healthy
subjects were given 25 g of the unmetabolizable pentose, xylose, in
test meals containing 12 g of various fiber preparations, including guar.
The urinary excretion of xylose from the guar meal was significantly
less than controls during an initial 2-hr collection period but was thereafter (up to 8 hr) more than in controls (Figure 4). These studies suggest that absorption of xylose was delayed, but also that there was little
if any impairment of overall absorption and excretion of the sugar.

The importance of viscosity of the fiber in delaying sugar absorption
has also been addressed in these studies (22). Using 50 g glucose
tolerance tests, in the presence of 12 g guar, there was a flattening of
the blood glucose response, accompanied by lower serum insulin levels.
However mild acid hydrolysis of the guar, which reduces viscosity, completely eliminated the flattening effect of the fiber on postprandial glycemia. Furthermore, there was an inverse relationship between the
measured viscosity of five different fiber preparations and their effects
on the peak rise in glucose under these test conditions. These studies

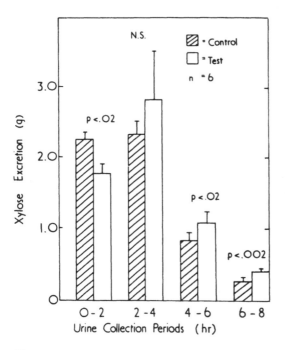

Figure 4 Urinary xylose excretion over an 8-hr period in subjects
given 25-g xylose in meals lacking or containing 12-g guar gum (22).
(Courtesy of D. J. A. Jenkins.)

suggest that viscous fibers can acutely influence the rate of glucose (and xylose) absorption from single test meals. This acute response also required that the fiber be present before the glucose tolerance test, emphasizing the importance of intimate mixing with the test meal (77).

Among the mechanisms to be considered in assessing the acute effects of food fiber preparations on nutrient bioavailability are: rates of gastric filling and emptying; effects on intestinal motility and transintestinal transit; altered pH in the stomach and intestine; interaction of the fiber with the intestinal surface; effects of entrapment of digestive enzymes and food nutrients; and effects on bulk-phase diffusion and accessibility of absorbable nutrients with the intestinal mucosa.

The role of dietary fiber, in general, and fiber constituents, in particular, on food intake (gastric filling) and on gastric emptying have been discussed earlier, and are undoubtedly important in modifying rates at which nutrients enter the small bowel. The effects of dietary fibers on pH profiles of the gastrointestinal tract have yet to be described. This is of considerable significance not only for assessing the mechanisms by which fiber influence nutrient availability but for potential therapeutic applications. In the stomach the regulation of gastrin release (78) has important implications in tissue growth responses to the hormone, as well as for secretory phenomena in the stomach, including secretion of hydrochloric acid and stomach pH. Fiber-dependent modifications of gastric pH, whether through alterations in gastrin responses or by physical "dampening" of the acid, may alter efficiency of peptic digestion of protein and subsequent proteolytic activity in the small intestine. Furthermore, there are important implications of potentially modified gastric pH on gastric and duodenal ulcers. It has been reported (79) that gastric pH is higher in patients with duodenal ulcers after a meal containing maize than after a meal containing corn flour.

The possibility that various dietary fibers could differentially modify gastrin and its responses is not without precedent. Secretion of other gastrointestinal hormones, such as gastric inhibitory peptide (GIP) (80-82) and intestinal glucagon (82,83), is significantly modified by dietary fiber. The effects on other important intestinal hormones have not been described. However delayed release of gastric contents, including acid, could well influence intestinal pancreozymin and cholesystokinin, which regulate pancreatic and biliary secretions, and such modifications have major implications in the digestibility of complex carbohydrates, proteins, and lipids. It is also important to assess whether these may be acute or adaptive responses of the intestine to fiber foods or supplements.

The rate of transit and the degree of mixing, or entrapment, are also determinants of the acute effects of dietary fibers on nutrient digestion and absorption. Rapid transintestinal transit might affect both the rate and extent of digestion by pancreatic and intestinal enzymes. Cellulose, for example, increases the rate of transit of the unabsorbable marker polyethylene glycol from stomach to cecum in experimental animals (Vahouny, unpublished data). In a similar model, Leeds (56) has suggested that transit in the mid-small intestine is slower with high-viscosity meals.

Another important aspect of nutrient availability is the possible interaction of enzymes and substrates with the fiber source, or entrapment in a hydrated fiber matrix. This interaction will limit the efficiency of digestion by pancreatic and/or intestinal enzymes and limit the diffusability of absorbable products to the absorptive surface. Evidence for the potential of this effect has been forwarded by Schneeman and her colleagues (84,85). During short-term in vitro incubations of human pancreatic juice with 5% levels of viscous fiber, the activities of various enzymes in the resulting supernatant were differentially affected (Figure 5). Incubations with pure cellulose, for example, resulted in a marked diminution in the activities of amylase, lipase, and the proteolytic enzymes. The viscous fibers, in contrast, were either without effect (e.g., oat bran) or actually caused an increase in some activities (pectin). In an extension of these studies in vivo (84,85), it was concluded that protein digestibility was reduced in rats given 20% levels of either cellulose or wheat bran (86). Studies by others (87,88) also

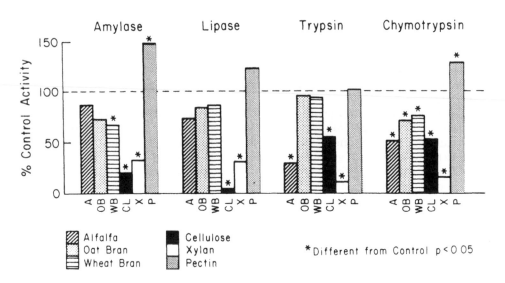

Figure 5 Activities of human pancreatic enzymes following a 5-min incubation with dietary fiber preparation (85). (Courtesy of B. O. Schneeman.)

suggest that dietary fiber may influence proteolytic activity. These
studies have not yet allowed definitive conclusions regarding effects
of dietary fibers on digestive enzyme activities in vivo but suggest that
certain fiber derivatives may play an acute role in protein bioavailability.
Similar conclusions have been forwarded regarding the digestion and
absorption of fat (75).

There is also evidence that viscous fibers, in particular, can influence accessibility of absorbable nutrients to the mucosal absorptive
surface. Studies on the uptake of sugars and amino acids by rat intestinal segments suggest that nutrient flux is inversely related to the
viscosity of the incubating solution (89,90). This response was evident
with various preparations of gums or with carboxymethylcellulose. These
effects, however, are completely dependent upon the immediate presence
of the fiber preparation since the inhibition of nutrient uptake was completely reversed either by washing or reincubating the tissue in the absence of fiber (90) or by using rinsed intestinal segments from fiber-
fed animals (91).

The various mechanisms by which dietary fibers may influence nutrient digestibility suggest an acute effect, which is dependent upon the
presence of the fiber in the intestine during these processes. However
these studies have not addressed two other potentially important aspects
of fiber action. The first is that fibers may transiently interact with
the intestinal surface, either through specific ionic interactions with
surface mucins or by nonspecific (e.g., hydrogen-bonding) processes.
The latter appears more likely based on the evidences just cited (90,91).
Such interactions however, could easily provide an explanation for a
variety of short-term and adaptive responses of the intestine with respect to altered transit, nutrient bioavailability, and various hormonal
responses. In addition, this type of interaction would effectively increase the thickness of the unstirred water layer barrier to nutrient
absorption, as has been suggested earlier (89,90).

E. Adaptive Responses to Dietary Fiber

It is well recognized (92-95) that intestinal structure and function can
be substantially altered by manipulation of the diet. There is also evidence that modifications of the enterohepatic circulation of bile acids,
and/or increased concentrations of the unconjugated bile acids, can result in morphological damage to the intestine (96-98) and cause abnormalities in water and salt transport (99-101). Thus chronic effects of
dietary fiber on gastric filling and emptying, on delayed or modified
nutrient bioavailability, and on bile acid sequesteration, might be expected to markedly alter the overall nutrition and characteristics of
the upper and lower small intestine and of the colon. Accordingly,
there is accumulating evidence that prolonged ingestion of dietary fibers and fiber derivatives can alter certain functions within the intestine,
as well as extraintestinally. In the intestine these include morphological

changes, modification of the activities of surface-associated digestive enzymes, and altered intestinal hormone secretion. Extraintestinal changes generally are not observed with acute administrations of fibers or fiber derivatives and suggest that dietary fibers may affect overall intestinal nutrition and consequently modify intestinal structure and function.

1. *Structural Modification*

There are numerous studies implying a relationship between dietary fiber intake and morphological changes in the small intestine. In the human the jejunal villi in the fetus are finger-like and regular, and this pattern is also typical of the adult in Western cultures (102-104). In contrast, the intestinal villi of men in developing countries (105) and in healthy vegetarians (104) are broad and leaf shaped with numerous ridges and convolutions. These patterns now have been reproduced in experimental animals. Tasman-Jones and co-workers (106, 107) have demonstrated that weanling rats fed a fiber-free diet maintain an immature finger-shaped villus pattern; rats on standard laboratory diets (containing fiber) or given diets containing pectin, developed mucosal ridges and a decrease in the total number of jejunal villi. We (M. M. Cassidy, F. G. Lightfoot, L. Fitzpatrick, and G. V. Vahouny) have obtained similar differences in weanling rats fed cellulose-containing diets or diets containing alfalfa or pectin. In the cellulose-fed animals the villi do not develop into mucosal ridges (Figure 6).

Studies by Brown et al. (108) suggested that dietary pectin caused an increase in total mucosal protein without affecting DNA content. Others (109) also have failed to detect a change in the crypt/villus cell ratio in animals fed pectin. Our recent studies (G. V. Vahouny, unpublished) suggest that a variety of either particulate fibers fed at 10% levels or viscous fibers at 5% levels, can increase the protein content of the proximal jejunum, without affecting crypt cell turnover (thymidine kinase). In a study comparing cellulose and pectin (107), morphological changes in dimensions of jejunal villi were not associated with changes in epithelial or goblet cell numbers. Diets containing wheat bran also had little effect on epithelial cell numbers but were reported to increase the numbers of goblet cells (85) and goblet cell secretory activity (110). These cells secrete mucins, which may be an important diffusion barrier to nutrient transport (111), and which are suggested to be involved in a variety of immunoprotective functions (112). These studies collectively suggest that specific dietary fiber components can express adaptive morphological changes in the small intestine, most likely related to the overall nutriture of that organ. Such morphological changes also expressly imply the possibility of modified cytokinetics (113) and functional capacity of the small intestine.

2. Modification of Surface-Associated Digestive Enzymes

Reports on the responses of intestinal enzyme activities to dietary fibers have not been entirely consistent. In the studies of Brown et al. (108), the activities of peptidase and alkaline phosphatase (surface markers of mature epithelial cells) were reduced in pectin-fed rats. The extent to which the tissues were freed of residual pectin was not demonstrated, however and as discussed earlier (Section III.D), this may be of importance is assessing changes in enzyme activities. Recently completed studies (G. V. Vahouny, unpublished) have compared the intestinal enzyme responses with a variety of particulate and viscous fiber preparations fed to rats for 4 weeks. Segments of the proximal, middle, and distal small intestine were carefully washed before analysis of villus and crypt cell marker enzymes. With all fiber preparations there was a significant and parallel increase in alkaline phosphatase and sucrase activities, but only in the proximal intestinal segment. Obviously, additional studies of this nature are required to assess fiber-related changes in the overall functional capacity of the intestinal tract.

3. Intestinal Hormone Responses

It now appears that at least one major gastrointestinal response to the intake of dietary fibers and specific fiber derivatives is a modification of the secretion of specific hormones intimately involved in glucose metabolism. As shown in Figure 7, intake of pectin is associated with a flattening of the glucose tolerance curve and lower levels of serum insulin (82). In addition, there are significantly lower levels of serum enteroglucagon and gastric inhibitory peptide (GIP). These gastrointestinal hormones are involved in regulation of insulin secretion (114), and their responses to diet modification may well explain the overall influence of dietary fiber on improved disposition of glucose in the body.

In a study by Morgan et al. (81), a test meal containing guar resulted in lower plasma GIP levels in both healthy volunteers and diabetics, but plasma glucagon (glucagon-like immunoreactivity, GL_1) levels were not

different. These effects were even more pronounced when guar was included in a carbohydrate meal. Other studies (82,83) however suggest that dietary fiber also modifies secretion of pancreas-specific intestinal glucogen levels in plasma.

These types of intestinal hormone responses, whether acute or adaptive, have major implications in the efficient metabolism and disposition of dietary carbohydrates and in the treatment of diabetes (see Section IV.D).

Figure 6 Scanning electron photomicrographs of the small intestine of weanling rats reared for 4 weeks on either (a) fiber-free or cellulose-containing diets or (b) standard laboratory chow.

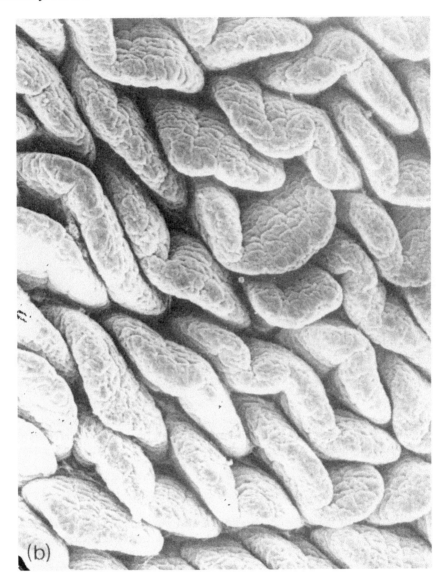

Figure 6   (continued)

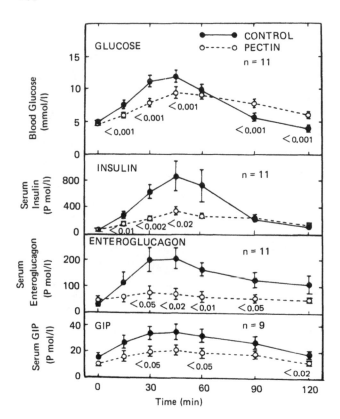

Figure 7 Responses of blood glucose and specific hormone levels to
50-g dose of glucose given with and without pectin (22). (Courtesy
of D. J. A. Jenkins.)

4. *Improved Sensitivity to Insulin*

The reductions in serum glucose and insulin responses, following fiber
supplements such as guar or pectin (82,115) or with high-carbohydrate
diets (116), suggest that dietary fiber may enhance tissue sensitivity
to insulin. Studies on insulin receptors of circulating monocytes (117,
118) has substantiated these reports. With either adult (117) or juve-
nile (118) diabetic patients given high-fiber diets, there is an increase
in the insulin binding and in the number of monocyte insulin receptors.
The mechanism(s) for these responses has yet to be explained, but a
role for bacterial metabolites of dietary fiber (i.e., short-chain fatty
acids) has been suggested (116).

5. *Metabolic Responses to Short-Chain Fatty Acids (SCFA)*

Among the major bacterial metabolites of dietary fiber in the colon are the SCFA, and these are primarily composed of acetate, propionate, and butyrate. It has been estimated (16) that the bacterial breakdown of 20 g of cell wall polysaccharides and other carbohydrates in the colon will result in approximately 200 mmol SCFA, of which 31% is acetate, 13% is propionate, and 8% is butyrate. The SCFA provide an important source of energy to the colonic mucosa (119) and will stimulate absorption of water and sodium (120,121). Although it appears that there is a substantial fecal excretion of SCFA (122), there is also substantial absorption from the colon (123) and subsequent hepatic metabolism of these metabolites.

Recent studies by Anderson and Bridges (124) on isolated hepatocytes suggest that low concentrations of propionate, in contrast to long-chain fatty acids, will stimulate glycolysis and inhibit gluconeogenesis. The effects of acetate, however, were similar to those of the long-chain fatty acids. Although these types of studies imply that propionate, in particular, can enhance glucose metabolism in the liver, other responses, such as effects on hepatic cholesterol and bile acid metabolism have not yet been investigated and are necessary to establish the physiological importance of these bacterial metabolites in the overall action of dietary fibers.

6. *Modifications in the Bile Acid Spectrum*

Modification of biliary bile acid content and composition by specific sources of dietary fiber has been investigated in both humans and experimental animals (125). In the rat, diets containing 20% bran decreased the overall concentration of colonic bile acids, but increased the levels of bacterial metabolites of the primary bile acids, cholic and chenodeoxycholic acids (126). In man wheat bran reduced the concentration of deoxycholate in gallbladder bile, suggesting an overall reduction in either bacterial metabolism of cholic acid or decreased resorption of the secondary bile acid (127). In general however, major alterations in bile acid pools and compositions have been observed only in patients with gallstones or in experimental models of gallstone production and dissolution. These are summarized in Section IV.F.

IV. THERAPEUTIC APPLICATIONS
A. Constipation

The therapeutic usefulness of dietary fibers in simple constipation is largely attributed to their effects on stool bulk and consistency and on lower-bowel motility. The apparent stool-bulking capacity of fibers and their derivatives from different sources varies considerably. There is considerable literature establishing the efficacy of wheat bran on these parameters of lower-bowel function (16), and this effect is greater than that provided, for example, by fruits and vegetables.

Brodtribb and Groves (24) reported that coarse wheat bran was more effective in increasing stool weights than fine bran, and Wyman et al. (128) demonstrated that cooking reduced the bulking capacity of bran. This is probably due to structural changes leading to increased bacterial degradation. These and other studies (129) suggest that the effect of bran may be related in part to water-holding capacity but also involves other mechanisms, such as bacterial digestibility, effects on bacterial growth, and production of osmotically active bacterial metabolites.

The early studies of Williams and Olmsted (130) initially addressed the question of the fraction of fiber that was active in stool-bulking responses. Of the three classes of fiber components that make up the indigestible residue, hemicelluloses were more effective than cellulose or lignins. These studies have been extended by Cummings et al. (60) with concentrated dietary fibers from a variety of sources. The overall results suggested that changes in fecal weight were directly correlated with the intake of pentose-containing polysaccharides in the fiber source and were mediated by their tendency to influence bacterial growth and excretion.

Bacteria represent a major component of stools (131) and their concentration appears to remain constant during changes in stool volume (132). Fiber components that are extensively metabolized by colonic microorganisms would be expected to cause increased bacterial growth and output. However this cause-and-effect relationship is not consistent with fibers such as pectin, which is extensively degraded but has little effect on stool bulk (68,133).

Despite this apparent paradox, fiber-containing foods and wheat bran, in particular, will increase the frequency and urgency of defecation (134,135) and increase overall stool volumes. Comparisons of average stool weights in populations ingesting high and low fiber-containing diets (136) suggest that a fecal output of 140-150 g/day can be used to establish the minimum daily intake of dietary fiber (137).

B. Diverticular Disease

A review of the available epidemiological data (138) suggests that diverticular disease is low or absent in populations consuming high levels of dietary fiber-containing foods and may be a "deficiency disease" of Western cultures (139). The disease is characterized by mucosal herniations of the sigmoid colon, often accompanied by muscle hypertrophy and increased intracolonic pressures (140,141). The known colonic responses to dietary fiber intake, and fiber-depletion studies in animals (142-144) are consistent with the concept (139) that low-fiber diets can induce characteristics of diverticular disease.

As with treatment of simple constipation, high-fiber diets, in particular wheat bran, have found general acceptance as the treatment of choice to relieve symptoms of uncomplicated diverticular disease (138). Bran has been reported to increase the rate of transit and stool bulk and to decrease the fecal streaming in the colon (145) that occurs in patients with diverticular disease (146). There has generally been a consistent relationship between the efficacy of coarse bran in relieving symptoms of the disease and the ability of bran to reduce intracolonic pressures (138). Finely ground wheat bran, however, was reported to have little effect on intracolonic pressures, while still relieving symptomatology (147).

C. Obesity and Weight Management

The hypothesis concerning the health attributes of dietary fiber has been extended to include its role in obesity (4). This relationship, and the potential mechanisms by which fiber can influence satiation and obesity, has been discussed extensively by others (10,13,148-150), and has been alluded to earlier.

The low-calorie bulking agents have found some success in weight reduction and management when used in combination with controlled caloric intake (151). The use of guar gum and methylcellulose at 10 g/day resulted in a reduction in voluntary food intake and increased weight loss, and the effect appeared to be more pronounced in obese subjects (152). Similar results were noted for female subjects given 15 g/day of guar gum (153), and for male subjects given high-fiber bread in their diets and compared with subjects on enriched white bread (154). Heaton and colleagues (155) have recently compared energy intakes on diets containing refined carbohydrates and diets in which refined carbohydrates were limited, but unrefined carbohydrate intakes were unrestricted. Subjects on the refined carbohydrate diet gained 1.6 kg over 14 weeks, while those on the unrefined carbohydrate diet lost 1.6 kg in the same period.

These types of results suggest that dietary fiber may have utility, when combined with other dietary modifications (e.g., decreased overall caloric intake), in reduction of body weights in obese subjects and in weight maintenance. The effect is unlikely to be a function of the dilution of caloric density of food, unless high levels of fiber are employed (156,157). Other mechanisms, such as gastric filling and emptying, satiation, and hormonal responses, may be of greater significance in predicting the efficacy of dietary fiber preparations.

Studies in animals have been less convincing in this regard (158-160). In our studies with rats (G. V. Vahouny, unpublished), addition of 10% levels of particulate fibers or 5% levels of viscous fibers to defined diets resulted in a compensatory increase in food intake and insignificant differences in weight gains with the various fiber diets.

D. Diabetes

The concept that the lack of dietary fiber may play a role in the etiology of diabetes mellitus is largely derived from a variety of epidemiological, clinical, and experimental evidence (161). This relationship has received additional attention with reports (162,163) that intake of dietary fiber has therapeutic effects in diabetic patients.

Two general approaches have been employed to improve glucose metabolism in both insulin-independent and insulin-dependent diabetics. One approach is to modify the diet to include high natural carbohydrate with high-fiber contents, and the second relies largely on the use of dietary fiber isolates or purified fiber components as supplements to the diet.

1. *Fiber Supplementation*

The favorable effects of fiber supplements on postprandial glucose concentrations has been clearly documented by Jenkins and co-workers (164). The addition of the viscous fiber preparations, guar or pectin, to a large breakfast significantly reduced the glycemic responses of lean diabetic patients (162). In an attempt to define the type of fiber supplement most effective in reducing this glycemia, the equivalent of 12 g of various fiber derivatives was added to 50 g glucose tolerance meals of healthy subjects (22). The greatest reductions in postprandial glycemia occurred with guar and tragacanth, both viscous fibers from the cluster bean. Other viscous agents (pectin and methylcellulose) and particulate bran were also effective. These studies also reported that serum insulin levels were depressed by the fiber supplements, suggesting that the glucose response was not due to an increase in insulin but more likely to increased insulin sensitivity (22). Other studies were designed to determine the importance of viscosity of the fiber preparation (22), the timing of the supplement and food intake (77,165), and if this effect was due to delayed glucose absorption rather than malabsorption (22,115). To improve palatability and acceptability of guar, it has been incorporated into crispbread preparations (166). Diabetic patients consuming guar in this form also exhibited reduced glycosurea and required less insulin. Other preparations of guar-containing foods have also been reported (167).

The efficacy of short-term fiber supplements on glucose or meal tolerance tests has generally been confirmed by others (see review 116). For example, noninsulin diabetics receiving a smaller (9 to 10 g) dose of guar and/or pectin in an oral glucose load also displayed a flattened tolerance curve, but in these patients there was no significant change in postprandial insulin responses (168).

The results of studies with insoluble or particulate fiber preparations, such as bran or cellulose, are not as clear. An acute response of control subjects to addition of bran to a glucose load was demonstrated by

Jenkins et al. (22). Similarly, cellulose added to an oral glucose load improved glucose tolerance in insulin-dependent diabetics, but there was no effect on the insulin response (169).

The long-term beneficial effects of fiber supplementation on glucose metabolism have been adequately demonstrated (83,170-173). For example, Miranda and Horwitz (83) have clearly shown the overall reduction in mean plasma glucose levels and insulin requirements in insulin-dependent diabetics given fiber-supplemented diets. Munoz and co-workers (173) reported that supplementation of diets with insoluble fiber preparation for up to 30 days in normal subjects resulted in improved glucose tolerance, despite the fact that the fiber was not included in the glucose load. These studies and those with high-carbohydrate-high fiber natural diets, discussed later, suggest that in addition to mechanisms related to acute viscous fiber intake, there are adaptive responses in the gastrointestinal tract to long-term fiber intake (see Section III.D and III.E).

2. *Natural High Fiber-Containing Diets*

Before the elegant studies of Anderson and co-workers (116) on the therapeutic value of the high-carbohydrate, high-fiber (HCF) diets for treatment of diabetes, there were several reports suggesting the utility of this approach (174-179). The therapeutic HCF diets employed by Anderson (180), provide 70% of the energy as carbohydrate, compared with 43% in control diets, and of this, three-fourths is complex carbohydrate providing 30 to 40 g dietary fiber per 1000 kcal. This is largely derived from vegetables, grains, and fruits.

Early studies with these diets (181) clearly demonstrated the beneficial responses of lean diabetic patients receiving moderate doses of insulin (25-40 U/day). A shift from the control to the HCF diet resulted in a reduced insulin requirement to half that required on the control diet, while maintaining plasma glucose levels constant. These studies have been extended, and as shown in Table 6, the HCF diet has been found effective in reducing insulin requirements of lean patients with either insulin-dependent or insulin-independent diabetes (116). For long-term outpatient therapy, the recommended high-fiber diets provide 55 to 60% of energy from carbohydrate and approximately 25 g dietary fiber per 1000 kcal (116). These diets are reported to have no significant negative implications on the mineral and vitamin status of these patients (182).

Since the majority of insulin-independent diabetics are overweight, studies have been conducted on the efficacy of the HCF diet coupled with moderate or severe restriction in calories (116,183). With moderate energy restriction, patients on high-fiber diets (40 g/1000 kcal) lost 2 lb/wk for 2 weeks and reduced their average insulin requirement to almost half. Severe caloric restriction (3.3 kcal/lb) with the high-fiber diet, resulted in a weekly loss of 6 lb and a discontinuation of insulin

Table 6 The Effect of High-Carbohydrate, High-Fiber (HCF) Diets
on Insulin Requirements of Lean Diabetic Patients

Patient group (number)	Insulin dose, U/day	Insulin dose, U/day	
		Control diet	HCF diet
Insulin-independent			
(12)	15-20	17 ± 2	0
(10)	25-38	31 ± 2	0
Insulin-dependent			
(10)	22-32	28 ± 2	15 ± 3
(6)	40-55	44 ± 2	31 ± 4

Source: Ref. 116.

therapy in five of eight patients studied. Thus the most pronounced
effects on plasma glucose levels and insulin requirements were observed
in obese patients demonstrating the largest weight reduction.

These studies have been largely confirmed by others (184-188) and
provide strong evidence for the long-term beneficial therapy of diabetes
with high dietary fiber intake.

E. Serum Lipid Response

Given the convincing epidemiological evidence suggesting an inverse
relationship between dietary fiber intake and coronary heart disease
(189,190), there have been extensive clinical and experimental studies
on the effects of various fibers and fiber components on circulating
lipid and lipoprotein levels. These are reviewed elsewhere (11,19,20,
42,43,62,63) and are summarized in the following section.

Clinical studies in man have involved three major approaches: effects
of high fiber-containing foods; effects of fiber concentrates such as
bran; and effects of purified fiber components.

1. *Fiber-Rich Foods*

The early studies of Keys and co-workers (191,192) demonstrated a
significant reduction in plasma cholesterol levels after various modifications of the diet including an increase in dietary fiber intake in the
form of vegetables, fruits, and legumes. Substitution of bread or sucrose with vegetables (40 g fiber) resulted in a similar decrease in
plasma cholesterol, while substitution with fruits (20 g fiber) was
without effect.

Other studies, however, have shown that diets rich in either fruits
or vegetables (192-194), or in legumes (195,196), can produce a small

but substantial decrease in plasma cholesterol levels, while cereal sources of dietary fiber are generally without effect (43). Problems relating to appropriate control diets and to diet adherence are discussed in detail elsewhere (62) and undoubtedly have contributed to the variability of results.

Recently, Kay and Truswell (43) have calculated the maximum change in plasma cholesterol that might be expected with a variety of dietary manipulations (191,192). These investigators have suggested that simply an increase in dietary fiber intake from 20 to 40 g/day could be expected to produce only a subtle lowering of plasma cholesterol (8 mg/dl, or about 4% of the original levels). Maximum lowering could only be expected in combination with other more dramatic modifications of diet (e.g., fat and cholesterol and vegetable protein intake). This result is perhaps best exemplified by the plasma lipid responses to the high-carbohydrate, high-fiber (HCF) diets employed by Anderson and colleagues (116,180). In diabetic patients exhibiting hyperlipidemia, the HCF diet produced significant reductions in both serum cholesterol and triglyceride levels. Other studies (184) suggest that similar diets (53% carbohydrate-high fiber) decrease both low-density lipoprotein (LDL) cholesterol and very low-density (VLDL) triglyceride levels. These types of reductions in plasma lipids also can be observed in control subjects (197).

2. Concentrated Fiber Sources

Among the fiber concentrates tested for effects on serum or plasma lipid levels, results with wheat bran have been consistently negative, even when given at levels of up to 72 g/day (11,19,20,43). Addition of other particulate fibers to human diets including those from corn bran, bagasse, soybean hulls, and corn hulls also have been without significant hypolipidemic effects, even in type II hypercholesteremic subjects (198).

Oat bran which, in contrast to wheat bran, contains mucilaginous glycans, has been reported to have hypocholesteremic effects in man (199-201), and this appears to be mediated through effects on plasma LDL levels (200,210). These effects of oat bran are most likely a response to its content of the mucilaginous gums present in concentrated fiber sources (202).

Guar gum, administered in various states of hydration or as crispbread, has been effective in reducing serum total and LDL cholesterol levels (203) and has been reported to be effective in patients with type II hyperlipoproteinemias (203,204). Metamucil (psyllium seed hydrocolloid) in moderate doses has also been reported to be effective in reducing plasma cholesterol levels (205).

Thus in general, the particulate fiber preparations have been ineffective in modifying serum lipids in man, while those preparations that have viscous components appear to be hypolipidemic, particularly in patients with elevated circulating lipoprotein levels.

3. *Isolated Fiber Components*

Among the isolated or purified fiber components, cellulose has generally
been found to be inert with respect to plasma lipid responses (11,19,20,
43). The effects with lignin have not been consistent (206). Viscous
pectins account for most consistent responses in lowering plasma lipids,
and in particular, cholesterol (11,19,43). This hypocholesteremic response has been observed within 1 to 3 weeks and appears to be approximately 15% using 15-45 g/day as a dietary supplement (11).

Studies in animals on the effects of high-fiber diets, supplementation
with concentrated fiber sources, and addition of purified fiber components largely complement the observations in humans. These are extensively documented elsewhere (11,19,43) and suggest that hypolipidemic responses are routinely observed with diets containing higher levels
of viscous fiber materials. The rat, unlike humans, has high levels of
circulating high-density lipoprotein (HDL) cholesterol. In this model,
hypocholesteremic responses to dietary fiber are not dramatic unless
the diet is initially modified to be hypercholesteremic. Under these
conditions, wheat bran is ineffective in preventing cholesterol-induced
hypercholesteremia (207), while pectin, guar gum, and oat bran not
only prevent diet-induced cholesteremia but also improve the degree to
which HDL contributes to circulating cholesterol levels (207,208).

Although these results suggest that the hypocholesteremic responses
may be, in large part, due to effects on cholesterol absorption, other
mechanisms (e.g., altered hormonal responses and improved glucose
metabolism) have yet to be elucidated.

F. Gallstones

The epidemiology, etiology, and experimental evidence linking diets
and gallstones have been reviewed extensively by Heaton (209,210).
Cholesterol-rich gallstones occur when gallbladder bile is supersaturated
with respect to cholesterol, and this disease is closely associated with
obesity and hypertriglyceridemia (type IV hyperlipoproteinemia). The
precipitation of cholesterol from bile is dependent upon a variety of
factors, among which are nucleating agents (89,211), the biliary bile
acid composition, and the lithogenic index (the molar relationship of
cholesterol, phospholipid, and bile salts in bile). Little is known concerning nucleating agents, but available evidence suggests that dietary
fiber can modify the biliary bile salt composition (212-215) and the
lithogenic index in gallstone patients (212-214).

Pomare et al. (212) reported that the intake of 30 g wheat bran daily
by gallstone patients reduced the lithogenic index and resulted in an
increased proportion of chenodeoxycholate to deoxycholate levels of
bile. The effect of bran has largely been reproduced by others (213,
214) but only in gallstone patients. Control subjects with lithogenic
indices at or near unity did not generally show a similar response to
dietary bran (214) or pectin (70). Data from experimental gallstone

models are limited, but suggest that diets rich in fiber or fiber components (e.g., lignin) may reduce lithogenicity and gallstone formation (74,216).

The mechanisms for the responses of gallstone subjects to dietary bran remains obscure since this fiber source has little or no affinity to bind bile acids. The overall reduction in the deoxycholate content of bile in these patients appears to be related to modified bile acid metabolism by colonic microorganisms in response to the fiber supplement. Further studies are required to clarify both the efficacy of fiber therapy for gallstone formation and dissolution and the mechanisms of the apparent biliary responses to dietary fibers.

V. UNDESIRABLE EFFECTS AND PATHOLOGICAL RESPONSES

A. Gastrointestinal Obstructions

Several types of gastrointestinal obstructions relating to dietary fiber intake have been described and require consideration, particularly with the increasing availability of commercial fiber-containing products.

Concentrated sources of hydrophilic fiber derivatives, such as the plant gums and pectins and the hydrophillic colloid laxatives, have become of concern because they will hydrate extensively, can swell in the gastrointestinal tract, and can cause bolus obstructions (217). Although not yet completely documented, there have been communications regarding the relationship of the increasing use of guar gum in Sweden and several cases of esophageal rupture (B. Jacobbsson, University of Grothenburg, personal communication). This problem is presumed to be caused by adherence of the gum to the esophagus, hydration at those sites, and the resultant distention and rupture of the esophagus.

Obstructions in the stomach (phytobezoirs) also have been described (218) as a result of the high intake of orange pith (pectins) with inadequate mastication. With the advent of newer commercial dry preparations of pectins and gums, it seems likely that the occurrence of similar concretions in the stomach can occur, particularly in subjects with gastric atony.

Volvulus involves a twisting of the sigmoid colon resulting in acute intestinal obstruction. This is associated with high intake of dietary fiber in specific population groups, particularly in children (219-222), but it is not generally associated with diets of westernized cultures.

B. Mineral and Vitamin Imbalances

Within the complex physical and chemical structure of plant cell walls there is a variety of potential binding sites for sequestering certain minerals, particularly the divalent cations such as calcium, magnesium, zinc, and ferrous iron. These include phenolic groups in lignin and the uronic acid residues in polysaccharides, such as the pectins and hemicelluloses. In addition, other materials associated with the cell walls, such as silicates, oxalates, and phytates, also chelate divalent cations.

Phytates (inositol hexaphosphate), in particular, have received considerable attention in this regard (223). This material is associated with plant seeds and storage tissues and usually occurs as salts of calcium, magnesium, and potassium. Although the phytates can provide a readily available source of phosphate to the body, their presence in foods, particularly when the phytate/metal cation ratio is high, may play a role in metal chelation and mineral balance (223). They are, however, lost during milling of flour and are readily susceptible to phytases and food preparation techniques.

The possible effect of unrefined foods, such as cereals, on calcium and iron balances in humans was emphasized in the early studies of McCance and co-workers (224-226), who attributed this effect to the phytate content of the foods. Reinhold and his colleagues (227-229) also demonstrated reduced bioavailability of zinc and magnesium during consumption of high-fiber bread, and originally attributed these mineral imbalances, at least in part, to phytates (227). However studies on the relationship between the breakdown of cereal phytate and bioavailability of zinc (228) and on the binding of cations to fiber in vitro (230), resulted in the conclusion that dietary fiber per se could be responsible for the observed mineral imbalances (231). This conclusion is supported by findings that purified cellulose reduced calcium and magnesium availability (232), and that cellulose and hemicellulose can increase fecal excretion of divalent cations (233).

These and other studies, reviewed elsewhere (11,223,234,235) strongly suggest that dietary fiber in its natural or derived forms have the capacity to sequester divalent cations and can play a role in overall micronutrient balances. It is also apparent that this effect is dependent upon the type and amount of fiber in the diet, by the levels of other macronutrients, such as proteins, and by the length of study.

Careful studies by Sanstead and colleagues (236) have demonstrated that modest increases in dietary fiber in the form of wheat bran, corn bran, or soybean hulls do not adversely affect zinc or copper balance when assessed after a 3-week period of adaptation. This protocol is in contrast to that of Reinhold et al. (227), in which zinc balances were determined at the time of diet change. Further studies by Sanstead et al. (237) have reemphasized the adaptive ability of the body to a change in diet, and suggest that reducing the protein levels in the face of increasing dietary fiber is associated with a more positive balance for zinc, calcium, and copper. This adaptive response, in part, may be due to colonic absorption of minerals as a mechanism for preserving mineral balance (234).

In general, it appears that modest increases in intake of dietary fiber may not have significant influence on mineral balance in human consuming a nutritionally adequate diet (238). However from studies in developing countries where the diet is largely of plant origin, it is apparent that mineral imbalances can occur, particularly where diets are marginal in micronutrients. This is of importance, for example, in children, whose zinc intake may be deficient (234) and in the elderly, who often are constipated and whose diets are likely to be marginal in vitamins and minerals. With the increasing emphasis on nutritional modifications and the greater availability of fiber-containing food additives and laxatives, it may become advisable to assure adequate intake of calcium, zinc, and iron in those diets that contain increasing levels of dietary fibers.

Information on dietary fiber and vitamin bioavailability is sketchy and has been summarized recently by Kelsay (235) and by Kasper (239). In general, the available studies suggest a lower bioavailability of riboflavin and nicotinic acid from diets containing cereal brans or wholewheat bread (240,241) and a lower bioavailability of vitamin B_6 from cereal than from nonfat dry milk (242). Rats fed cellulose or pectin had significantly greater excretions of vitamin B_{12} than rats on fiber-free diets (243). In contrast, folate availability appears to be unaffected by dietary fiber (244).

With the exception of a limited literature on vitamin A and carotene (235,239) there is little information on the effect of fiber-containing diets on fat-soluble vitamins. This is of importance because the current data suggest that viscous fibers can interfere with the normal digestability and absorbability of lipids (75), and these mechanisms may be of importance in absorption of fat-soluble vitamins. Limited studies in rats (235) suggest that 10% levels of various fibers caused a decrease in liver and serum levels of vitamin A. A study in humans (235) suggested that a greater amount of vitamin A was excreted with a fiber diet containing fruits and vegetables than on a low-fiber diet which was supplemented with beta-carotene to provide equivalent dietary levels.

Again, the available evidence suggests that dietary fiber may have only a subtle influence of vitamin bioavailability and may be of concern only in populations on a marginal intake of certain vitamins.

C. Fecal Energy Loss

There is now a considerable literature to suggest that high intake of dietary fiber is associated with increased fecal energy loss (245-248). This response was initially demonstrated in young women on low, medium, and high intake of fiber, who respectively excreted 83, 127, and 210 kcal/day (245). Such an effect can be expressed by fruit, vegetable, or cereal fibers (248) and will vary from 58-321 kcal/day in fecal energy loss depending upon the amount and type of dietary fiber ingested (245-248).

The increased energy loss is largely in the form of protein and fat, but it is unclear whether this represents largely bacterial sources of these materials or may be, in part, a result of altered nutrient metabolism and absorption. For example, pectins and guar in large doses cause a significant increase in fecal fat (68,249) and are believed to interfere with normal physiological processes of fat digestion, micellar solubilization, and absorption of fat digestion products. It is unclear whether these products are subsequently absorbed in distal portions of the small intestine or, in addition to the fiber, provide an energy source to colonic bacteria. In the latter case, energy is lost to the body, but in either case, the fecal fat composition would reflect that of the bacteria rather than that of the diet. Thus the increased fecal fat loss in response to dietary fiber intake is largely a result of increased bacterial proliferation and excretion (250).

The effect of dietary fiber on protein metabolism also has not been clarified. Studies in which fecal nitrogen has been fractionated suggest that the majority of fecal nitrogen loss is from nondietary sources. These include bacterial products, mucosal cell debris, and unabsorbed intestinal secretions. These studies however do not consider the extent to which dietary protein metabolism is impaired and therefore provides, together with fiber, energy sources to colonic bacteria. There is evidence to suggest that increased fiber intake increases fecal nitrogen but does not alter urinary nitrogen levels (248). The implication is that there is no major nitrogen imbalance under these conditions. Nevertheless, it has been reported that fiber may contain inhibitors of proteolytic enzymes that can alter protein utilization (251) and that certain fiber derivatives can alter the spectrum of pancreatic digestive enzyme activities (85).

Overall, fecal energy loss can occur in varying extents during increases in dietary fiber consumption, but the extent of losses are relatively small and may be inconsequential, except in population groups on marginal energy balance.

D. Persorption

The process of persorption involves the passage of particulate material through the intestinal wall directly into blood or lymph. This may be of greater significance in areas of the intestine where desquamation of intestinal epithelium occurs or where the epithelial cell layer of the mucosal surface is limited (252-254). Volkheimer and his colleagues (252-254) initially reported on the persorption of raw starch granules administered as a suspension to experimental animals and humans. Within minutes, particulate matter was detected in venous blood. Similar results were reported in experimental animals administered microcrystalline cellulose (255) and humans given oatmeal containing methylene blue or creamed corn stained with fuchsin (256). In the latter study, dyed plant fibers up to 500 μm in length were observed in venous blood from 4 hr to 6 days after ingestion of the test food. The evidence suggests that persorption of large-molecular-weight, complex carbohydrates can occur (257), and there has been concern expressed over the long-term and possibly toxic effects of these small amounts of persorbed materials (257).

It is obvious that this area requires additional consideration to establish the extent and toxicological importance of this process. The relationship between the chemical nature of the complex polysaccharide and the extent of persorption, the possible role of plant lectins on modifying permeability of the intestinal epithelium, the relationship between observed morphological responses to certain fiber components (160) and the persorption phenomenon, and the distribution and tissue effects of the presorbed materials, are all areas that require further study.

E. Other Responses Attributed to Fiber Supplements

The rapid introduction of fiber supplements, particularly insoluble fibers such as bran, can cause abdominal distension and considerable discomfort. This effect may be alleviated with continued intake, but some patients may not tolerate these supplements.

There have been recent reports (258-260) that diet supplementation with alfalfa seed and sprouts can produce symptoms of systemic lupus erythematosus (SLE) in monkeys and humans. Systemic lupus erythematosus is an autoimmune disorder that affects connective tissue and expresses symptoms such as anemia, arthritis, fever, hair loss, and disorders of the renal, cardiovascular, and central nervous systems. In monkeys (*Macaca fascicularis*) fed defined diets containing 45% ground alfalfa seeds, symptoms of SLE were evident within 5 months. Similar findings were reported using alfalfa sprouts, and it appeared that the syndrome was reversible by removal of the dietary alfalfa supplement. In fact, it has been reported (261) that consumption of 80 to 160 g of alfalfa seeds daily for 5 months by one of the investigators resulted in hemolytic anemia and other subclinical symptomatology compatible with SLE. Although the levels of the supplements employed in these studies are far in excess of those which humans might consume, the studies do suggest the presence of lectins or other toxins in alfalfa products, which if consumed over long periods, might cause adverse symptoms in humans.

In this connection, it should be emphasized that plants as well as other organisms produce glycoproteins, termed lectins, which play a role in cell surface recognition. These lectins are inserted into cell membranes and selectively bind to specific sugar residues of membrane glycoproteins and glycolipids, resulting in agglutination (262,263). Although lectins have been employed to investigate cell surface properties of epithelial cells (264), the extent to which the lectins of fiber preparations influence intestinal pathophysiology has not been studied in any detail. It has been speculated (265) that the lectins may play a role in some of the intestinal responses currently attributed to dietary fiber per se. These include effects on mucin production, mitotic rates of epithelial cells, local stimulation of immune phenomena, as well as effects on bacterial growth and fecal bulk. Obviously this important aspect of fiber pharmacology and pathophysiology has been largely neglected.

VI. FIBER, BILE ACIDS, AND COLON CANCER

A variety of demographic studies on the prevalence of bowel cancer (266-269), together with studies on metabolic epidemiology (270-171), have provided convincing evidence that dietary factors are important in the etiology of colon cancer (for reviews, see 272-274). There is however less agreement on the component or components of the diet that are responsible, since fat (275,276), animal protein (276-277), meat (268,278), refined carbohydrate (279), and lack of dietary fiber (266,280), all have been implicated. As summarized in Table 7, population studies suggest a strong correlation between the prevalence of colon cancer and intake of fat and animal protein; a weaker correlation between this disease and refined sugar; and no correlation or a negative correlation with dietary fiber. This epidemiological "jigsaw puzzle" (281) largely serves to emphasize the complexity of these types of studies in unraveling the well-recognized interactions of nutritional

Table 7 Correlations Between Incidence of Colon Cancer and Dietary Components

Dietary component	Correlation coefficient
Fat	
Total	0.81
Animal	0.84
Protein	
Total	0.70
Animal	0.87
Refined sugar	0.32
Fiber	
Total	0.02
Fruits	0.22
Cereals	0.32
Nuts	0.07
Starchy foods	0.07

Source: Ref. 268.

components, but they do, in fact, implicate nutritional factors in the etiology of colon cancer.

Important advances in the search for causative agents in colon cancer were contained in the reports by Hill and associates (282, 283), who suggested that the epidemiological correlation between diet and colon cancer might be explained by an involvement of bile steroids. These investigators reported a strong correlation between fecal concentrations of deoxycholic acid (and microorganisms possessing 7-alpha-dehydroxylase activity) and frequency of colon cancer. They also demonstrated a strong correlation between fat intake and the fecal concentrations of bile acids and neutral steroids (284). Subsequent studies have confirmed these findings (270,285–287). Although this relationship may be more complicated than originally proposed (287), the purported role of bile acids as promoters of colon cancer has also received considerable experimental support.

A. Bile Acids and Experimental Carcinogenesis

1. *The Animal Models for Colon Cancer*

The chemical induction of intestinal neoplasia in experimental animals (usually rodents) can be accomplished by administration of any one of several carcinogenic agents (288). These agents appear to provide a high degree of specificity and tumor yield, and the overall pathology appears largely compatible with the human disease (288). The most commonly used agents are either indirect- or direct-acting carcinogens, and include 1,2-dimethylhydrazine (DMH), azoxymethane (AOM), and methylazoxymethanol (MAM), which are closely related chemically, and N-methyl-N'-nitro-N-nitroguanidine (MNNG) or N-methylnitrosurea (MNU). Dimethylhydrazine requires metabolic activation to azomethane and azoxymethane, which are direct acting carcinogens, and these chemically related analogues have been effectively administered either orally or subcutaneously. The nitroguanidine and nitrosourea derivatives are effective by rectal instillation, suggesting direct action on the colonic mucosa.

Following administration of DMH, for example, tumors will occur in both the small and large intestine of the rodent. In the small intestine, these are localized proximal to the bile duct and are rarely observed in the ileum (288); in the large bowel, tumors tend to be localized distally, paralleling the distribution in the human. Although extrapolation of chemically induced tumor production in the rodent to the human disease is open to criticism, studies with these models have allowed important findings on aspects of nutrition and colon cancer that are not possible to explore in man.

2. *Bile Acids as Promoters of Experimental Colon Cancer*

The involvement of bile acids as promoters of experimentally induced colonic tumors has been demonstrated by several approaches. These

Table 8 Relationship Between Fecal Bile Acid Concentration and
Experimental Colon Carcinogenesis

Experimental manipulation	Effect on fecal bile acid concentration	Effect on experimental colon cancer
Bile diversion to cecum	Increase	Increase
Diversion of fecal stream	Decrease	Decrease
Rectal instillation of bile acids	Increase	Increase
Administration of cholestyramine	Increase	Increase
Diet manipulations		
Increased fat	Increase	Increase
Added meat	Increase	Increase
Added bran	Decrease	Decrease
Added cellulose	Decrease	Decrease
Added pectin	Increase	Increase (?)
Added Metamucil	Increase	Increase
Added guar gum	Increase	Increase (?)

Source: Ref. 287.

include: effects of direct rectal instillation of bile acids; surgical
displacement of the bile duct; modification of the fecal stream; effects
of bile acid sequestering resins; and dietary-induced increases in
fecal excretion of bile acids (see Table 8).

Narisawa et al. (289) reported that multiple rectal instillations of
lithocholic or taurodeoxycholic acids alone produced no tumors in rats;
however both bile acids doubled the frequency of MNNG-induced neoplasms over that induced by the carcinogen alone, suggesting that
these agents were acting as promoters rather than inducers of neoplasia. Similar results were obtained in germ-free rats using deoxycholic acid (290), implying that further bacterial modification of the
bile acid was not responsible for the promoting action of the steroid.

Surgical approaches also have been employed to demonstrate the
promoting action of bile acids in experimental carcinogenesis. Chomchai
et al. (291) surgically displaced the bile duct to the mid-small intestine,
resulting in an increased fecal output of bile acids. The subcutaneous
administration of AOM to these animals resulted in markedly increased
tumor formation in the large bowel with the largest effect being in the
distal colon. The importance of the fecal stream was demonstrated by
Campbell et al. (292), who reported that colostomy in rats decreased
by half the number of colonic tumors observed in control animals.

Cholestyramine is a nonabsorbable anion-exchange resin, which se-
questers bile acids and increases fecal output of these acidic steroids

(293). The inclusion of this agent as 2% of the diet to rats, which were also administered either DMH, AOM, or MAM, produced a striking increase in malignant intestinal tumors, largely in the distal large intestine (294). Similar results have been reported by others (292,295). Furthermore, the cholestyramine-induced increase in tumors has been reported to be independent of colonic microorganisms (295) and is not dependent upon a release of bile acids from the resin in the lower bowel (295).

Finally and perhaps more appropriate to the human correlative studies, have been the effects of diet modification on bile acid excretion and experimental carcinogenesis. In early studies of Reddy et al. (296), DMH-induced tumorgenesis was investigated in rats receiving semipurified diets containing 0.5, 4, and 20% corn oil. Rats given the high-fat diet excreted higher levels of bile acids (primarily deoxycholic and beta-muricholic acids) and developed significantly more tumors than those on low- or "normal"-fat diets. These observations have been confirmed by others (297-299), and at least in one case (300), there were no differences between animal and vegetable fats on tumor formation when fed at 20% levels. These studies, in general, tend to support the epidemiological data relating fat intake and prevalence of human colorectal cancer but have not yet adequately demonstrated the importance of several other nutritional modifications and interactions.

3. *Dietary Fiber and Experimental Carcinogenesis*

The results of various studies on dietary fibers and chemically induced carcinogenesis have been extensively reviewed elsewhere (274,288,301) and have not been entirely consistent. The protective effects of wheat bran have been reported by several investigators (300,302-305), while others have reported no differences (306,307) with this fiber supplement. Similarly, the effects of cellulose (308-310), pectin (305,311, 312), and guar (311,312) supplementation also have been inconsistent. These differences appear to be, in part, a function of the chemical carcinogen employed (305), of the levels of fiber supplementation included in the diets (274,301), of the levels of dietary fat (313), and of differences in food intake and nutritional status of the animals (309).

Despite these potentially confounding factors, there does appear to be a developing pattern relating the type of fiber supplementation, the dilution of fecal bile acid concentration, and the extent of chemically induced tumorigenesis. This correlation however is by no means conclusive and should be interpreted with considerable reservation. Dietary fiber supplements that generally increase colonic transit, dilute colonic contents, and increase fecal volume are also those that appear to reduce experimental tumorigenesis, particularly when dietary fat levels are low. These include the particulate fibers such as wheat bran and cellulose, and these fibers, perhaps coincidentally, also do not increase fecal bile acid output and have little effect on plasma lipid levels.

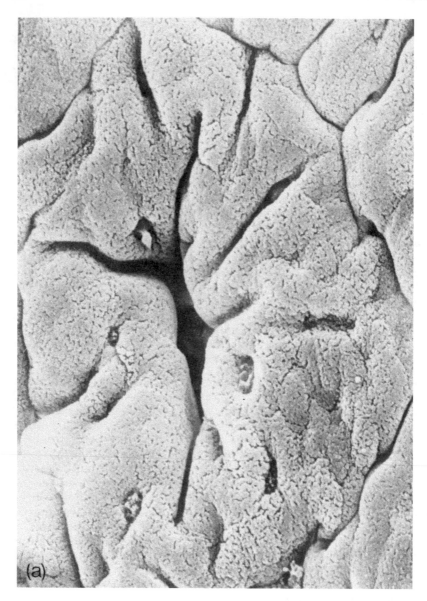

(a)

Figure 8 Scanning electron photomicrographs of the colon of mature rats fed for 4 weeks on defined diets containing (a) no fiber or (b) 5% alfalfa. Diets containing 10% cellulose or bran gave results comparable with control, while inclusion of bile acid-sequestering fibers or resins in the diet caused a variety of morphological anomalies.

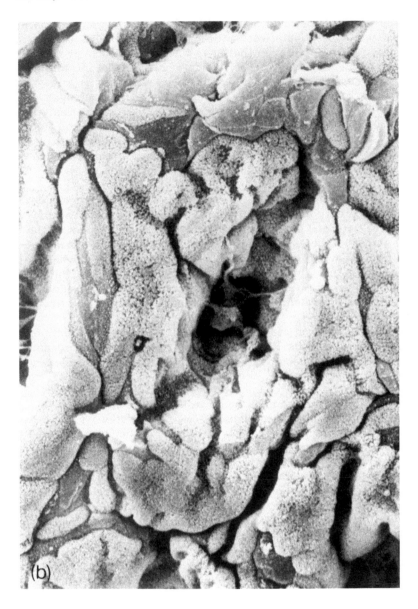

(b)

Figure 8 (continued)

When dietary fat levels are high (e.g., 20% of the diet), there appears to be little if any protective effect of these fiber supplements (313).

In contrast, fiber supplements that have bile acid-binding properties in vitro, or increase the fecal excretion of bile acids (bile acid concentrations) in vivo, are either without protective effects, or in several cases have been reported to enhance the activity of the carcinogen. Included in this group of inactive or "promoting" fiber supplements are agar (314), alfalfa (305), pectin (311,312), guar gum (311), and Metamucil (311). This relationship between the fecal bile acid concentrations, resulting from various experimental approaches, and effects on colon carcinogenesis have been summarized by Hill (287) and is shown in Table 8.

Obviously, these generalizations are to be interpreted with considerable caution since they are derived from experimental models and in some case have not been given sufficient attention with respect to the nutritional status of the animal. In addition, they largely reflect aspects of fiber pharmacology that may have little direct relationship to human nutrition and disease. They are, nevertheless, difficult to discount and require more highly controlled and standardized experimental approaches to evaluate the intestinal responses to these fiber supplements.

In addition to studies on the effects of dietary fiber supplements on experimental carcinogenesis, there have been recent reports suggesting morphological responses in rats to certain preparations of dietary fibers (160). In these studies, chronic ingestion of defined diets, containing either wheat bran or cellulose, had little effect on the topography and morphology of either the small or large bowel but did appear to demonstrate increased goblet cell activity (mucin secretion) (110). In contrast, as shown in Figure 8, those preparations that demonstrated bile acid-binding properties (36,39), such as alfalfa and pectin and the resin, cholestyramine, also produced the greatest morphological deviations from animals fed the control fiber-free diet (315,316). These studies suggest that there was a strong quantitative correlation between the bile acid-sequestering properties of the various dietary supplements and their effects on inducing morphological abnormalities in the intestine and colon (160). However other possibilities, such as the lectin content of the preparations, were not addressed.

Irrespective of the mechanisms by which these dietary supplements "induce" mucosal injury, it is well recognized (113) that various types of irritant agents, including bile acids (317), which evoke a similar necroses of villus cells, will result in a "feedback" stimulation of crypt cell production, migration, and villus repair rates. These increased cell proliferative rates are also regulated by locally produced mitotic inhibitors (318-320) and by trophic regulatory peptides of the gastrointestinal tract (321).

It is also believed that tumor promoters may act by increasing intestinal cell proliferation in tissue previously exposed to carcinogens (322). Thus it appears possible that the relationship between colonic bile acids and colorectal cancer is a reflection of the hyperplastic effects of increased steroid concentrations in the large bowel, and that nutritional or pharmacological modifications of bile acid excretion (concentration) may not be of overall benefit with respect to lower bowel pathology. Studies underway in several laboratories (321,322) may further elucidate these conjectures.

VII. CONCLUSIONS

The revival of interest in dietary fibers during this decade has been largely attributed to the hypothesis forwarded by Burkitt and Trowell, which contains two primary statements (18):

1. Diets containing high levels of foods of plant origin are protective against a wide range of disorders common to Western cultures, e.g., constipation, diverticular disease, appendicitis, colorectal cancer, obesity, gallstones, diabetes, and coronary heart disease.
2. In certain cases, diets low in fiber content may be a causative factor in the etiology of Western culture disorders.

Thus it is not surprising that both the lay and scientific communities have considered dietary fiber as a magical nutritional bullet for both the therapeutic and the prophylactic control of disease. However epidemiological and demographic evidence, as well as the proliferation of experimental evidence, clearly indicates that this type of simple relationship between dietary fiber and disease is not easily demonstrable. This could be anticipated from a nutritional standpoint, since the hypothesis implies that various modifications of the diet must occur during the shift to high intake of plant materials, and that the overall composition of the diet and the interactions between dietary components must play a role in health and disease. Trowell and others have advanced the view that the hypothesis is based on consumption of plant foods in their natural state, containing intact cellular structures and other cellular components.

Undoubtedly, the supplementation of diets with specific fiber derivatives, particularly wheat bran, has been useful in the therapeutic relief of simple constipation and associated disorders and of the symptomatology of diverticular disease. Other types of supplementation, particularly with viscous fiber products, have been effective in control of sugar metabolism in diabetes and in reducing the requirements for oral hypoglycemic agents or for insulin. These therapeutic approaches, however, represent aspects of fiber pharmacology rather than nutrition and are not included in the original nutritional hypothesis. The demonstrations of the beneficial effects of total dietary modification on plasma lipids and lipoproteins and on overall sugar metabolism, are more compatible with the fiber hypothesis.

This chapter has attempted to provide a broad perspective of the definition, the physicochemical properties of, and the physiological and pharmacological responses to dietary fibers and their isolated components. The accepted and potential therapeutic applications of diet modification and dietary fiber supplementation has been briefly reviewed to demonstrate the important impact of these plant-derived materials in medicine. Throughout the chapter, an attempt has been made to distinguish the nutritional and pharmacological aspects of dietary fiber applications and to emphasize the importance of this difference.

The potential undesirable responses to increased dietary fiber intake have also been considered in some detail. This section, in particular, is in no way meant to minimize the nutritional importance and therapeutic utility of dietary fibers but largely to review areas that require attention, particularly among specific population groups and with certain fiber supplements.

If this review has provided a basis for establishing that dietary fiber can no longer be considered an inert component of the diet, then its original intent has been served

ACKNOWLEDGMENTS

The author wishes to acknowledge: the collaborative efforts of Drs. Marie M. Cassidy, David Kritchevsky, Jon Story, and Fred Lightfoot; the technical assistance of Mr. S. Satchithanandam, Ms. Sally Haas-Smith, and Ms. Lauretta Grau; and Ms. A. Jackson and Ms. J. Brereton for the typing and editing of the manuscript.

This work was supported by U.S.D.A. Grant 82-CRCR-1-1001.

VIII. GLOSSARY OF TERMS (From Ref. 17)

Acid detergent fiber: The residue from heating food with hot, dilute sulfuric acid containing cetyl trimethylammonium bromide and taken as a measure of cellulose and lignin.

Agar: A cell wall constituent of red marine algae that is extracted without water. It is a viscous mixture of polysaccharides of which the main constituent is agarose.

Available carbohydrates: Those carbohydrates digestible by human secretions in the digestive tract, including glucose, fructose, lactose, sucrose, dextrins, starch, and glycogen.

Carageenan: An algal polysaccharide extractable from an edible seaweed (carragheen, Irish moss) that contains, among other constituents, polymerized sulfated D-galactopyranose units.

Crude fiber: A residue remaining after boiling defatted food in dilute base and dilute acid. This "classical" method does not allow complete recovery of plant cell wall constituents (e.g., 20% of hemicelluloses) and is no longer an acceptable method or term for dietary fiber.

Cutin: A lipid component of the covering and cuticle of the outer cellulose walls of plants.

Galactomannans: Polysaccharides, containing galactose and mannose and usually in the hemicellulose fraction of plant cell walls, e.g., guar gum.

Glycans: A generic name for polysaccharides. The major beta-glucan in nature is cellulose [beta-(1-4)-glucopyranoside].

Guar gum: A neutral polysaccharide isolated from the ground endosperm of the Indian cluster bean.

Heteroglycans: Polysaccharides containing two or more monosaccharides or uronic acid species, e.g., hemicelluloses, pectins.

Neutral detergent fiber: Residue remaining after extraction of food with a hot neutral solution of the detergent sodium lauryl sulfate. It is a measure of cell wall constituents of vegetable foods.

Noncellulosic polysaccharides: Another term for hemicelluloses.

Nonnutritive fiber: A term originally used to designate crude fiber or acid-detergent fiber and currently considered ambiguous.

Pectic substances: Mixtures of acidic and neutral polysaccharides that are rich in galacturonic acid and are water-soluble.

Plantix: A term coined from plant matrix and suggested as a replacement for the term dietary fiber.

Unavailable carbohydrates: Originally applied to distinguish available carbohydrate from dietary fiber.

Uronic acids: Commonly, glucuronic and galacturonic acids which are present in the hemicelluloses and pectic substances. Hexose derivatives in which the terminal primary hydroxyl is oxidized to a carboxylic acid, allowing formation of alkyl and methyl esters, amides, and metal salts.

Water-holding capacity: The amount of water than can be taken up by a unit weight of dry fiber.

REFERENCES

1. T. L. Cleave, The neglect of natural principles in current medical practice. *J. R. Nav. Med. Serv.* 42:55 (1956).
2. T. L. Cleave and G. D. Campbell, *Diabetes, Coronary Thrombosis and the Saccharine Disease.* John Wright, Bristol, 1966.
3. T. L. Cleave, *The Saccharine Disease.* John Wright, Bristol, 1974.

4. D. P. Burkitt and H. C. Trowell, *Refined Carbohydrate Foods and Disease—Some Implication of Dietary Fibre*. Academic Press, London, 1975.

5. A. R. P. Walker, Dietary fiber and pattern of diseases. *Ann. Intern. Med. 80*:663 (1974).

6. H. Trowell, Definition of dietary fiber and hypothesis that it is a protective factor in certain diseases. *Am. J. Clin. Nutr. 29*: 417 (1976).

7. G. V. Vahouny and D. Kritchevsky, Preface, in *Dietary Fiber in Health and Disease* (G. V. Vahouny and D. Kritchevsky, eds.). Plenum Press., New York, 1982, p. ix.

8. H. P. Roth and M. A. Mehlman, *Symposium on Role of Dietary Fiber in Health. Am. J. Clin. Nutr. 31*(Suppl.) (1978).

9. G. A. Spiller and R. M. Kay, *Medical Aspects of Dietary Fiber*. Plenum Press, New York, 1980.

10. Royal College of Physicians of London, *Medical Aspects of Dietary Fiber*, Summary of a report. Pitman Medical Limited, Kent, 1980.

11. R. M. Kay, Dietary fiber. *J. Lipid Res. 23*:221 (1982).

12. G. V. Vahouny and D. Kritchevsky, eds., *Dietary Fiber in Health and Disease*. Plenum Press, New York, 1982.

13. K. W. Heaton, Dietary fibre in perspective. *Hum. Nutr. Clin. Nutr. 37C*:151 (1983).

14. G. Wallace and L. Bell, *Fibre in Human and Animal Nutrition*. Royal Society of New Zealand, Wellington, 1983.

15. H. C. Trowell, D. A. T. Southgate, T. M. S. Wolever, A. R. Leeds, M. A. Gassul, and D. J. A. Jenkins, Dietary fibre redefined. *Lancet 1*:967 (1976).

16. J. H. Cummings, Consequences of the metabolism of fiber in the human large intestine, in *Dietary Fiber in Health and Disease* (G. V. Vahouny and D. Kritchevsky, eds.). Plenum Press, New York, 1982, p. 9.

17. D. A. T. Southgate and M. White, Glossary, in *Medical Aspects of Dietary Fiber* (G. A. Spiller and R. Kay, eds.). Plenum Press, New York, 1980, p. 285.

18. D. A. T. Southgate, Definitions and terminology of dietary fiber, in *Dietary Fiber in Health and Disease* (G. V. Vahouny and D. Kritchevsky, eds.). Plenum Press, New York, 1982, p. 1.

19. R. M. Kay and S. M. Strasberg, Origin, chemistry, physiological effects and clinical importance of dietary fiber. *Clin. Invest. Med. 1*:9 (1978).

20. G. V. Vahouny, Dietary fiber, lipid metabolism and atherosclerosis. *Fed. Proc. 41*:2801 (1982).

21. J. A. Robertson and M. A. Eastwood, An examination of factors which may affect the water holding capacity of dietary fibre. *Br. J. Nutr. 45*:83 (1981).

22. D. J. A. Jenkins, T. M. S. Wolever, A. R. Leeds, M. A. Gassull, P. Hiasman, J. Dilawari, D. V. Goff, G. L. Metz, and K. G. Alberti, Dietary fibres, fibre analogues and glucose tolerance, importance of viscosity. *Br. Med. J. 1*:1392 (1978).

23. J. A. Robertson, M. A. Eastwood, and M. M. Yeoman, An investigation into the physical properties of fibre prepared from several carrot varieties at different stages of development. *J. Sci. Food Agric. 31*:633 (1980).

24. A. J. M. Brodtribb and C. Groves, Effect of bran particle size on stool weight. *Gut. 19*:60 (1978).

25. S. N. Heller, L. R. Hackler, J. M. Rivers, P. J. Van Soest, D. A. Roe, B. A. Lewis, and J. Robertson, Dietary fiber: The effect of particle size of wheat bran on colonic function in young adult men. *Am. J. Clin. Nutr. 33*:1734 (1980).

26. A. M. Stephen and J. H. Cummings, Mechanisms of action of dietary fibre in the human colon. *Nature 284*:283 (1980).

27. A. A. McConnell, M. A. Eastwood, and W. D. Mitchell, Physical characteristics of vegetable foodstuffs that could influence bowel function. *J. Sci. Food Agric. 25*:1457 (1974).

28. W. P. T. James, W. J. Branch, and D. A. T. Southgate, Calcium binding by dietary fibre. *Lancet 1*:638 (1978).

29. F. Ismail-Beigi, B. Faraji, and J. G. Reinhold, The binding of zinc and iron to wheat bread, wheat bran and their components. *Am. J. Clin. Nutr. 30*:1721 (1977).

30. M. A. Eastwood and G. S. Boyd, The distribution of bile salts among the small intestine of rats. *Biochim. Biophys. Acta 137*: 393 (1967).

31. M. A. Eastwood and D. Hamilton, Studies on the adsorption of bile salts to non-absorbed components of the diet. *Biochim. Biophys. Acta 152*:165 (1968).

32. D. Kritchevsky and J. A. Story, Binding of bile salts in vitro by non-nutritive fiber. *J. Nutr. 104*:458 (1974).

33. N. J. Birkner and F. Kern, Jr., In vitro adsorption of bile salts to food residues, salicylazosulfapyridine and hemicellulose. *Gastroenterology 67*:237 (1974).

34. J. Balmer and D. B. Zilversmit, Effects of dietary roughage on cholesterol absorption, cholesterol turnover and steroid excretion in the rat. *J. Nutr. 104*:1319 (1974).

35. J. A. Story and D. Kritchevsky, Comparison of the binding of various bile acids and bile salts in vitro to several types of fiber. *J. Nutr. 106*:1292 (1976).

36. G. V. Vahouny, R. Tombes, M. M. C. Cassidy, D. Kritchevsky, and L. L. Gallo, Binding of bile salts, phospholipids and cholesterol from mixed micelles by bile acid sequestrants and dietary fibers. *Lipids 15*:1012 (1980).

37. M. A. Eastwood and L. Mowbray, The binding of components of mixed micelles to dietary fibers. *Am. J. Clin. Nutr.* 29:146 (1976).

38. G. A. Leveille and N. E. Sauberlich, Mechanisms of the cholesterol depressing effect of pectin in the cholesterol-fed rat. *J. Nutr.* 88:209 (1966).

39. G. V. Vahouny, R. Tombes, M. M. Cassidy, D. Kritchevsky, and L. L. Gallo, Dietary fibers. VI. Binding of fatty acids and monolein from mixed micelles containing bile salts and lecithin. *Proc. Soc. Exp. Biol. Med.* 166:12 (1981).

40. J. Nagyvary and E. L. Bradbury, Hypocholesterolemic effect of Al^{3+} complexes. *Biochim. Biophys. Acta* 77:592 (1977).

41. I. Furda, Interaction of pectinaceous dietary fiber with some metals and lipids, in *Dietary Fibers—Chemistry and Nutrition* (G. Inglett and L. Falkehag, eds.). Academic Press, New York, 1979, p. 31.

42. J. A. Story, Dietary fiber and lipid metabolism: An update, in *Medical Aspects of Dietary Fiber* (G. A. Spiller and R. M. Kay, eds.). Plenum Press, New York, 1980, p. 137.

43. R. M. Kay and A. S. Truswell, Dietary fiber: Effects on plasma and biliary lipids in man, in *Medical Aspects of Dietary Fibers* (G. A. Spiller and R. M. Kay, eds.). Plenum Press, New York, 1980, p. 153.

44. G. H. Bornside, Stability of human fecal flora. *Am. J. Clin. Nutr.* 31:S141 (1978).

45. R. D. Williams and W. H. Olmstead, The manner in which food controls the bulk of feces. *Ann. Intern. Med.* 10:717 (1936).

46. C. J. Prynne and D. A. T. Southgate, The effects of a supplement of dietary fibre on faecal excretion by human subjects. *Br. J. Nutr.* 41:495 (1979).

47. R. A. McCance, K. M. Prior, and E. M. Widdowson, A radiological study of the rate of passage of brown and white bread through the digestive tract of man. *Br. J. Nutr.* 7:98 (1953).

48. B. K. Anand, Neurological mechanisms regulating appetite, in *Obesity Symposium* (W. L. Burland, P. D. Samuel, and J. Yudkin, eds.). Churchill Livingston, Edinburg, 1974, p. 116.

49. I. L. MacGregor, P. Martin, and J. H. Mayer, Gastric emptying of solid food in normal man after subtotal gastrectomy and truncal vagotomy with pyloroplasty. *Gastroenterology* 72:206 (1977).

50. J. D. Davis and B. J. Collins, Distention of the small intestine, satiety, and the control of food intake. *Am. J. Clin. Nutr.* 31:S255 (1978).

51. G. B. Haber, K. W. Heaton, D. Murphy, and L. Burroughs, Depletion and disruption of dietary fibre. Effects on satiety, plasma glucose and serum insulin. *Lancet* 2:679 (1977).

52. K. W. Heaton, Food intake and regulation by fiber, in *Medical Aspects of Dietary Fiber* (G. A. Spiller and R. M. Kay, eds.). Plenum Press, New York, 1980, p. 223.
53. S. Holt, R. C. Heading, D. C. Carter, L. F. Prescott, and P. Tothill, Effect of gel fiber on gastric emptying and absorption of glucose and paracetamol. *Lancet 1*:636 (1979).
54. P. Wilmshurst and J. C. W. Crawley, The measurement of gastric transit time in obese subjects using [24]Na and the effects of energy content and guar gum on gastric emptying and satiety. *Br. J. Nutr. 44*:1 (1980).
55. A. R. Leeds, D. N. L. Ralphs, F. Ebied, G. Metz, and J. B. Dilawari, Pectin and the dumping syndrome: Reduction of symtoms and plasma volume changes. *Lancet 1*:1075 (1981).
56. A. R. Leeds, Modification of intestinal absorption by dietary fiber and fiber components, in *Dietary Fiber in Health and Disease* (G. V. Vahouny and D. Kritchevsky, eds.). Plenum Press, New York. 1982, p. 53.
57. J. Winreich, O. Pederson, and K. Dinesein, Role of bran in normals: Serum levels of cholesterol, triglycerides, calcium and total 3-hydroxycholanic acid, and intestinal transit time. *Acta Med. Scand. 202*:125 (1977).
58. J. B. Wyman, K. W. Heaton, A. P. Manning, and A. C. B. Wicks, Variability of colon function in healthy subjects. *Gut 19*:146 (1978).
59. A. N. Smith and M. A. Eastwood, The measurement of intestinal transit time, in *Medical Aspects of Dietary Fiber* (G. A. Spiller and R. M. Kay, eds.). Plenum Press, New York, 1980, p. 27.
60. J. H. Cummings, W. Branch, D. J. A. Jenkins, D. A. T. Southgate, H. Houston, and W. P. T. James, Colonic response to dietary fibre from carrot, cabbage, apple, bran and guar gum. *Lancet 1*:5 (1978).
61. J. H. Cummings, Physiological effects of dietary fibre in man, *Topics in Gastroenterology*, Vol. 6 (S. C. Truelove and M. F. Hayworth, eds.). Blackwell, Oxford, 1978, p. 49.
62. J. L. Kelsay, A review of research on effects of fiber intake on man. *Am. J. Clin. Nutr. 31*:141 (1978).
63. D. Kritchevsky, J. A. Story, and G. V. Vahouny, Influence of fiber on lipid metabolism, in *Nutrition and Food Sciences*, Vol. 3. (W. Santos, N. Lopes, J. J. Barbosa, D. Chaves, and J. C. Valante, eds.). Plenum Press, New York, 1980, p. 461.
64. S. A. Hyun, G. V. Vahouny, and C. R. Treadwell, Effects of hypocholesterolemic agents on intestinal cholesterol absorption. *Proc. Soc. Exp. Biol. Med. 112*:496 (1963).
65. G. V. Vahouny, T. Roy, L. Gallo, J. A. Story, D. Kritchevsky, M. M. Cassidy, B. Grund, and C. R. Treadwell, Dietary fiber and intestinal absorption of cholesterol in the rat. *Am. J. Clin. Nutr. 31*:S208 (1978).

66. G. V. Vahouny, T. Roy, L. Gallo, J. A. Story, D. Kritchevsky, and M. M. Cassidy, Dietary fibers. III. Effects of chronic intake on cholesterol absorption and metabolism in the rat. *Am. J. Clin. Nutr.* *33*:2182 (1980).

67. D. J. A. Jenkins, A. R. Leeds, M. A. Gassull, H. Houston, D. V. Goff, and M. J. Hill, The cholesterol lowering properties of guar and pectin. *Clin. Sci. Mol. Med.* *51*:8 (1976).

68. R. M. Kay and A. S. Truswell, Effects of citrus pectin on blood lipids and fecal excretion in man. *Am. J. Clin. Nutr.* *30*:171 (1977).

69. P. A. Judd, R. M. Kay, and A. S. Truswell, The cholesterol lowering properties of guar and pectin. *Nutr. Metab.* *21*:85 (1977).

70. T. A. Miettinen and S. Tarpila, Effect of pectin on serum cholesterol, fecal bile acids and biliary lipids in normolipidemic and hyperlipidemic individuals. *Clin. Chem. Acta* *79*:471 (1977).

71. D. T. Foreman, J. E. Garvin, J. E. Forestner, and C. B. Taylor, Increased excretion of fecal bile acids by a hydrophylic colloid. *Proc. Soc. Exp. Biol. Med.* *127*:1060 (1968).

72. M. M. Stanley, D. Paul, D. Gacke, and J. Murphy, Effect of cholestyramine, Metamucil and cellulose on fecal bile acid excretion in man. *Gastroenterology* *65*:889 (1973).

73. R. W. Kirby, J. W. Anderson, B. Sieling, E. D. Rees, W. J. L. Chen, R. E. Miller, and R. M. Kay, Oat bran intake selectivity lowers serum low density liporpotein concentration: Studies of hypercholesterolemic men. *Am. J. Clin. Nutr.* *34*:824 (1981).

74. O. D. Rotstein, R. M. Kay, M. Wayman, and S. M. Strasberg, Prevention of cholesterol gallstone by lignin and lactulose. *Gastroenterology* *81*:1098 (1981).

75. G. V. Vahouny, Dietary fibers and intestinal absorption of lipids, in *Dietary Fiber in Health and Disease* (G. V. Vahouny and D. Kritchevsky, eds.). Plenum Press, New York, 1982, p. 203.

76. M. A. Eastwood and R. M. Kay, A hypothesis for the action of fiber along the gastrointestinal tract. *Am. J. Clin. Nutr.* *32*: 364 (1979).

77. D. J. A. Jenkins, R. Nineham, C. Craddock, P. Graig-McFeely, K. Donaldson, T. Leigh, and J. Snook, Fiber and diabetes. *Lancet* *1*:434 (1979).

78. L. M. Lichtenberger, Importance of food in the regulation of gastrin release and formation. *Am. J. Physiol.* *243*:G429 (1982).

79. J. E. Lennard-Jones, J. Fletcher, and D. G. Shaw, Effect of different foods on the acidity of the gastric contents in patients with duodenal ulcer. III. Effect of altering the proportions of protein and carbohydrate. *Gut* *9*:177 (1968).

80. T. J. Goulder, L. M. Morgan, V. Marks, T. Smythe, and L. Hinks, Fiber and diabetes. *Diabetologia* *14*:235 (1978).

81. L. M. Morgan, T. J. Goulder, D. Tsiolakis, V. Marks, and K. G. M. M. Alberti, The effect of unabsorbable carbohydrate on gut hormones. *Diabetologia* 17:85 (1979).

82. D. J. A. Jenkins, S. R. Bloom, R. H. Albuquerque, A. R. Leeds, D. L. Sarson, G. L. Metz, and K. G. M. M. Alberti, Pectin and complication after gastric surgery: Normalization of postprandial glucose and endocrine responses. *Gut* 21:574 (1980).

83. P. M. Miranda and D. L. Horwitz, High fiber diets in the treatment of diabetes mellitus. *Ann. Intern Med.* 88:482 (1978).

84. G. Dunaif and B. O. Schneeman, The effect of dietary fiber on human pancreatic enzyme activity in vitro. *Am. J. Clin. Nutr.* 34:1034 (1981).

85. B. O. Schneeman, Pancreatic and digestive function, in *Dietary Fiber in Health and Disease* (G. V. Vahouny and D. Kritchevsky, eds.). Plenum Press, New York, 1982, p. 73.

86. B. O. Schneeman and D. Gallagher, Changes in small intestinal digestive enzyme activity and bile acids with dietary cellulose in rats. *J. Nutr.* 110:584 (1980).

87. J. C. Acton, L. Breyer, and L. D. Satterlee, Effect of dietary fiber constituents on the in vitro digestibility of casein. *J. Food Sci.* 47:556 (1982).

88. C. M. Gagne and J. C. Acton, Fiber constituents and fibrous food residue effects on the in vitro enzymatic digestion of protein. *J. Food Sci.* 48:734 (1983).

89. I. T. Johnson and J. M. Gee, Inhibitory effect of guar gum on the intestinal absorption of glucose in vitro. *Proc. Nutr. Soc.* 39:52A (1980).

90. B. Elsenhans, U. Sufke, R. Blume, and W. F. Caspary, The influence of carbohydrate gelling agents on rat intestinal transport of monosaccharides and neutral amino acids in vitro. *Clin. Sci.* 59:373 (1980).

91. S. Sigleo, M. J. Jackson, and G. V. Vahouny, Effect of dietary fiber constituents on intestinal morphology and nutrient transport. *Am. J. Physiol.* 246:G 34 (1984).

92. B. Creamer, Intestinal structure in relation to absorption. *Biomembranes* 4A:1 (1974).

93. L. D. Dworkin, G. M. Levine, J. J. Farber, and M. H. Spector, Small intestinal mass of the rat is partially determined by indirect effects on intraluminal nutrition. *Gastroenterology* 71:626 (1976).

94. R. F. Hageman and J. Stragand, Fasting and refeeding: Cell kinetic response of jejunum, ileum and colon. *Cell Tissue Kinet.* 10:3 (1977).

95. D. V. Muadsley, J. Lief, and Y. Kobayashi, Ornithine decarboxylase in rat small intestine: Stimulation with food or insulin. *Am. J. Physiol.* 231:1557 (1976).

96. A. M. Dawson and K. J. Isselbacher, Studies on lipid metabolism in the small intestine with observations on the role of bile salt. *J. Clin. Invest.* *39*:730 (1960).

97. T. S. Low-Beer, R. E. Schneider, and W. O. Dobbins, Morphological changes of the small intestinal mucosa of guinea pig and hamster following incubation in vitro and perfusion in vivo with unconjugated bile salts. *Gut* *11*:486 (1970).

98. M. V. Teem and S. F. Phillips, Perfusion of the hamster jejunum with conjugated and unconjugated bile acids: Inhibition of water absorption and effects on morphology. *Gastroenterology* *62*:261 (1972).

99. H. S. Mekhijian and S. F. Phillips, Perfusion of the canine colon with unconjugated bile acids: Effect on water and electrolyte transport, morphology and bile acid absorption. *Gastroenterology* *59*:120 (1970).

100. H. S. Mekhijian, S. F. Phillips, and A. F. Hofmann, Colonic secretion of water and electrolytes induced by bile acids: Perfusion studies in man. *J. Clin. Invest.* *50*:1569 (1971).

101. G. E. Sladen and J. T. Harries, Studies on the effects of unconjugated dihydroxy bile salts on rat small intestinal function in vivo. *Biochim. Biophys. Acta* *288*:443 (1972).

102. S. J. Baker, M. Ignatino, V. L. Mathan, S. K. Waich, and C. C. Chacko, Intestinal biopsy in tropical sprue, in *Intestinal Biopsy* (G. E. W. Wolstanholme and M. P. Cameron, eds.). Churchill, London, 1962, p. 82.

103. G. C. Cook, S. K. Kajubi, and F. D. Lu, Jejunal morphology of the African in Uganda. *J. Pathol.* *98*:157 (1969).

104. C. J. G. Chacko, K. A. Paulson, V. I. Mathan and S. J. Bahu, The villus architecture of the small intestine in the tropics. A necroscopy study. *J. Pathol.* *98*:146 (1969).

105. R. L. Owen and L. L. Brandborg, Jejunal morphologic consequences of vegetarian diet in humans. *Gastroenterology* *72*: A88 (1977).

106. C. Tasman-Jones, A. L. Jones, and R. L. Owen, Jejunal morphological consequences of dietary fibre in rats. *Gastroenterology* *74*:1102 (1978).

107. C. Tasman-Jones, Effects of dietary fiber on the structure and function of the small intestine, in *Medical Aspects of Dietary Fiber* (G. A. Spiller and R. M. Kay, eds.). Plenum Press, New York, 1980, p. 67.

108. R. C. Brown, J. Kelleher, and M. S. Losowsky, The effect of pectin on the structure and function of the rat small intestine. *Br. J. Nutr.* *42*:357 (1979).

109. S. E. Schwartz and G. D. Levine, Effect of dietary fiber on intestinal glucose absorption and glucose tolerance in rats. *Gastroenterology* *79*:833 (1980).

110. G. V. Vahouny, T. Le, F. G. Lightfoot, and M. M. Cassidy, Stimulation of intestinal mucin by dietary fiber, in *A Decade of Achievements and Challenges in Large Bowel Cancer Research*. National Large Bowel Cancer Project, 1983.

111. F. Nimmerfall and J. Rosenthaler, Significance of the goblet-cell mucin layer, the outermost barrier through the gut wall. *Biochem. Biophys. Res. Commun.* *94*:960 (1980).

112. M. M. Cassidy, F. G. Lightfoot, and G. V. Vahouny, Structural-functional modulation of mucin secretory patterns, in *Structural-Functional Correlations in Epithelia* (M. Dinno, ed.). Alan R. Liss, Inc., New York, 1981, p. 97.

113. H. Sprinz, Factors influencing intestinal cell renewal. *Cancer* *28*:71 (1971).

114. V. Marks and D. S. Turner, The gastrointestinal hormones with particular reference to their role in the regulation of insulin secretion. *Essays Biochem.* *3*:109 (1977).

115. D. J. A. Jenkins, A. R. Leeds, M. A. Gassull, B. Cochet, and K. G. M. M. Alberti, Decrease in postprandial insulin and glucose concentration by guar and pectin. *Ann. Intern. Med.* *86*:20 (1977).

116. J. W. Anderson, Dietary fiber and diabetes, in *Dietary Fiber in Health and Disease* (G. V. Vahouny and D. Kritchevsky, eds.). Plenum Press, New York, 1982, p. 151.

117. J. W. Anderson and K. Ward, High carbohydrate, high fiber diets for insulin treated men with diabetes mellitus. *Am. J. Clin. Nutr.* *32*:2312 (1979).

118. O. Pederson, E. Hollund, H. O. Lindskov, and N. S. Sorensen, Increased insulin receptors on monocytes from insulin-dependent diabetes after a high starch-high fiber diet. *Diabetologia* *19*:306 (1980).

119. W. E. W. Roediger, The colonic epithelium in ulcerative colitis: An energy-deficiency disease. *Lancet* *2*:712 (1980).

120. M. H. Crump, R. A. Argenzio, and S. C. Whipp, The effect of acetate on the absorption of solute and water from the pig colon. *Am. J. Vet. Res.* *41*:1565 (1980).

121. W. E. W. Roediger and A. Moore, Effect of short chain fatty acid on sodium absorption in isolated human colon perfused through the vascular bed. *Dig. Dis. Sci.* *26*:100 (1981).

122. J. H. Cummings, M. J. Hill, E. S. Bone, J. W. Branch, and D. J. A. Jenkins, The effect of meat protein and dietary fiber on colonic function and metabolism. II. Bacterial metabolites in feces and urine. *Am. J. Clin. Nutr.* *32*:2094 (1979).

123. N. I. McNeil, J. H. Cummings, and W. P. T. James, Short chain fatty acid absorption by the human large intestine. *Gut* *19*:819 (1978).

124. J. W. Anderson and S. Bridges, Short chain fatty acids alter the rates of glycolysis and gluconeogenesis in isolated rat hepatocytes. *Clin. Res. 28*:776A (1980).

125. J. A. Story and J. N. Thomas, Dietary fiber and lipoproteins, in *Dietary Fiber in Health and Disease* (G. V. Vahouny and D. Kritchevsky, eds.). Plenum Press, New York, 1982, p. 193.

126. W. G. Brydon, K. Tadesse, and M. A. Eastwood, The effect of dietary fibre on bile acid metabolism in rats. *Br. J. Nutr. 43*:101 (1980).

127. A. C. B. Wicks, J. Yeates, and K. W. Heaton, Bran and bile: Time course of changes in normal young men given a standard dose. *Scand. J. Gastroenterol. 13*:289 (1978).

128. J. B. Wyman, K. W. Heaton, A. P. Manning, and A. C. B. Wicks, Variability of colonic function in healthy adults. *Am. J. Clin. Nutr. 29*:1474 (1976).

129. M. A. Eastwood, W. G. Brydon, and K. Tadesse, Effect of fiber on colon function, in *Medical Aspects of Dietary Fiber* (G. A. Spiller and R. M. Kay, eds.). Plenum Press, New York, 1980, p. 1.

130. R. D. Williams and W. H. Olmstead, The effect of cellulose, hemicellulose and lignin on the weight of the stool. A contribution to the study of laxation in man. *J. Nutr. 11*:433 (1936).

131. A. E. Williams, M. A. Eastwood, and R. Cregeen, SEM and light microscope study of the matrix structures of human feces. *Scanning Electron Microsc. 2*:707 (1978).

132. G. H. Bornside, Stability of human flora. *Am. J. Clin. Nutr. 31*:S141 (1978).

133. P. N. Durington, C. H. Bolton, A. P. Manning, and M. Hartog, Effect of pectin on serum lipids and lipoprotein, whole-gut transit time, and stool weight. *Lancet 2*:394 (1976).

134. G. R. Cowgill and W. E. Anderson, Laxative effects of wheat bran and "washed" bran in healthy men. *J. Am. Med. Assoc. 98*:1866 (1932).

135. E. M. Dimock, The prevention of constipation. *Br. Med. J. 2*: 906 (1937).

136. R. C. Rendtorff and M. Kashgarian, Stool patterns of healthy adult males. *Dis. Colon Rectum 10*:222 (1967).

137. G. A. Spiller, M. C. Cernoff, E. A. Shipley, M. A. Bergher, and G. A. Briggs, Can fecal weight be used to establish a recommended intake of dietary fiber (plantix)? *Am. J. Clin. Nutr. 30*:659 (1977).

138. A. J. M. Brodiff, Dietary fiber in diverticular disease of the colon, in *Medical Aspects of Dietary Fiber* (G. A. Spiller and R. M. Kay, eds.). Plenum Press, New York, 1980, p. 43.

139. N. S. Painter and D. P. Burkitt, Diverticular disease of the colon. *Br. Med. J. 2*:450 (1971).

140. S. Arfwidsson, Pathogenesis of multiple diverticular of the sigmoid colon in diverticula disease. *Acta Chir. Scand. Suppl. 342*:1 (1964).
141. N. S. Painter and S. C. Truelove, The intraluminal pressure patterns in diverticulosis of the colon. *Gut 5*:201 (1964).
142. D. M. Lubbock, W. Thomson, and R. G. Garry, Epithelial overgrowths and diverticula in the gut. *Br. Med. J. 1*:1252 (1937).
143. A. J. Carlson and F. Hoelzel, Relationship of diet to diverticulosis of the colon in rats. *Gastroenterology 12*:108 (1949).
144. A. J. M. Brotribb, R. E. Condon, V. Cowles, and J. J. DeCosse, Effect of dietary fiber on intraluminal pressure and myoelectrical activity in the left colon in monkeys. *Gastroenterology 77*:70 (1979).
145. J. A. Ritchie, S. C. Truelove, and G. M. Ardran, Propulsion and retropulsion in the human colon demonstrated by time lapse cinefluorography. *Gut 9*:735 (1968).
146. J. M. Findlay, W. D. Mitchell, M. A. Eastwood, A. J. B. Anderson, and A. N. Smith, Intestinal streaming patterns in cholerrhoeic enteropathy and diverticular diseases. *Gut 15*:207 (1974).
147. A. N. Smith, Effect of bulk additives on constipation and in diverticular diseases, in *Dietary Fiber: Current Developments of Importance to Health* (K. W. Heaton, ed.). T. Libbey Co., London, 1978, p. 97.
148. D. A. T. Southgate, Has dietary fiber a role in the presentation and treatment of obesity? *Bibl. Nutr. Dieta 26*:70 (1978).
149. T. B. Van Itallie, Dietary fiber and obesity. *Am. J. Clin. Nutr. 31*;S43 (1978).
150. R. Ali, H. Staub, G. A. Leveille, and P. C. Boyle, Dietary fiber and obesity, in *Dietary Fiber in Health and Disease* (G. V. Vahouny and D. Kritchevsky, eds.). Plenum Press, New York, 1982, p. 139.
151. J. J. Beereboom, Low calorie bulking agents. *Crit. Rev. Food Sci. Nutr.* p. 401 May (1979).
152. E. Evans and D. S. Miller, Bulking agents in the treatment of obesity. *Nutr. Metab. 18*:199 (1975).
153. J. Tuomilehto, E. Voutilainen, J. Huttemen, S. Vinni, and K. Homen, Effect of guar gum on body weight and serum lipids in hypercholesterolemic females. *Acta Med. Scand. 208*:45 (1980).
154. O. Mickelson, D. P. Makdani, R. H. Cotton, S. T. Titcomb, J. C. Colmey, and R. Gatty, Effect of a high fiber bread on weight loss in college-age males. *Am. J. Clin. Nutr. 32*:1703 (1979).
155. K. W. Heaton, P. M. Emmett, C. L. Henry, J. R. Thornton, A. Manhire, and M. Hertog, Not just fibre: The nutritional consequences of refined carbohydrate foods. *Hum. Nutr. Clin. Nutr. 37*:31 (1983).

156. R. G. Campbell, S. A. Hashim, and T. B. Van Itallie, Studies on food intake regulation in man: Responses to variation in nutritive density in lean and obese subjects. *N. Engl. J. Med. 285*:1402 (1971).

157. T. S. Spiegel, Caloric regulation of food intake in man. *J. Comp. Physiol. Psychol. 84*:24 (1973).

158. K. Keim and C. Keis, Effect of dietary fiber on nutritional status of weanling mice. *Cereal Chem. 56*:73 (1979).

159. A. V. Rao and E. Bright-See, Effect of graded amounts of dietary pectin on growth parameters of rats. *Nutr. Rep. Int. 19*:411 (1979).

160. M. M. Cassidy, F. G. Lightfoot, and G. V. Vahouny, Dietary fiber, bile acids and intestinal morphology, in *Dietary Fiber Health and Disease* (G. V. Vahouny and D. Kritchevsky, eds.) Plenum Press, New York, 1982, p. 239.

161. H. C. Trowell, Dietary fiber hypothesis of the etiology of diabetes mellitus. *Diabetes 24*:762 (1975).

162. D. J. A. Jenkins, A. R. Leeds, M. A. Gassull, T. M. S. Wolever, D. V. Goff, K. G. M. M. Alberti, and T. D. R. Hockaday, Unabsorbable carbohydrates and diabetes: Decreased post-prandial hyperglycemia. *Lancet 2*:172 (1976).

163. T. G. Kiehm, J. W. Anderson, and K. Ward, Beneficial effects of a high carbohydrate, high fiber diet on hyperglycemic diabetic men. *Am. J. Clin. Nutr. 29*:895 (1976).

164. D. J. A. Jenkins, Dietary fiber and carbohydrate metabolism, in *Medical Aspects of Dietary Fiber* (G. A. Spiller and R. M. Kay, eds.). Plenum Press, New York, 1980, p. 175.

165. T. M. S. Wolever, D. J. A. Jenkins, R. Nineham, and K. G. M. M. Alberti, Guar gum and the reduction of post-prandial glycemia: Effect of incorporation into solid food, liquid food and both. *Br. J. Nutr. 41*:505 (1979).

166. D. J. A. Jenkins, T. M. S. Wolever, R. Nineham, R. Taylor, G. L. Metz, S. Bacon, and T. D. R. Hockaday, Guar crispbread in the diabetic rat. *Br. Med. J. 2*:1744 (1978).

167. J. Tredgar and J. Ramsley, Guar gum—its acceptability to diabetic patients when incorporated into baked food products. *J. Hum. Nutr. 32*:427 (1978).

168. N. A. Levitt, A. I. Vinik, A. A. Sive, P. T. Child, and W. P. U. Jackson, The effect of dietary fiber on glucose and hormone responses to a mixed meal in normal subjects and diabetic subjects with and without autonomic neuropathy. *Diabetes Care 3*:515 (1980).

169. L. Monnier, T. C. Pham. L. Aguirre, A. Orsetti, and J. Mirouze, Influence of indigestible fibers on glucose tolerance. *Diabetes Care 1*:83 (1980).

170. O. Bosello, R. Ostuzzi, F. Armellini, R. Micciolo, and L. A. Scuro, Glucose tolerance and blood lipids in bran-fed patients with impaired glucose tolerance. *Diabetes Care 3*:46 (1980).

171. A. J. M. Brodtribb and D. M. Humphreys, Diverticular disease: Three studies. *Br. Med. J. 2*:424 (1976).

172. Y. Kanter, N. Eitan, G. Brook, and D. Barzilai, Improved glucose tolerance and insulin. *Isr. J. Med. Sci. 16*:1 (1980).

173. J. M. Munoz, H. H. Sanstead, and R. A. Jacob, Effect of dietary fiber on glucose tolerance of normal mice. *Diabetes 28*: 496 (1979).

174. J. C. Patel, A. B. Metha, M. D. Dhirawani, V. J. Juthani, and L. Airyer, High carbohydrate diet in the treatment of diabetes mellitus. *Diabetologia 5*:243 (1969).

175. P. D. Gulati, M. B. Rao, and H. Vaishnava, Diets for diabetes. *Lancet 2*:397 (1974).

176. K. M. West. Prevention and therapy of diabetes mellitus. *Nutr. Rev. 33*:193 (1975).

177. E. L. Bierman and R. Nelson, Carbohydrates, diabetes and blood lipids. *World Rev. Nutr. Diet. 22*:280 (1975).

178. J. M. Douglass, Raw diet and insulin requirements. *Ann. Intern Med. 82*:61 (1975).

179. J. J. Albrink, P. C. Davidson, and T. Newman, Lipid-lowering effect of a very high carbohydrate high fiber diet. *Diabetes 26* (Suppl. 1):324A (1976).

180. J. W. Anderson, Dietary fiber and diabetes, in *Medical Aspects of Dietary Fiber* (G. A. Spiller and R. M. Kay, eds.). Plenum Press, New York, 1980, p. 183.

181. J. W. Anderson and K. Ward, Long-term effects of high carbohydrate, high fiber diets on glucose and lipid metabolism. A preliminary report on patients with diabetes. *Diabetes Care 1*: 77 (1978).

182. J. W. Anderson, S. K. Ferguson, D. Karonous, L. O'Malley, B. Sieling, and W. L. Chen, Mineral and vitamin status in high-fiber diets: Long-term studies on diabetic patients. *Diabetes Care 3*:38 (1980).

183. J. W. Anderson and B. Sieling, High fiber diets for obese diabetic patients. *J. Obes. Bar. Med. 9*:109 (1980).

184. A. Rivellese, A. Giacco, S. Genovese, G. Riccardi, D. Pacioni, P. L. Mattioli, and M. Mancini, Effect of dietary fiber on glucose control and serum lipoproteins in diabetic patients. *Lancet 1*: 447 (1980).

185. R. W. Simpson, J. L. Mann, J. Eaton, R. A. Moore, R. Carter, and T. D. R. Hockaday, Improved glucose control in maturity-onset diabetes treated with high carbohydrate-modified fat diet. *Br. Med. J. 1*:1753 (1979).

186. R. W. Simpson, J. I. Mann, J. Eaton, R. Carter, and T. D.R. Hockaday, High carbohydrate diets and insulin-dependent diabetics. *Br. Med. J. 2*:523 (1979).

187. H. C. R. Simpson, S. Lousley, M. Geekie, R. W. Simpson, R. D. Carter, T. D. Hockaday, and J. I. Mann, A high carbohydrate leguminous fibre diet improves all aspects of diabetic control. *Lancet 1*:1 (1981).

188. R. M. Kay, W. Grobin, and N. S. Tract, Diets rich in natural fibre improve carbohydrate tolerance in maturity-onset, non-insulin dependent diabetics. *Diabetologia 20*:18 (1981).

189. S. L. Malhotra, Serum lipids, dietary factors and ischemic heart disease. *Am. J. Clin. Nutr. 20*:462 (1967).

190. H. Trowell and D. R. Burkitt, Dietary fibre and cardiovascular disease. *Artery 3*:107 (1977).

191. A. Keys, J. T. Anderson, and F. Grande, Diet-type and blood lipids in man. *J. Nutr. 70*:257 (1960).

192. F. Grande, J. T. Anderson, and A. Keys, Sucrose and various carbohydrate-containing foods and serum lipids in man. *Am. J. Clin. Nutr. 27*:1043 (1974).

193. D. J. A. Jenkins, D. Reynolds, A. R. Leeds, A. L. Waller, and J. H. Cummings, Hypocholesterolemic action of dietary fiber unrelated to fecal bulking effect. *Am. J. Clin. Nutr. 32*: 2430 (1979).

194. T. A. Miettinen, Effects of dietary fibers and ion-exchange resins on cholesterol metabolism in man, in *Atherosclerosis* (A. M. Gotto, L. C. Smith, and B. Allen, eds.). Springer-Verlag, New York, 1980, p. 311.

195. P. Luyken, N. A. Pikaar, H. Palman, and H. Schippers, The influence of legumes on the serum cholesterol level. *Voeding 23*:447 (1962).

196. F. Grande, J. T. Anderson, and A. Keys, Effect of carbohydrates of leguminous seeds, wheat and potatoes on serum cholesterol concentrations in man. *J. Nutr. 86*:313 (1965).

197. M. J. Albrink and I. H. Ullrich, Effect of dietary fiber on lipids and glucose tolerance of healthy young men, in *Dietary Fiber in Health and Disease* (G. V. Vahouny and D. Kritchevsky, eds.). Plenum Press, New York, 1982, p. 169.

198. P. J. Palumbo, E. R. Broines, and R. A. Nelson, Sucrose sensitivity of patients with coronary artery disease. *J. Am. Med. Assoc. 240*:223 (1978).

199. J. W. Anderson and W. J. L. Chen, Plant fiber: Carbohydrate and lipid metabolism. *Am. J. Clin. Nutr. 32*:346 (1976).

200. J. W. Anderson, R. W. Kirby, and E. D. Rees, Oat-bran selectivity lowers serum low-density lipoprotein concentrations in man. *Am. J. Clin. Nutr. 33*:914 (1980).

201. R. W. Kirby, J. W. Anderson, R. Sieling, E. D. Rees, W. J. L. Chen, R. E. Miller, and R. M. Kay, Oat bran intake selectively lowers serum low density lipoprotein cholesterol concentration. *Am. J. Clin. Nutr. 34*:824 (1981).

202. W. J. L. Chen, J. W. Anderson, and M. R. Gould, Cholesterol lowering effects of oat bran and oat gum. *Fed. Proc. 40*:853 (1981).

203. D. J. A. Jenkins, D. Reynolds, B. Slavin, A. R. Leeds, A. L. Jenkins, and E. M. Jepson, Dietary fiber and blood lipids: Treatment of hypercholesterolemia with guar crispbread. *Am. J. Clin. Nutr. 33*:575 (1980).

204. D. J. A. Jenkins, A. R. Leeds, B. Slavin, J. Mann, and E. M. Jepsom, Dietary fiber and blood lipids: Reduction of serum cholesterol in type II hyperlipidemia by guar gum. *Am. J. Clin. Nutr. 32*:16 (1979).

205. C. Kies and H. M. Fox, Dietary hemicellulose interactions influencing serum lipid patterns and protein nutritional status of adult man. *J. Food Sci. 42*:440 (1977).

206. P. Linder and B. Moller, Lignin: A cholesterol lowering agent. *Lancet 2*:1259 (1973).

207. W. J. L. Chen and J. W. Anderson, Effect of guar gum and wheat bran on lipid metabolism of rats. *J. Nutr. 109*:1028 (1979).

208. W. J. L. Chen and J. W. Anderson, Effects of plant fiber in decreasing plasma total cholesterol and increasing high-density lipoprotein cholesterol. *Proc. Soc. Exp. Biol. Med. 162*:310 (1979).

209. K. W. Heaton, The epidemiology of gallstones and suggested aetiology. *Clin. Gastroenterol. 2*:67 (1973).

210. K. W. Heaton, Diet and gallstones, in *Gallstones, Hepatology: Research and Clinical Issues* (E. A. Shaffer and S. M. Strasberg, eds.). Plenum Press, New York, 1979, p. 371.

211. J. W. Marks, G. G. Bonorris, and L. J. Schoenfield, Pathophysiology and dissolution of cholesterol gallstones, in *The Bile Acids* (P. P. Nair and D. Kritchevsky, eds.). Plenum Press, New York, 1976, p. 81.

212. E. W. Pomare, K. W. Heaton, T. S. Low-Beer, and H. J. Espiner, The effects of wheat bran upon bile salt metabolism and upon the lipid composition of bile in gallstone patients. *Am. J. Dig. Dis. 21*:521 (1976).

213. J. M. Watts, P. Jablonski, and J. Toorili, The effect of added bran to the diet on the saturation of bile in people without gallstones. *Am. J. Surg. 135*:321 (1978).

214. R. M. McDougall, L. Yakymyshyn, K. Walker, and O. G. Thurston, The effect of wheat bran on serum lipoproteins and biliary lipids. *Can. J. Surg. 21*:422 (1978).

215. S. Tarpila, T. A. Miettinen, and L. Metsaranta, Effects of bran on serum cholesterol, faecal mass, fat, bile acids, and neutral steroids and biliary lipids in patients with diverticular disease of the colon. *Gut 19*:137 (1978).

216. T. Osuga, O. W. Portman, N. Tanaka, M. Alexander, and A. J. Ochsner, The effect of diet on hepatic bile acid metabolism in squirrel monkeys with and without cholesterol gallstones. *J. Lab. Clin. Med. 88*:649 (1976).

217. W. A. Souter, Bolus obstruction of gut after use of hydro-phyllic colloid laxatives. *Br. Med. J. 1*:166 (1965).

218. J. C. McLoughlin, A. H. G. Love, A. A. J. Adgey, A. D. Gough, and M. P. S. Varma, Intact removal of phytobezoir using fiberoptic endoscope in patients with gastric atony. *Br. Med. J. 2*:1466 (1979).

219. N. S. Painter, Below the belt. *Lancet 2*:381 (1971).

220. D. P. Burkitt, C. L. Nelson, and E. H. Williams, Some geo-graphical variations in disease pattern in East and Central Africa. *East Afr. Med. J. 40*:1 (1963).

221. A. Ghavarni and F. Saidi, Patterns of colonic disorders in Iran. *Dis. Colon Rectum 12*:462 (1969).

222. F. Saidi, High incidence of intestinal volvulus in Iran. *Gut 10*: 838 (1969).

223. N. T. Davies, Effect of phytic acid on mineral availability, in *Dietary Fiber in Health and Disease* (G. V. Vahouny and D. Kritchevsky, eds.). Plenum Press, New York, 1982, p. 105.

224. R. A. McCance and E. M. Widdowson, Mineral metabolism of healthy adults on white and brown bread dietaries. *J. Physiol. 101*:44 (1942).

225. R. A. McCance and C. M. Walsham, The diegestibility and absorption of the calories, protein, fat and calcium on whole-meal wheaten bread. *Br. J. Nutr. 2*:26 (1948).

226. R. A. McCance and E. M. Glaser, The energy value of oatmeal and the digestibility and absorption of its protein, fat and calcium. *Br. J. Nutr. 2*:221 (1948).

227. J. G. Reinhold, K. Nasr, A. Lehimgarzedeh, and H. Hedayati, Effect of purified phytate and phytate-rich bread upon metabo-lism of zinc, calcium, phosphorus, and nitrogen in man. *Lancet 1*:283 (1973).

228. J. G. Reinhold, A. Parsa, N. Karimian, J. W. Hammide, and F. Ismail-Beigi, Availability of zinc in leavened and unleavened wholemeal wheaten breads as measured by solubility and uptake by rat intestine in vitro. *J. Nutr. 104*:976 (1974).

229. J. C. Reinhold, B. Faraji, P. Abadi, and F. Ismail-Beigi, De-creased absorption of calcium, magnesium, zinc and phosphorus by humans due to increased fiber and phosphorus consumption as wheat bread. *J. Nutr. 106*:493 (1976).

230. J. G. Reinhold, R. Ismail-Beigi, and B. Faradji, Fiber vs. phytate as determinant of the availability of calcium, zinc and iron in breadstuffs. *Nutr. Rep. Int. 12*:75 (1975).

231. J. G. Reinhold, Rickets in Asian immigrants. *Lancet 2*:1132 (1976).

232. F. Ismail-Beigi, J. G. Reinhold, B. Faradji, and P. Abadi, The effects of cellulose added to diets of low and high fibre content upon the metabolism of calcium, magnesium, zinc and phosphorus by man. *J. Nutr. 107*:510 (1979).

233. L. M. Drews, C. Kies, and H. M. Fox, Effect of dietary fiber on copper, zinc and magnesium utilization by adolescent boys. *Am. J. Clin. Nut. 32*:1893 (1979).

234. W. P. T. James, Dietary fiber and mineral absorption, in *Medical Aspects of Dietary Fibers* (G. A. Spiller and R. M. Kay, eds.). Plenum Press, New York, 1980, p. 239.

235. J. L. Kelsay, Effects of fiber on mineral and vitamin bioavailability, in *Dietary Fiber in Health and Disease* (G. V. Vahouny and D. Kritchevsky, eds.). Plenum Press, New York, 1982, p. 91.

236. H. H. Sanstead, J. M. Munoz, R. A. Jacob, L. M. Klevay, S. J. Reck, G. M. Logan, Jr., F. R. Dintzis, G. E. Inglett, and W. C. Shuey, Influence of dietary fiber on trace element balance. *Am. J. Clin. Nutr. 31*:S180 (1978).

237. H. H. Sanstead, L. M. Klevay, R. A. Jacob, J. M. Munoz, G. M. Logan, Jr., and S. Reck, Effects of dietary fiber and protein level on mineral element metabolism, in *Dietary Fibers: Chemistry and Nutrition* (G. E. Inglett and S. I. Falkehag, eds.). Academic Press, New York, 1979, p. 147.

238. *Carbohydrates in Human Nutrition*, A joint FAO/WHO report. Food and Nutrition paper 15. FAO, Rome 1980, p. 55.

239. H. Kasper, Influence of dietary fiber substances on vitamin absorption, in *Fiber in Human and Animal Nutrition* (G. Wallace and L. Bell, eds.). Royal Society New Zealand, Wellington, 1983, p. 195.

240. W. I. M. Holman, Medical Research Council Special Report Series, No. 287, HMSO, London, 1954, p. 92.

241. K. J. Carpenter, E. Kodicek, and P. W. Wilson, The availability of bound nicotinic acid to the rat: The effect of boiling maize in water. *Br. J. Nutr. 14*:25 (1960).

242. J. F. Gregory, Bioavailability of vitamin B_6 in nonfat dry milk and a fortified breakfast cereal fiber. *J. Food Sci. 45*:84 (1980).

243. R. W. Cullen and S. M. Oace, Methylmalonic acid and vitamin B_{12} excretion of rats consuming diet varying in cellulose and pectin. *J. Nutr. 108*:640 (1978).

244. S. Babu and S. G. Srikantia, Availability of folates from some foods. *Am. J. Clin. Nutr.* 29:376 (1976).

245. D. A. T. Southgate and J. V. G. A. Durnin, Calorie conversion factors: An experimental reassessment of the factors used in the calculation of the energy value of human diets. *Br. J. Nutr.* 24:517 (1970).

246. T. F. Macrae, J. C. D. Hutchinson, J. O. Irwin, J. S. D. Bacon, and E. T. McDougall, Comparative digestibility of whole meal and white breads and the effect of the degree of fineness of grinding on the former. *J. Hyg.* 42:423 (1942).

247. D. J. Farrell, L. Girle, and J. Arthur, Effect of dietary fibre on the apparent digestibility of major components and on blood lipids in man. *Aust. J. Exp. Biol. Med. Sci.* 56:469 (1978).

248. J. L. Kelsay, K. M. Behall, and E. S. Prather, Effect of fiber from fruits and vegetables on metabolic responses of human subjects. I. Bowel transit time, number of defecations, fecal weight, urinary excretions of energy and nitrogen and apparent digestibilities of energy, nitrogen and fat. *Am. J. Clin. Nutr.* 31:1149 (1978).

249. D. J. A. Jenkins, A. R. Leeds, M. A. Gassull, H. Houston, D. V. Goff, and M. J. Hill, The cholesterol lowering properties of guar and pectin. *Clin. Sci. Mol. Med.* 51:8P (1976).

250. D. A. T. Southgate, W. J. Branch, M. J. Hill, B. S. Draser, R. L. Walters, P. S. Davies, and I. McLean Baird, Metabolic responses to dietary supplements of bran. *Metabolism* 25:1129 (1976).

251. T. Mistunaga, Some properties of protease inhibitors in wheat grain. *J. Nutr. Sci. Vitaminol.* 20(2):153 (1974).

252. T. G. Volkheimer and F. H. Schulz, The phenomenon of persorption. *Digestion* 1:213 (1968).

253. G. Volkheimer, F. H. Schulz, I. Aurich, S. Strauch, K. Beuthin, and H. Wendlandt, Persorption of particles. *Digestion* 1:78 (1968).

254. G. Volkheimer, F. H. Schulz, H. Lehmann, I. Aurich, R. Hubner, M. Hubner, A. Hallmaiyer, F. Munch, H. Opperman, and S. Strauch, Primary portal transport of persorbed starch granules from the intestinal wall. *Med. Exp.* 18:103 (1968).

255. G. Pahlke and R. Friedrich, Persorption von milcrokristalliner Cellulose. *Naturwissenschaften* 61:35 (1974).

256. G. Schreiber, Ingested dyed cellulose in the blood and urine of man. *Arch. Environ. Health* 29:39 (1974).

257. P. Grasso, Persorption—a new long-term problem. *Br. Ind. Biol. Res. Assoc. Bull.* 15(3):22 (1976).

258. M. R. Malinow, E. J. Bardana, Jr., and B. Pirofsky, Systemic lupus erythematosus-like syndrome induced by diet in monkeys. *Clin. Res.* 29:626A (1981).

259. M. R. Malinow, E. J. Bardana, Jr., and S. H. Goodnight, Jr., Pancytopenia during ingestion of alfalfa seeds. *Lancet 1*:615 (1981).
260. M. R. Malinow, E. J. Bardana, Jr., B. Pirofsky, S. Craig, and P. McLaughlin, Systemic lupus erythematosus-like syndrome in monkeys fed alfalfa sprouts: Role of a nonprotein amino acid. *Science 216*:415 (1982).
261. Research Resources Reporter, National Institutes of Health *VI No. 9*:12 (1982).
262. L. Sequeria, Lectins and their role in host-pathogen specificity. *Annu. Rev. Phytophatol. 16*:453 (1978).
263. T. J. Williams, N. R. Pleases, I. J. Goldstein, and J. Lonngran, A new class of model glycolipids: Synthesis, characterization and interaction with lectins. *Arch. Biochem. Biophys. 195*:145 (1979).
264. M. E. Etzler, Lectins as probes in studies of intestinal glycoproteins and glycolipids. *Am. J. Clin. Nutr. 32*:133 (1979).
265. D. L. J. Freed and F. Y. H. Green, Do dietary lectins protect against colon cancer? *Lancet 2*:1261 (1975).
266. D. P. Burkitt, Epidemiology of cancer of the colon and rectum. *Cancer 28*:3 (1971).
267. R. Doll, The geographic distribution of cancer. *Proc. R. Soc. Med. 65*:49 (1972).
268. W. Haenszel, J. W. Berg, M. Segi, M. Kurihara, and F. B. Locke, Large bowel cancer in Hawaiian Japanese. *J. Nat. Cancer Inst. 51*:1765 (1973).
269. E. L. Wynder, The epidemiology of large bowel cancer. *Cancer Res. 35*:3388 (1975).
270. B. S. Reddy and E. L. Wynder, Fecal constituents of populations with diverse incidence rates of colon cancer. *J. Nat. Cancer Inst. 50*:1437 (1973).
271. M. J. Hill, Metabolic epidemiology of dietary factors in large bowel cancer. *Cancer Res. 35*:3398 (1975).
272. *Diet, Nutrition and Cancer.* National Academy Press, 1982.
273. J. Cummings, Dietary fiber and large bowel cancer. *Proc. Nutr. Soc. 40*:7 (1981).
274. B. S. Reddy, dietary fiber and colon carcinogenesis. A critical review, in *Dietary Fiber in Health and Disease* (G. V. Vahouny and D. Kritchevsky, eds.). Plenum Press, New York, 1982, p. 265.
275. E. L. Wynder and T. Shigematsu, Environmental factors and cancer of the colon and rectum. *Cancer 20*:1520 (1967).
276. B. S. Draser and D. Irving, Environmental factors and cancer of the colon and breast. *Br. J. Cancer 27*:167 (1973).
277. O. Gregor, R. Toman, and F. Prusova, Gastrointestinal cancer and nutrition. *Gut 10*:1031 (1969).

278. B. Armstrong and R. Doll, Environmental factors and cancer incidence and mortality in different countries with special reference to dietary patterns. *Int. J. Cancer 15*:616 (1975).
279. T. L. Cleave, *The Saccharine Diseases.* Wright, Bristol, 1974.
280. A. R. P. Walker, B. F. Walker and B. D. Richardson, Bowel motility and colonic cancer. *Br. Med. J. 3*:238 (1969).
281. D. P. Burkitt, Large bowel carcinogenesis. An epidemiologic jigsaw puzzle. *J. Nat. Cancer Inst. 54*:3 (1975).
282. M. J. Hill, B. S. Draser, V. Aries, J. S. Crowther, G. Hawksworth, and R. E. O Williams, Bacteria and aetiology of cancer of the large bowel. *Lancet 1*:95 (1971).
283. M. J. Hill and V. C. Aries, Fecal steroid composition and its relation to cancer of the large bowel. *J. Pathol. 104*:129 (1971).
284. M. J. Hill, The effect of some factors on the fecal concentration of acid steroids, neutral steroids and urobilins. *J. Pathol. 104*: 239 (1971).
285. E. L. Wynder and B. S. Reddy, Metabolic epidemiology of colorectal cancer. *Cancer 34*:801 (1974).
286. M. J. Hill, B. S. Draser, R. E. O. Williams, T. W. Meade, A. G. Cox, J. E. P. Simpson, and B. C. Morson, Faecal bile acids and clostridia in patients wtih cancer of the large bowel. *Lancet 1*:535 (1975).
287. M. J. Hill, Bile acids and human colorectal cancer, in *Dietary Fiber in Health and Disease* (G. V. Vahouny and D. Kritchevsky, eds.). Plenum Press, New York, 1982, p. 299.
288. H. J. Freeman, Experimental animal studies in colonic carcino-genesis and dietary fiber, in *Medical Aspects of Dietary Fibers* (G. A. Spiller and R. M. Kay, eds.). Plenum Press, New York, 1980, p. 83.
289. T. Narisawa, N. E. Magadia, J. H. Weisburger, and E. L. Wynder, Promoting effect of bile acids on colon carcinogenesis after intrarectal instillation of N-methyl-N'-nitro-N-nitro-soguanidine in rats. *J. Nat. Cancer Inst. 53*:1093 (1974).
290. B. S. Reddy, T. Narisawa, J. H. Weisburger, and E. L. Wynder, Promoting effect of sodium deoxycholate on colon adeno-carcinomas in germfree rats. *J. Nat. Cancer Inst. 56*:441 (1976).
291. C. Chomchai, N. Bhadrachari, and N. D. Nigro, The effect of bile on the induction of experimental intestinal tumors in rats. *Dis. Colon Rectum 17*:310 (1974).
292. R. L. Campbell, D. V. Singh, and N. D. Nigro, Importance of the fecal stream in the induction of colon tumors by azoxy-methane in rats. *Cancer Res. 35*:1369 (1975).
293. D. Kritchevsky, in *The Bile Acids* (P. P. Nair and D. Kritchevsky, eds.). Plenum Press, New York, 1973, p. 273.

294. N. D. Nigro, N. Bhadrachari, and C. Chomchai, A rat model for studying colonic cancer: Effect of cholestyramine on induced tumors. *Dis. Colon Rectum 16*:438 (1973).

295. T. Asano, M. Pollard, and D. C. Madsen, Effects of cholestyramine on 1,2-dimethylhydrazine-induced enteric carcinoma in germfree rats. *Proc. Soc. Exp. Biol. Med. 150*:780 (1975).

296. B. S. Reddy, J. H. Weisburger, and E. L. Wynder, Effect of dietary fat level and dimethylhydrazine on fecal acid and neutral sterol excretion and colon carcinogenesis in rats. *J. Nat. Cancer Inst. 52*:507 (1974).

297. N. D. Nigro, D. V. Singh, R. L. Campbell, and M. S. Pak, Affect of dietary beef fat on intestinal tumor formation by azoxymethane in rats. *J. Nat. Cancer Inst. 54*:439 (1975).

298. A. W. Bull, B. K. Soullier, P. S. Wilson, M. T. Hayden, and N. D. Nigro, The promotion of azoxymethane-induced intestinal cancer by high-fat diet in rats. *Cancer Res. 39*:4956 (1979).

299. P. M. Newberne, Dietary fat, immunological response, and cancer in rats. *Cancer Res. 41*:3783 (1981).

300. R. B. Wilson, D. P. Hutcheson, and L. Wideman, Dimethylhydrazine-induced colon tumors in rats fed diets containing beef fat or corn oil with and without wheat bran. *Am. J. Clin. Nutr. 30*:176 (1977).

301. H. J. Freeman, Studies on the effects of single fiber sources in the dimethylhydrazine-rodent model of human bowel neoplasia, in *Dietary Fiber in Health and Disease* (G. V. Vahouny and D. Kritchevsky, eds.). Plenum Press, New York, 1982, p. 287.

302. J. A. Barbolt and R. Abraham, The effect of bran on dimethylhydrazine-induced colon carcinogenesis in the rat. *Proc. Soc. Exp. Biol. Med. 157*:656 (1978).

303. W. F. Chen, A. S. Patchefsky, and H. S. Goldsmith, Colonic protection from dimethylhydrazine by a high fiber diet. *Surg. Gynecol. Obstet. 147*:503 (1978).

304. D. Fleiszer, D. Murray, H. MacFarlane, and R. Brown, A protective effect of dietary fiber against chemically induced bowel tumors in rats. *Lancet 2*:522 (1978).

305. K. Watanabe, B. S. Reddy, J. H. Weisburger, and D. Kritchevsky, Effect of dietary alfalfa, pectin and wheat bran on azoxymethane- or methylnitrosourea-induced colon carcinogenesis in F344 rats. *Nat. Cancer Inst. 63*:141 (1979).

306. J. P. Cruse, M. R. Lewin, and C. G. Clark, Failure of bran to protect against experimental colon cancer in rats. *Lancet 2*: 1278 (1978).

307. H. G. Bauer, N-G. Asp, R. Oste, A. Dahlqvist, and P. E. Fredlund, Effect of dietary fiber on the induction of colorectal tumors and fecal β-glucuronidase activity in the rat. *Cancer Res. 39*:3752 (1979).

308. J. M. Ward, R. S. Yamamoto, and J. H. Weisburger, Cellulose dietary bulk and azoxymethane-induced intestinal cancer. *J. Nat. Cancer Inst. 51*:713 (1973).

309. H. J. Freeman, G. A. Spiller, and Y. S. Kim, A double-blind study on the effect of purified cellular dietary fiber on 1,2-dimethylhydrazine-induced rat colonic neoplasia. *Cancer Res. 38*:2912 (1978).

310. N. D. Nigro, A. W. Bull, B. A. Klopfer, M. S. Pak, and R. L. Campbell, Effect of dietary fiber on azoxymethane-induced intestinal carcinogenesis in rats. *J. Nat. Cancer Inst. 62*:1097 (1979).

311. W. M. Castledon, Prolonged survival and decrease in intestinal tumors in dimethylhydrazine-treated rats fed a chemically defined diet. *Br. J. Cancer 35*:491 (1977).

312. H. G. Bauer, N-G. Asp. A. Dahlqvist, P. E. Fredlund, M. Nymen, and R. Oste, Effect of two kinds of pectin and guar gum on 1,2-dimethylhydrazine initiation of colon tumors and on fecal β-glucuronidase activity in the rat. *Cancer Res. 41*:2518 (1981).

313. N. D. Nigro, Animal studies implicating fat and fecal steroids. in intestinal cancer. *Cancer Res. 41*:3769 (1981).

314. H. P. Glauert, M. R. Bennick, and C. H. Sander, Enhancement of 1,2-dimethylhydrazine-induced colon carcinogenesis in mice by dietary agar. *Food Cosmet. Toxicol. 19*:281 (1981).

315. M. M. Cassidy, F. G. Lightfoot, L. Grau, J. Story, D. Kritchevsky, and G. V. Vahouny, Effect of chronic intake of dietary fibers on the ultrastructural topography of the rat jejunum and colon. *Am. J. Clin. Nutr. 34*:218 (1981).

316. M. M. Cassidy, L. Grau, D. Kritchevsky, F. Lightfoot, J. Story, and G. V. Vahouny, Effect of bile-salt binding resins on the morphology of the rat jejunum and colon. *Dig. Dis. Sci. 25*:504 (1980).

317. C. C. Ray, G. Laurendeau, G. Doylon, L. Chartrand, and M. R. Rivest, The effect of bile and of sodium taurocholate on the epithelial cell dynamics of the rat small intestine. *Proc. Soc. Exp. Biol. Med. 149*:1000 (1975).

318. P. J. M. Tutton, Control of epithelial cell proliferation in the small intestinal crypt. *Cell Tissue Kinet. 6*:211 (1973).

319. P. Sassier and M. Bergeron, Existence of an endogenous inhibitor of DNA synthesis in rabbit small intestine specifically effective on cell proliferation in adult mouse intestine. *Cell Tissue Kinet. 13*:251 (1980).

320. R. J. May, A. Quaroni, K. Kirsch, and K. J. Isselbacher, A villous cell derived inhibitor of intestinal cell proliferation. *Am. J. Physiol. 241*:G520 (1981).

321. L. R. Jacobs and F. A. White, Modulation of mucosal cell proliferation in the intestine of rats fed a wheat bran diet. *Am. J.
Clin. Nutr. 37*:945 (1983).

322. L. Diamond, T. G. O'Brian, and W. M. Baird, Tumor promoters
and the mechanism of tumor promotion. *Adv. Cancer Res. 32*:
1 (1980).

323. H. P. Glauert and M. R. Bennick, Influence of diet or intrarectal bile acid injections on colon epithelial cell proliferation
in rats previously injected with 1,2-dimethylhydrazine. *J. Nutr.
113*:475 (1983).

7

Pathology of Choline Deficiency

Hisashi Shinozuka and Sikandar L. Katyal

University of Pittsburgh School of Medicine
Pittsburgh, Pennsylvania

I. INTRODUCTION

Some 50 years ago Best et al. (1) reported that fatty degeneration of
the liver in diabetic dogs kept on a lean meat and sugar diet could be
alleviated by the administration of lecithin. The observation led these
investigators to identify the lipotropic action of choline, a component
of lecithin (2). The term lipotropic was applied to dietary substances
that either prevent the deposition of abnormal amounts of lipids in the
liver or accelerate their removal into the circulation (3,4). Within a
relatively short time thereafter, the importance of choline, in addition
to its lipotropic action, was recognized in the general nutritional requirement for experimental animals. Thus choline was essential for
weanling rats, not only for the prevention of fatty liver but also for
the maintenance of tissue structure and even for their growth and
survival (5,6). The absence of choline in the diet induces characteristic lesions not only in rats, but also in species as varied as mice,
monkeys, ducklings, and mosquitoes (7). During the subsequent
years, it became evident that choline serves as a donor of a labile
methyl group in the biological system (8,9). Further biochemical
studies of methyl metabolism and methyl requirements in a wide variety of biological functions have made it clear that natural sources of
labile methyl are largely limited to choline, betaine, and methionine
and that de novo synthesis of methyl groups is dependent upon the

dietary provision of adequate supplies of water-soluble vitamins, especially folic acid and vitamin B_{12} (7,10).

Considering the number of important functions that choline performs, it can be readily appreciated that its deficiency will lead to a variety of pathological manifestations. Indeed, over the past 50 years, the subject of choline deficiency has been investigated extensively, many pathological conditions have been described, and their pathogenetic mechanisms have been explored with varying degrees of intensity. Signs of deficiency would be expected whenever the demands for choline or other methyl-containing metabolites exceed the dietary supply of labile methyl and the capacity of the organism to carry out adequate de novo synthesis. The complexities of the interrelationship between a deficiency of choline per se and of methionine, vitamin B_{12}, or folic acid are considerable. In this chapter, we will review the basic pathology of choline deficiency and how the deficiency state alters responses of organisms to selected exogenous stimuli. No attempt will be made to provide a comprehensive and exhaustive review of all aspects of choline deficiency. Rather, efforts will be directed to highlighting pathogenetic mechanisms of selected pathological conditions. The subjects of choline and choline deficiency have been reviewed periodically in the past, and excellent reviews are available in the literature (7,11-15).

II. BIOCHEMISTRY OF CHOLINE

Choline, (beta-hydroxyethyl)trimethylammonium, is a quaternary ammonium compound and has the structural formula: $HOCH_2CH_2N^+(CH_3)_3$. It is widely distributed in living organisms as free choline; as acetylcholine, an important neurotransmitter; as phospholipids (phosphatidylcholine and sphingomyelin); and as the intermediates involved in the biosynthesis and catabolism of phospholipids. Choline functions as a methyl group donor owing to the presence of biologically labile methyl groups in its structure. As moiety of phosphatidylcholine, choline is required for the synthesis of serum lipoproteins and cellular membranes.

Biosynthesis of phosphatidylcholine occurs through two distinct pathways. In the first pathway (the Kennedy pathway) (16), preformed choline is converted to phosphatidylcholine (Figure 1). The first reaction of this pathway is carried out by choline kinase, a cytosolic enzyme present in several tissues. The phosphorylcholine thus formed is converted to CDP-choline by the action of cytidyl transferase, a cytosolic as well as microsomal enzyme that appears to be the rate-limiting enzyme in the pathway (17). The CDP-choline is subsequently

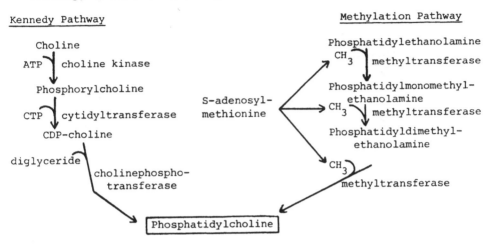

Figure 1 Pathways for the biosynthesis of phosphatidylcholine. De novo synthesis of choline moiety of phosphatidylcholine occurs via the methylation pathway.

combined with diglyceride by a microsomal enzyme, choline phosphotransferase, to form phosphatidylcholine. Phosphatidylcholine is synthesized in tissues preponderantly by the Kennedy pathway. Hepatic choline kinase activity is substantially increased by feeding polycyclic aromatic hydrocarbons, inducers of hepatic drug-metabolizing enzymes (18).

In addition to the dietary supply, choline is also synthesized endogenously as phosphatidylcholine, chiefly in the liver and brain, by sequential methylation of phosphatidylethanolamine (the methylation pathway) (19-21). The methyl donor in each of the three methylation steps is S-adenosylmethionine (Figure 1). Two enzymes appear to be involved in the methylation pathway: one that transfers the methyl group from S-adenosylmethionine to phosphatidylethanolamine, and the second that transfers the methyl groups to phosphatidylmonomethylethanolamine and phosphatidyldimethylethanolamine, eventually synthesizing phosphatidylcholine. The activity of the first enzyme appears to be rate limiting (22). Activity of the methylation pathway is substantially increased in the livers of choline-deficient rats (23). A twofold increase in the activity of phosphatidylethanolamine methyltransferase occurs as a result of choline deficiency (24).

Free choline found in tissues, either that derived from the diet or that released from phosphatidylcholine by the action of phospholipases and glycerophosphorylcholine diesterase, is converted to acetylcholine by the enzyme choline acetyltransferase, or it is phosphorylated or

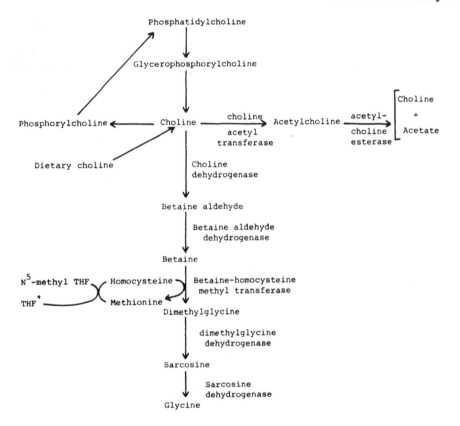

Figure 2 Pathways for the utilization and catabolism of choline. *THF, tetrahydrofolate.

oxidized (Figure 2). Oxidation of choline is several times more rapid than its phosphorylation (25). During oxidation of choline, its methyl groups are removed and reutilized. Betaine, an intermediate in the catabolic pathway of choline, is able to transfer its methyl group to homocysteine with the formation of methionine and dimethylglycine. Further oxidation of dimethylglycine to sarcosine, and then to glycine, results in the removal of the two remaining methyl groups, which enter the 1-carbon pool and are reutilized in the transmethylation reactions as folic acid derivatives (26).

III. PATHOLOGICAL LESIONS INDUCED BY CHOLINE DEFICIENCY

Choline exerts extremely important functions in maintaining the structural as well as the functional integrity of cells, organs, and bodies.
It is evident that dietary deficiency of choline will affect almost every
organ in the body, particularly when animals are in the state of high
metabolic activity and growth. In the absence of choline from the
diet, many species of animals develop a variety of pathological conditions. The rat, dog, mouse, rabbit, hamster, calf, pig, chicken,
duckling, and even the monkey exhibit fatty livers when the diet lacks
sufficient lipotropic agents, choline or its precursors (7,27). Despite
the unlikeliness that pathological conditions related to the dietary deficiency of choline per se exist in man, inadequacy of lipotrope uptake
may occur in conjunction with other types of nutritional deficiency,
and lipotrope deficiency may play a role in the development of certain
types of fatty liver (12,28,29). Signs and pathological lesions induced
by choline deficiency in one particular species do not necessarily du-
plicate those seen in other species. For instance, in guinea pigs it is
rather difficult to produce fatty liver by feeding diets low in choline
(30,31). A crippling skeletal deformity known as perosis is a characteristic lesion in poultry caused by choline deficiency, but it is not found
in other species (32,33). Duodenal ulcers as a result of long-term
feeding of a choline-deficient diet are recorded only in dogs (34,35).
The animal's age and sex also influence the severity of the pathological lesions; young animals are more susceptible than old ones. Females
are less susceptible to adverse effects of choline deficiency than males.
     An important consideration in assessing the significance of various
pathological lesions is the nature of the choline-deficient diets. The
composition of the deficient diets varied considerably from one experimental study to another. Differences in the quality and quantity of
protein and/or of fat contained in the deficient diets have marked effects on creating the conditions of choline deficiency (27,36). When
caloric intake is low, even severely hypolipotropic diets do not produce
signs of choline deficiency, such as fatty liver. The level of methionine in the dietary protein mixtures is one of the critical factors in
creating a lipotrope-deficient state. It is recognized that complete removal of methionine from a diet leads to loss of appetite, and thus a
diet low in methionine seems more suitable for lipotropic studies than
one completely free of it (27). Lipids and lipotropes in the diet are
closely interrelated, since lipotropes are required for the normal metabolism of lipids, and the dietary lipid contents partially determine
the lipotrope requirements. As will be discussed later, the types of
fat in the deficient diet will modify the severity of fatty liver.

Furthermore, in the earlier studies of choline deficiency, many diets used to induce choline deficiency in animals were also deficient in folic acid and vitamin B_{12}. Since these vitamins are intimately involved in the metabolism of lipotropic compounds, their omission certainly modifies the nature and severity of the pathological lesions induced by choline deficiency. Difficulties in the preparation of choline-deficient diets that sustain good animal growth and create proper signs of lipotrope deficiency have been discussed by several investigators (27,36).

Pathology related to choline deficiency can be divided into two broad categories: the lesions resulting from choline deficiency per se and the lesions related to altered reactivity of the organs to exogenous stimuli because of the deficiency. The pathological lesions that belong in the first category are numerous, and those other than in the liver are listed in Table 1. Many lesions related to choline deficiency were described in the period between the late 1930s and the early 1950s, the era before the introduction of modern techniques of cell biology and the concept of biochemical pathology. As a consequence, most of the pathological lesions listed were limited to the gross and light microscopic descriptions of the lesions without clear insight into their pathogenetic mechanisms. Notable exceptions to this are the lesions in the livers and kidneys. The liver is the most extensively studied organ affected by a dietary deficiency of choline, and the pathogenesis of choline-deficiency fatty liver has been clarified in considerable depth. Table 2 lists the functional and morphological pathology of the liver induced by choline deficiency.

The pathology of the lesions that belong to the second category as just described, appears to be extremely important in the appreciation of the pathological consequences of choline deficiency. However this field has been explored in depth only in three areas: modified liver carcinogenesis, acute hemorrhagic pancreatitis of mice fed a choline-deficient diet containing ethionine, and altered immunological states. The subject of choline deficiency and liver carcinogenesis will be discussed later, and acute pancreatitis of mice due to choline deficiency and ethionine is discussed in Chapter 3.

Some 40 years ago thymic atrophy was described as one of the pathological lesions resulting from choline deficiency (37). However its implication for the functional alterations of the host had not been experimentally tested until the 1970s when Newberne and his associates investigated the possible effects of lipotrope deficiency on the animal's immune systems (38-41). Their studies showed that maternal nutritional deficiency of lipotropes adversely affected T-cell functions of rat offspring, and predisposed them to infections much later in life (38,39). A number of reports indicate that folate or methionine deficiency influences the immunocompetence of animals (42). It would be extremely interesting to determine how choline deficiency modifies the immunoregulatory systems of animals, and how altered immunological competence of the host creates secondary pathological lesions.

Table 1 Pathology of Choline Deficiency in Organs Other Than the Liver

Organ	Pathological lesions	References
Cardiovascular	Atheromas, medial calcification (aorta, coronary arteries)	43,44,45,46
	Myocarditis	46,47,48
	Hemangioendothelioma (mesenteric and subcutaneous fat)	49
Nervous system	Cerebellar hemorrhage	50
	Decrease in neurotransmitter	See Section IV, (4)
Eyes	Intraocular hemorrhage	51,52,53
Lungs	Squamous metaplasia of bronchi	49
	Bronchogenic carcinomas	49
Intestine	Inhibition of chylomicron secretion	54,55,56
	Duodenal ulcers	34,35
Kidneys	Hemorrhagic kidney	See Section IV, (2)
Pancreas	Centroacinar cell hyperplasia	49
	Adenocarcinomas	49
	Acute hemorrhagic pancreatitis	57,58
Adrenals	Cortical hyperplasic	59
Thymus	Reversible involution	37
Skeletal	Perosis (slipped tendon disease)	32,33

Table 2 Pathology of Choline Deficiency in the Liver[a]

Functional changes	Morphological findings
Disturbance in lipoprotein secretion	Fatty liver
Bile flow and bile composition	Cirrhosis
Alterations in cell membrane	Hepatomas

[a] Details in Section IV.A.

IV. PATHOGENETIC MECHANISMS

A. Fatty Liver

Fatty liver may be defined as an accumulation of lipids, consisting principally of triglycerides, which exceed 5% of the liver weight. Fatty liver is a common and nonspecific response of the liver to different forms of acquired injury or inherited metabolic derangements. Various degrees of fatty infiltration of the liver are found in association with many human disease states (60,61). In man, some cases of fatty liver may be of little clinical consequence, while in others (i.e., fatty liver of pregnancy and of Rye's syndrome, etc.), fatty liver may lead to hepatic failure and death. In experimental animals, fatty liver can be induced by many means and by a large number of agents. Therefore, studies concerned with the pathogenesis of fatty liver have been pursued by many investigators of different disciplines.

The sequence of pathological events, which begins almost as soon as animals are placed on the low-choline dietary regimen, has now been studied in a large number of animal species. Most of the early studies, which resulted in the discovery by Best and his associates (1,2,62) of the lipotropic action of choline, were carried out on dogs. However, since the end of World War II, many studies have been carried out using the rat. It is now well established that species of animals other than the dog and rat are similarly affected (7). Within 24 hr after rats are placed on a diet low in choline and its precursors, droplets of stainable lipids appear in the hepatic cells around hepatic veins. Large intracytoplasmic masses of fat are formed from smaller droplets during the first week and the hepatic parenchymal cells are expanded to twice their original size. The fat content of the liver in this stage may be increased 10 times, but it can rapidly disappear if choline is administered to the rat. If rats are continued on the choline-deficient diet for 1-2 months, distended hepatic cell walls rupture and large fatty masses, termed lipodiastemata, are formed by fusion of several cells (63,64). Lipodiastemata and fatty liver cells compress the sinusoids in the center of the lobules where the intravascular pressure is lowest, resulting in interference with the oxygen supply. The partial anoxia is believed to be an important factor in the ensuing atrophy and regression of liver cells, followed by disappearance of the fat, and by the growth of fibrous tissue leading to cirrhosis of the liver (64).

The quantity and quality of the fat in a choline-deficient diet influence the severity of the fatty liver. However, the quantity of fat in the diet is not a determining factor because accumulation of liver lipids occurs when animals are fed low-choline diets that are low in fat (65, 66). Concerning the quality of fat in a choline-deficient diet, the proportion of saturated fatty acids in the diet that contain 14 to 18 carbons appears to increase the deposition of fat more than the unsaturated fatty acids, including a solid unsaturated acid, elaidic acid, which is solid at physiological temperature (67). The severest degree of fatty liver was observed in rats fed butter fat, while the mildest degree occurred in those fed olive oil, and the severity of fatty liver for those fed coconut oil was intermediate. Benton et al. found that more choline was required to prevent fatty livers from saturated than from unsaturated dietary fats (68,69). Stetten and colleagues, on the basis of their experiments of feeding ethyl esters of fatty acids in low-choline diets, demonstrated that the degree of fat accumulation increased as the chain length decreased from 18 to 14 (70,71). The severity of the fatty livers was low if the dietary fatty acid contained less than 12 carbons. Carroll and Williams (72) also noted less severe fatty infiltration in livers of rats fed low-choline diets containing safflower oil as compared with that of rats fed low-choline diets containing blend or tallow. They suggested a possible protective effect of linoleic acid, the predominant fatty acid in safflower oil, on fatty liver. Perhaps the presence of substantial amounts of a more saturated fatty acid in the tallow and blend enhanced lipid accumulation in the liver of choline-deficient rats.

The rapidity and extent of the increase in liver lipids with choline deficiency, as well as their removal after the administration of choline, suggest that the effectiveness of choline as a lipotropic agent depends upon some function of choline-containing phospholipids in fat metabolism. The fact that in almost all fatty livers, despite the multiplicity and heterogeneity of etiologic factors, the lipid that accumulates in the liver is predominantly triglyceride suggests that common mechanisms exist in the pathogenesis of fatty livers. It has been postulated that the underlying mechanism of any fatty liver is related to one of a few basic mechanisms, and that different etiological factors of fatty liver may share common mechanisms in triggering the pathological conditions (73). Studies dealing with the pathogenesis of choline-deficiency fatty liver have contributed greatly to creating a unified concept of the mechanism of fatty liver. Fatty liver is the result of an imbalance between the rate of synthesis and the rate of utilization of hepatic triglyceride. Three basic conditions have been considered; (a) the rate of synthesis of hepatic triglycerides is normal, but there is a block in their utilization; (b) the rate of synthesis of triglycerides is increased without impairment of their utilization; and (c) there is both an increase in the rate of triglyceride synthesis and a block in their utilization.

To elucidate the mechanisms by which choline-deficient diets interfere with one of these processes, it is necessary to evaluate the effects of the deficient diets on the phospholipid metabolism of hepatocytes. Indeed, the most extensive work regarding the effects of choline deficiency on phospholipid metabolism pertains to the liver, especially in relation to explaining the mechanisms of development of the fatty liver. Early effects of choline-deficient diets on the liver are: (a) a decrease in the content of phosphatidylcholine in the whole liver, microsomes, and mitochondria (23,74,75); (b) a shift toward an increase in the content of polyunsaturated species of phosphatidylcholine (23); (c) an increase in the synthesis of phosphatidylcholine by the methylation pathway and a decrease via the Kennedy pathway (23,76); (d) a decrease in the activity of choline phosphotransferase (24) and an increase in the activity of ethanolamine methyltransferase (77); and (e) a decrease in the hepatic phosphatidylcholine/phosphatidylethanolamine ratio (74,78). No doubt, these metabolic alterations of phospholipids lead to alterations in the structure and function of cell membranes and serum lipoproteins. It has been shown that the membranes of microsomes isolated from the livers of choline-deficient rats contain not only less lecithin per unit of membrane protein (74) but also a different proportion of lecithin species (23). The membranes of smooth endoplasmic reticulum showed a greater depletion of lecithins than those of the rough endoplasmic reticulum (74). Similarly, an insufficient availability of phospholipids or a lack of specific phospholipids could result in a failure in the conjugation of the various moieties of plasma lipoproteins and, thereby, impair their secretion into plasma. A widely held view concerning the mechanisms of fatty liver development in choline deficiency is that it occurs as a result of an impairment of the synthesis and the secretion of very low-density lipoproteins, caused either by alterations in the cell membrane components involved in their synthesis and/or the nonavailability of sufficient phosphatidylcholine for incorporation into lipoproteins. Functional alterations of liver cell membranes are suggested by the study of Leelavathi et al. (75), who reported that a depletion in mitochondrial lecithin caused by choline deficiency resulted in the lowering of a membrane-bound, phosphatidylcholine-dependent enzyme, beta-hydroxybutyrate dehydrogenase. Partial restoration of the enzyme activity was observed by the treatment of mitochondria with sonicated preparations of lecithins.

The disturbances of phospholipid metabolism in the liver induced by choline deficiency are also reflected on bile metabolism. Biliary secretion of phosphatidylcholine, cholesterol, and conjugated bile acids is also reduced in choline-deficient animals (79). Balint et al. (80) showed that the synthesis of bile lecithins preferentially occurs by incorporation of choline by the Kennedy pathway rather than by the methylation pathway. Again, a probable explanation for the altered bile compositions is that less phosphatidylcholine is available in the livers of choline-deficient rats for the formation of bile salt micelles to be transported in the bile. Robins and Armstrong (81) fed rats either a choline-deficient diet or the same diet supplemented with 0.5 or 5% choline and studied ^{32}P incorporation into bile lecithins. Synthesis was at least 50% greater in animals fed the diet with 5% choline than in deficient animals. However changes in dietary choline produced no alterations in the bile salt pool size or in the bile secretion rate. With each increment of dietary choline, biliary lecithin and cholesterol secretion rates were significantly increased. Their results also indicated that biliary cholesterol secretion is more closely linked to lecithin than to bile salt secretion. The pathological consequences caused by altered bile compositions induced by a choline-deficient diet have not been carefully analyzed.

Alterations in liver phosphatidycholines as a result of choline deficiency are also reflected in the blood. For instance, low levels of plasma phospholipids were observed in rats fed a choline-deficient diet (82). The incorporation of radiolabeled leucine into plasma proteins, both lipoprotein and nonlipoprotein fractions, in choline-deficient rats was considerably lower than that in the choline-supplemented animals (82). However there were no changes in protein synthesis in the liver (82). These results indicate that there is a general inhibition or delay in the release of proteins synthesized by the liver into plasma in rats fed a choline-deficient diet. Thus the diminished secretion of the lipoproteins from the liver appears to be caused by less availability of the phosphatidylcholine moiety of the lipoproteins as well as by derangement of factors that control the secretion of proteins, both of lipoproteins and nonlipoprotein types. The synthesis of secretory proteins, including lipoproteins, takes place in the rough endoplasmic reticulum, and secretory products are passed to the smooth endoplasmic reticulum and to the Golgi apparatus, where glycosylation of lipoproteins occurs, and finally to the plasma membranes (83). Although the precise qualitative or quantitative changes of lipids in the membrane systems of the organelles are not known, it is reasonable to assume that changes in lipid composition occur in these organelles resulting in functional disturbances of the secretory processes. Electron micrographs of the liver cells in the early stages of choline deficiency showed inactive and atrophic Golgi apparatus and an increased number of osmiophilic droplets (liposomes) in the lumen of the rough and smooth endoplasmic reticulum (84). Liposomes may represent the morphological manifestation of a block in the release of hepatic triglycerides to the plasma, which occurs as a result of interference with the synthesis and/or the secretion of low-density lipoproteins. These morphological observations are compatible with the biochemical evidence of faulty secretory processes of proteins and lipoproteins, which serve as the basic mechanism of the development of fatty liver by choline deficiency.

B. Hemorrhagic Kidneys

The acute renal lesion, characterized by a fulminating hemorrhagic necrosis, which results from a deficiency of choline in the diet of weanling rats was first described by Griffith and Wade (6). Subsequent studies by a number of investigators clearly established that the renal lesion represents one of the principal effects of choline deficiency (85-92). In typical experiments, when weanling rats are fed a

diet low in choline the renal changes develop over 24 to 48 hr during
which time the animals become noticeably sick with an elevation of
blood nonprotein nitrogen and proteinuria. Renal cortical and tubular
necrosis with extensive hemorrhages occur between the sixth and
eighth day, and many animals die (6). Surviving rats develop chronic
lesions consisting of interstitial nephritis and scarring (93). As with
choline-deficiency fatty liver, the severity of the lesion depends upon
the sex and age of animals, male and younger animals being more susceptible than female or older animals (94). Intake restriction of the
deficient diet completely prevents the hemorrhagic kidneys (95). In
one study, the blood pressure of rats was unaltered during the acute
and chronic stages of renal involvement induced by choline deficiency
(96), while in other studies sustained elevation of blood pressure was
reported (97-99).

Detailed histological analyses of the sequence of events leading to
renal tubular necrosis and hemorrhages were made by Christensen
(100) and by Hartroft (101). Two cellular elements were involved in
the early phase of renal necrosis, namely the cortical capillaries and
the proximal tubular epithelial cells. Glomeruli were generally unaffected in the process. There have been considerable debates as to
whether the vascular or the tubular epithelial cell changes represent
the primary event in initiating renal necrosis. Hartroft (101), on the
bases of morphological observations, postulated that the renal necrosis was a result of obstruction of the circulation resulting from com-
pression of capillaries by swollen fatty tubular cells. This observation
formed the basis for the tubular obstruction of capillaries (TOC) mechanism, which was proposed to explain the capillary ischemia, tubular
necrosis, and hemorrhage of the kidney. It is, however, debatable
whether or not tubular fatty changes occur in the prenecrotic stages,
since extensive tubular necrosis was frequently observed in the presence of very little visible fat (102). Furthermore, biochemical analyses of total lipids indicated that there was no increase in renal lipids
up to the approximate time of onset of renal necrosis (88,89,103-105).
The ultrastructural analyses of the tubular epithelial cells were consistent with the findings of the biochemical study (102). Although an
increase in the number and size of lysosomes in the tubular epithelial
cells was found during the early stages of tubular necrosis by the
ultrastructural investigation (102), it does not appear to be the primary event in the causation of tubular cell necrosis.

In contrast to the view that the primary event of the renal lesions
was degeneration of tubular epithelial cells, several investigators
stressed the importance of vascular spasm as a cause of tubular damage
(59,105-107). It was postulated that feeding choline-deficient diets
results in an imbalance in the vasoactive mediators, such as acetylcholine and catecholamine, leading to periods of ischemia and subsequent
tubular necrosis and hemorrhage in the kidney. This view was

supported by the experimental observations of Nagler et al. (108,109), who demonstrated decreased concentrations of acetylcholine in the brain, gut, and kidneys of choline-deficient weanling rats but no changes in kidney cholinesterase levels. They demonstrated further that the mesoappendicial circulation in the choline-deficient rats was indeed ischemic compared with control rats and that the responses of microvessels to topical application of epinephrine were altered in the choline-deficient rats (110).

Despite the controversy concerning the primary cause of ischemia in the kidneys by choline deficiency, it appears that temporary ischemia to areas of the proximal tubules may play a pivotal role in inducing cell necrosis and subsequent hemorrhages. In the preceding section, evidence was presented that indicates that feeding a choline-deficient diet leads to structural and functional alterations of the cellular membranes of the liver. It is possible that similar alterations can be induced in the kidneys through disturbances of phospholipid metabolism caused by choline deficiency. Although studies specifically aimed at determining biochemical alterations of the membrane fractions of the kidneys are not yet available, there is ample evidence that a decrease in the renal content of phospholipids and the changes in fatty acid composition of the kidneys occur during choline deficiency (104,111-114). The importance of cell membrane alterations in triggering ischemic cell necrosis was recently reemphasized (115). Thus it is reasonable to assume that choline deficiency induces membrane alterations in tubular cells, and such alterations may make membranes much more sensitive to ischemic insults.

Recently, there has been a renewed interest in implicating lipid peroxidation as a fundamental pathogenetic mechanism of cell necrosis occurring under a number of circumstances (115,116). There are reports that suggest that lipid peroxidation plays an important role in the pathogenesis of renal necrosis induced by choline deficiency (117, 118). It has been shown that the additions of the antioxidants, NN^1-diphenyl-p-phenylenediamine, butylated hydroxyanisole, and butylated hydroxytoluene, to the choline-deficient diets effectively reduce the frequency and severity of the renal lesions (117). Monserrat et al. demonstrated an elevated diene conjugate formation of the total renal lipids in rats fed a choline-deficient diet before the appearance of morphological evidence of necrosis (118). Further studies are warranted to confirm these observations.

One of the characteristic features of the renal lesions caused by choline deficiency is extensive hemorrhage associated with tubular necrosis. Wells (92,119) postulated that the extensive hemorrhage seen in choline-deficient rats may be due to another choline-related pathological alteration —a decrease in the concentration of factor V of the blood-clotting system. He emphasized that two choline-associated defects, a decrease in acetylcholine and a decrease in factor V of the clotting system, would be operative in the development of hemorrhagic kidney. As Table 1 shows, the occurrence of hemorrhages is seen in many other organs than the kidney as pathological manifestations of choline deficiency. The defect in the blood-clotting system may, in part, contribute to such an occurrence (92). There are wide variations in plasma factor V concentrations among different species of animals, and it has been postulated that the different degrees of reduction of factor V in response to choline deficiency may be responsible for the species-specificity of the hemorrhagic kidney of choline deficiency (92).

C. Fetal Lung Maturation

The effects of maternal choline deficiency on fetal development have been investigated by several groups. Although the subject is extremely important, only two systems have been explored in sufficient depth; lung maturation and the immune system. In an earlier section, we briefly mentioned the pathological consequences of maternal choline deficiency on the immune responses of the offspring. Development of the fetus and the placenta creates an extra demand for choline during pregnancy. Gwee and Sim (120) reported a reduction in the choline content of livers in rats during late pregnancy as compared with the content in adult nonpregnant rats. Liver phosphatidylethanolamine-N-methyltransferase activity was increased by 24% during pregnancy. Fetal liver, on the other hand, had only 7% of the activity found in maternal liver. These results suggest that during pregnancy, the maternal liver supplies choline to the placenta and fetal organs, and that the demand for maternal liver choline far exceeds the dietary supply of choline. Therefore it is possible to create a deficiency of choline in the fetal organs by restricting the supply of choline in the diet. Superimposition of choline deficiency with a lower supply of methionine would further curtail the methylation pathway for the de novo synthesis of choline in maternal liver.

Significant effects upon phospholipid and surfactant metabolism in the lungs of newborns have been observed after feeding a combined choline-deficient and low-methionine diet to pregnant rats (121). The lungs synthesize and secrete a surface-active material, pulmonary surfactant, which in large part is composed of disaturated species of phosphatidylcholine (122). Pulmonary surfactant maintains the alveolar stability in the newborn, and its deficiency causes them respiratory distress. It was considered important to determine if maternal dietary deficiency of choline resulted in alterations in lung phosphatidylcholine and surfactant metabolism in the newborn. Although newborn rats *from choline-deficient mothers apparently showed no adverse effects in their respiration and survival*, significant reduction of the following occurred in their lungs: (a) disaturated and total phosphatidylcholine; (b) choline phosphotransferase and cytidyltransferase activities, and (c) the amount of surfactant recovered from their lungs by differential

and density-gradient centrifugations (121). A further decrease occurred in the amount of surfactant recovered from the newborn of
pregnant mothers fed 1% dimethylaminoethanol in addition to the above
regimen. Most of the newborn from this group died within 36 hr after
birth. A large proportion of phosphatidyldimethylaminoethanol accumulated in the lungs of these newborn as well as in the surfactant isolated from them (121).

Lung phosphatidylcholine is almost exclusively synthesized by the
Kennedy pathway, and the relative activity of the methylation pathway
is negligible (123). However in choline-deficient adult rats, the methylation pathway shows a significant increase in activity (124). The
adult lung appears to be less prone to the effects of choline deficiency
since it can reutilize choline secreted in the form of surfactant phosphatidylcholine (125). Choline deficiency has no effect on the phosphatidylcholine content of the lung of adult rats (124). In the neonatal
lung, on the other hand, the synthesis of surfactant and of cell membrane phospholipids compete with each other for the available choline
(121).

D. Central Nervous System (CNS) Function

In recent years, the regulation of cholinergic nerve transmission by
the dietary supply of choline has attracted much attention, primarily
because a number of disease conditions related to the CNS may be due,
in part, to disturbances of cholinergic neurotransmission. As mentioned earlier, choline serves as a precursor of acetylcholine: choline
+ acetyl CoA → acetylcholine + CoA. The reaction is catalyzed by choline acetyltransferase. Conditions that alter brain choline or acetyl
CoA concentrations affect the rates at which neurons accumulate
acetylcholine.

Several workers have reported that choline availability in the brain
directly influences the rates at which the cholinergic neurons synthesize and release acetylcholine (126-129). Choline acetyltransferase,
the enzyme that converts choline to acetylcholine, is not saturated
with choline under physiological conditions (129). Thus increases in
the amount of choline available to the enzyme accelerate acetylcholine
formation. Choline can be generated in the brain by the breakdown
of phosphatidylcholine, that derived from the circulation, or that from
the synthesis in the brain by the Kennedy pathway or by the methylation pathway (130). Circulating free choline is also available to the
brain. Blood choline, on the other hand, is acquired from the liver
and from the diet, primarily in the form of lecithin. Cohen and Wurtman (126,127) and Haubrich et al. (128) have reported that the administration of choline or lecithin to rats results in increased levels of
blood and brain choline and of brain acetylcholine. Thus higher levels
of acetylcholine may be released at the synapses when neurons are
depolarized. Whether or not increased levels of acetylcholine, resulting

from choline administration, also increase the concentration of acetylcholine receptors was examined by Morley et al. (131). These workers reported an increase in the number of acetylcholine receptors in the brain of choline-supplemented compared with choline-deficient rats. It must be pointed out that the effects of administration of choline or of choline deficiency on the levels of choline and acetylcholine in the brain and other organs have not yet been established. Pepeu et al. (132) and Pedata et al. (133), in contrast with the studies by Cohen and Wurtman (126,127) and Haubrich et al. (128), observed no increase in rat brain acetylcholine following the administration of choline. Zahniser et al. (134) observed no differences in the choline and acetylcholine levels in the brains of pups or of their mothers who were maintained on a choline-deficient diet throughout pregnancy. Nagler et al. (108,109), on the other hand, reported that lower levels of choline and acetylcholine are present in the brain, kidney, and intestine of adult choline-deficient rats. Haubrich et al. (135) also reported that the concentration of acetylcholine is not reduced in the brain, duodenum, heart, kidney, or stomach of 21-week-old rats raised from birth on a choline-deficient diet. The discrepancy in the results could be due to the age and the strain of the rats or to the differences in methods used for sacrificing the animals and those used for the determination of choline and acetylcholine.

Choline and lecithin have been used to treat several human diseases of the CNS. Choline was first used to treat tardive dyskinesia, a disorder caused by prolonged treatment with several neuroleptic drugs (136,137). Inadequate neurotransmission at striatal cholinergic interneurons appears to be involved in the pathophysiology of this syndrome. Choline and lecithin have been used, with partial success, in the therapy for several psychiatric disorders (138), Huntington's disease (139,140), Friedreich's ataxia, (141) and Tourette's syndrome (142).

There is considerable evidence that the activity of the cholinergic neurons is essential for maintaining normal memory. For example, healthy humans treated with anticholinergic drugs, exhibit a loss of short-term memory (143). In such patients, an enhancement of cholinergic function by the administration of choline or lecithin could improve short-term memory. In experimental animals, age-related diminution of memory can be improved by a dietary supplement of choline (144). Choline and lecithin are currently being tested in patients with Alzheimer's disease who suffer short-term memory losses as a result of deficient hippocampal neurons. The activity of choline acetyltransferase is reduced in the brains of patients with Alzheimer's disease (145, 146). Thus decreased levels of acetylcholine may exist and contribute to cholinergic dysfunction. Administration of choline may therefore have beneficial effects.

E. Choline Deficiency as Carcinogen, Cocarcinogen, and Tumor Promoter

1. *Carcinogen*

In 1946, Copeland and Salmon published a paper demonstrating the carcinogenic effects of a choline-deficient diet (49). In their study, feeding a diet low in choline and methionine to rats for 8 months, or longer, induced a variety of neoplasms, including hepatic tumors (hepatomas and adenocarcinomas), hemangioendotheliomas of peritoneal and subcutaneous fat, lung carcinomas (type not specified) and retro-peritoneal fibrosarcomas. It is important that not a single neoplasm developed in littermate controls receiving the same diet supplemented with adequate amounts of choline. Their subsequent studies (35,147,148) repeatedly confirmed the original observation and added several other types of neoplasms (carcinomas of the pancreas and urinary bladder) that developed in rats after long-term feeding of a choline-deficient diet. Chickens were equally susceptible to the deficient diet: they developed liver tumors (148,149). In contrast to the rat, the liver tumors in the chicken were seldom associated with cirrhosis of the liver and were preponderantly of ductal origin, cholangiocarcinomas (149).

The original findings attracted widespread attention among investigators interested in chemical carcinogenesis because the tumors were induced by a deletion of the dietary components rather than by an addition of chemical carcinogens to the diet. However the real significance of this early observation remains unclear because the diets used in these experiments were not nutritionally complete, namely they were free of vitamin B_{12} or folic acid (148). Later, Salmon and Copeland noted that no tumors developed in the liver of rats receiving choline-deficient diets supplemented with vitamin B_{12} or riboflavin (148, 150). In addition, the studies by Newberne et al., in the early 1960s, suggested a possible contamination by aflatoxin B_1 (*Aspergillus flavus*) of the peanut meal present in the lipotrope-deficient diets (151, 152). It was speculated that the aflatoxin that contaminated the choline-deficient diets used in the earlier studies acted as the causative agent of the neoplastic development. It must be stressed, however, that the animals fed choline-supplemented diets containing the same batch of peanut meal developed no, or only occasional, tumors. Thus regardless of a possible contamination of the diets by a carcinogen, these earlier studies indicated that choline-deficient (lipotrope-deficient) diets appear to exert profound effects on the development of neoplasms.

Recently, the idea that a choline-deficient diet acts as a tumor promoter of hepatoma induction has gained considerable support (see the subsequent discussion). To verify this concept, the carcinogenicity of choline-deficient diets per se has been reinvestigated under strict control to avoid diet contamination by aflatoxin or by other chemical carcinogens. Two recent studies (abstracts presented at a 1983 meeting) (153,154) demonstrated that either a semisynthetic diet devoid of choline and low in methionine or a complete amino acid-defined diet devoid of methionine and choline induced a high frequency of hepatomas in Fisher 344 rats. However, caution must be exercised in interpreting these results because Fisher 344 rats have a relatively high background toward development of presumptive preneoplastic lesions and even frank neoplasia in the liver (155,156). Nevertheless, the possibility must remain open that a choline-deficient diet may have a complete carcinogenic effect, and exploration of its possible underlying mechanism is an interesting area of future research.

2. Cocarcinogen

An increasing awareness that diet strongly influences the response of experimental animals to carcinogen, and that nutritional factors may play a key role in the development of human cancers, has generated revived interest in studies of the dietary effects on carcinogenesis (157). Particularly, the effects of lipids and lipotropes on cancer causation have been extensively studied by Rogers and Newberne (158, 159). This group, using a marginally lipotrope-deficient diet that is deficient in choline, methionine, and folic acid, has performed a series of investigations in which they compared the carcinogenicity of chemicals of different classes on several target organs of rats. The results of their published papers are summarized in Table 3. By using final tumor frequency and cumulative probability of death with tumor as an assay method, they have shown that the effects of a lipotrope-deficient diet on carcinogenicity depend upon the specific carcinogen, the target organ, and the strain and sex of the experimental animals. It is interesting that when confined to the liver as the target organ, aflatoxin B_1 (AFB_1), diethylnitrosamine (DEN), and acetylaminofluorene

(AAF), all induced liver tumors earlier, or with a higher prevalence, or both in rats fed the marginally lipotrope-deficient diets compared with rats fed diets with adequately balanced lipotropes.

In our laboratory, we have begun investigations to determine if a diet devoid of only choline is able to modify hepatocarcinogenesis. We decided to use such a dietary regimen rather than the combined deficiency of lipotropic agents employed by Rogers and Newberne. We have demonstrated that feeding a diet devoid only of choline showed three striking effects on tumor induction by three well-known carcinogens, ethionine, AAF, and azaserine (169-171). The choline-devoid diet sensitizes animals to extremely low doses of carcinogens, reduces the latent periods of hepatoma induction, and increases tumor frequency. In our experiments, as well as those of Rogers and Newberne, carcinogens were given either mixed with the deficient diets or given

Table 3 Effects of Lipotrope-Deficient Diet on Chemical Carcinogenesis

Carcinogen tested	Target organ	Tumor frequency			Ref.
		Enhanced	No change	Inhibited	
Aflatoxin B$_1$	Liver	+			160,161,162
Diethylnitrosamine	Liver	+			163,164,165
	Esophagus		+		
Nitrosobutylamine	Liver	+			164
	Esophagus		+		
	Bladder and kidney		+		
	Lung		+		
Dimethylnitrosamine (DMN)	Liver		+		164
	Kidney		+		
Acetylaminofluorene (AAF)	Liver	+			166,167
	Mammary gland		+		
N-[4,(5 nitro-2-furyl)-2-thiazolyl]formamide (FANFT)	Bladder		+		160
7,12-Dimethylbenzanthracene (DMBA)	Mammary gland			+	166,167
1,2-Dimethylhydrazine (DMH)	Colon	+			161,168
	Small bowel		+		

Table 3 (Continued) Effects of Lipotrope-Deficient Diet on Chemical Carcinogenesis

Carcinogen tested	Target organ	Tumor frequency			Ref.
		Enhanced	No change	Inhibited	
3,3-Diphenyl-3-dimethyl-carbamoyl-1-propyne (DDCP)	Liver Forestomach	+	+		160
N-Methyl-*N*'-nitro-*N*-nitrosoguanidine (MNNG)	Forestomach		+		160

while the animals were fed the deficient diet, and thus the enhancing effects of the diet on carcinogenesis can be best characterized as co-carcinogenic effects.

It has been postulated that the most likely explanation for the mechanisms by which choline (lipotrope) deficiency influences the carcinogenic responses is through an alteration in the metabolism of the carcinogen. In rats fed a choline (lipotrope)-deficient diet, hepatic microsomal oxidases, NAD-NADH and cytochrome P450 content were reduced (162,165,172,173). The in vitro Ames mutagenesis test showed that the S9 fraction from the livers of rats fed a choline (lipotrope)-deficient diet produced less mutagen from AFB_1 or AAF than did the fraction from the livers of rats fed a balanced diet (174,175). These findings, indicating reduced rather than increased activities for hepatic metabolism of carcinogens, are difficult to correlate with the in vivo enhancing effects of the deficient diet on carcinogenesis. The complexity of metabolic pathways of carcinogens, particularly of the interrelationship between carcinogen activation and detoxification, may be one of the reasons for the difficulty in correlating carcinogen metabolism with carcinogenesis. Indeed, little information is currently available regarding the possible metabolic defects of the carcinogen-detoxifying capacity of the livers of rats fed a choline-deficient diet. The extent of carcinogen-DNA interaction and the subsequent repair capabilities of the damaged DNA have been implicated as critical factors in the carcinogenic processes (176). Again, in this area, no comprehensive studies have been reported regarding the effect of feeding a choline-deficient diet on DNA damage and repair induced by carcinogens.

Feeding a lipotrope-deficient diet has been shown to reduce cellular levels of S-adenosylmethionine (177,178), and it is possible that such reductions may interfere with important methyl transfer reactions in cells. Reduction of S-adenosylmethionine and folate was also induced by several carcinogens (177,179) and by several known liver tumor promoters (180,181), thus it may have causal relationship to carcinogenesis. Recent studies by several investigators indicate the important roles played by the 5-methylcytosine of DNA in maintaining proper states of cell differentiation (182-184). In the carcinogen-treated cells, newly synthesized DNA at the repair sites appears to be hypomethylated (185). Several carcinogens have been shown to inhibit DNA methylase activity (186,187). Taking these observations together, the reduction of hepatic S-adenosylmethionine induced by lipotrope deficiency may lead cells to become more vulnerable to the creation of hypomethylated DNA. Whether or not such a condition favors enhancement of carcinogenesis remains to be established. Methylation of other cellular macromolecules, such as RNAs and proteins, also plays a critical role in maintaining proper cellular functions. If and how choline- or lipotrope-deficient diets modify the state of methylation of these cellular macromolecules is not known.

Another important consideration in relation to the mechanisms of the enhancing effect of choline-deficient diets on liver carcinogenesis is the diet-induced liver cell proliferation. Feeding a choline-deficient diet or a combined lipotrope-deficient diet increases hepatocyte mitosis and DNA synthesis (188-190). Enhanced liver cell proliferation occurs relatively early after beginning the dietary feeding and is maintained during chronic feeding (190). As will be discussed subsequently, enhanced cell proliferation plays a critical role both in the initiation and in the promotion phases of liver carcinogenesis. It is not clear however how a choline- or a combined lipotrope-deficient diet induces liver cell proliferation. As already discussed, choline deficiency is known to induce structural and functional properties of cell membranes and may alter the responsiveness of cells to a variety of growth-stimulating factors, such as hormones or newly discovered and yet incompletely characterized hepatocyte growth-stimulating factors (191,192). Alternatively, the enhanced cell proliferation may represent a regenerative reponse of the liver to the necrogenic effects of the deficient diet. Using the loss of prelabeled DNA as an index of cell necrosis, it was demonstrated that a choline-deficient diet induces considerable liver cell necrosis (193).

3. *Promoter*

The concept of the two-stage theory of carcinogenesis, originally developed for the skin (194), has now been adopted for many organ systems including the liver (195). The importance of separating the stages of carcinogenesis into *initiation* and *promotion* in understanding the development of malignant neoplasms has been reviewed repeatedly (196,197). The question of whether the choline deficiency-associated enhancement of liver tumor induction is mediated through the initiation or through the promotion phases of carcinogenesis was addressed by us several years ago. Using the experimental model of rapid induction of enzyme-altered, presumptive preneoplastic foci in the liver (198), we demonstrated that the modifying effects of choline deficiency on liver carcinogenesis were primarily through the promotion process (199, 200). Subsequent study has shown that choline deficiency not only promotes the induction of enzyme-altered hepatocytic foci in the DEN-initiated rats, but it also promotes the progression of enzyme-altered foci to hepatomas (201). Thus it is clear that feeding a choline-deficient diet is an efficient promoter in the induction of hepatomas. These findings have now been confirmed by studies of several other laboratories (202,203). Even though large numbers of chemicals of diverse structures and actions have been identified as promoters of liver carcinogenesis during the past several years (195), choline deficiency stands unique as the only strictly dietary means of promoting liver tumor induction.

Somewhat to our surprise, the initial experiments revealed that feeding of choline-devoid diet for 1 to 2 weeks exerted no significant effects on initiation of liver target cells by a single injection of DEN (199). As already discussed, the livers of rats fed a choline-deficient diet exhibit marked alterations in carcinogen-metabolizing enzymes, and the deficient diet induces sustained proliferation of liver cells. These changes can be important modifying factors of initiation. An increasing number of studies over the past few years have shown the dependence on cell proliferation for initiation of liver carcinogenesis (204-206). Subsequent studies indicate that the effects of choline deficiency on the initiation phase of carcinogenesis appear to be dependent upon the types of carcinogens used as initiators; Ghoshal and Farber have demonstrated that a choline-deficient diet is indeed capable of enhancing the initiating potency of 1,2-dimethylhydrazine, benzo-[a]pyrene, and ethionine (207). We have also demonstrated a similar effect of the deficient diet using methapyrilene as an initiator of liver tumor induction (208).

The mechanisms of action of a choline-deficient diet as a promoter of liver tumor induction have not been explored in detail, and only fragmentary information is presently available. Most investigations concerning the mechanisms of tumor promotion have dealt with the analysis of a well-defined tumor promoter, such as 12-O-tetradecanoylphorbol-13-acetate (TPA), on the skin of mice and more recently on cells in culture. Compounds with tumor-promoting activity induce many different biological and biochemical effects (196,197), and it is difficult to determine which of the many responses are essential components of the promotion process. Among the many effects of promoters, the role of cell proliferation has been extensively studied in the past. Even though there is a good correlation between the promoting abilities of a series of phorbol esters and their ability to stimulate epidermal hyperplasia, it is generally accepted that the induction of cellular proliferation is possibly a necessary, but not sufficient, condition for tumor promotion (209). A similar conclusion was also derived from studies dealing with the effects of tumor promoters on cultured cells (210). A recent development of considerable interest is the increasing body of evidence supporting the theory that skin tumor promotion has at least two separate stages. This concept is based on the observation that only a few applications of a weak or nonpromoter are sufficient to promote tumor development when followed by long-term treatment with the irritant skin mitogen (211,212). Furstenburger et al. (213) recently synthesized a TPA analogue that induces epidermal hyperplasia, but has no promoting activity, and demonstrated that the agent acts as a typical second-stage promoter. Obviously, the mechanisms underlying the distinct stages of promotion may be different (214). It remains to be clarified if such a division of the promotion stage can be applicable to liver tumor promotion.

In contrast with many studies on the skin and culture cells, very few studies have been directed toward an understanding of the mechanism of the action of promoting agents effective in liver carcinogenesis. Similar to skin tumor promotion, enhanced cell proliferation induced by a promoter appears to play an important role in liver tumor promotion. There are currently two views regarding the roles of cell proliferation on liver tumor promotion. Schulte-Hermann and co-workers (215,216) analyzed the effects of various promoters on proliferations of normal and putative preneoplastic hepatocytes and demonstrated that cells in preneoplastic foci are much more sensitive to the mitogenic effects of tumor promoters than are normal unaltered hepatocytes. This selective stimulation of the initiated cells by promoters provides a clue to explaining the rapid growth of preneoplastic lesions. Farber and his associates (217), based on their experiments of the rapid induction of foci of GGT positive hepatocytes by feeding AAF coupled with a partial hepatectomy in DEN-initiated rats, postulated that differential inhibition of cell proliferation in noninitiated cells may exert a maximal proliferative effect on initiated cell populations. The creation of a differential growth environment, leading to amplification of the initiated cells, may play a critical role in tumor promotion. We recently demonstrated that the addition of phenobarbital or pentobarbital to the choline-deficient diet inhibits liver cell proliferation induced by the deficient diet, while such an addition actually enhances the promoting efficacy of the diet (190,218). This observation supports the idea that differential responses of initiated and noninitiated cells to cell proliferative stimuli may be an important factor in tumor promotion.

One area of the rapidly progressing research in carcinogenesis during the past several years is the investigation into a possible role of cellular oncogenes on cancer induction (219,220). It has been shown that cell proliferation in rat liver following partial hepatectomy or after carbon tetrachloride administration activates certain types of cellular oncogenes (c-alb, ras^H and myc) (221). The expression of certain classes of oncogenes also increases during early stage of liver carcinogenesis in rats fed a choline-deficient diet containing ethionine (222). The significance of these findings in relation to the mechanism of initiation and/or promotion of liver carcinogenesis is not at all clear at the present time. However these observations obviously raise a number of interesting questions, the answers to which may provide better and perhaps critical understanding of the roles of cell proliferation during liver carcinogenesis.

It is generally thought that promoters act on the expression of genomes without the necessity of a covalent interaction of the promoter with the genetic component of the cells. The term epigenetic carcinogen with promoter type has been introduced to designate the mode of action of promoters (223). Alterations in cell-to-cell communication

caused by membrane changes induced by promoters are considered by
some as the basic mechanism of the promoting action (224,225). Whether or not such functional alterations of the liver cell membranes occur
because of choline deficiency remains to be investigated. Recently,
we demonstrated a number of experimental conditions that modulate
the promoting efficacy of the choline-deficient diet (226). For instance, changes in both quantity and quality of fat in the diet influence the promoting activity of the diet. While an addition of barbiturates to the choline-deficient diet enhances the promoting activity of
such a diet, an addition of a hypolipidemic agent inhibits it. These
experimental manipulations would be extremely useful to analyze the
critical mechanisms of liver tumor promotion by a choline-deficient diet.
There is increasing awareness that in human tumor development many
of the environmental factors play the role of promoters rather than of
genuine initiators. Thus clear understanding of the mechanisms underlying each step of carcinogenesis is a prerequisite to establish rational
approaches towards treatment and/or prevention of cancer.

V. CONCLUSIONS

Studies over the past 50 years have generated a wealth of new information concerning the chemistry, biochemistry, and biology of choline
and its closely related lipotropic substances. The pathological lesions
associated with its deficiency have been extensively characterized and
they have been briefly reviewed in this chapter. Lack of lipotropic
factors in the diet results in a variety of anatomical lesions in many
organs of a wide variety of experimental animals. However direct evidence of disease in man caused by choline deficiency is lacking. Altered cellular structures and functions induced by choline deficiency
also lead to modified responsiveness of the organs to external noxious
stimuli. Although this aspect was explored in considerable depth in
choline deficiency fatty liver, which shows a marked sensitivity to
chemical carcinogens, similar studies on other organ systems are lacking. There is increasing evidence indicating that nutritional factors
play a role in the regulation of the host immune system. However only
limited information is available about how dietary choline deficiency
modifies the immune regulation of the host, and thus this area has to
be explored further.

Many of the pathological lesions caused by choline deficiency are
the consequences of the disturbances of three major biological functions of choline. It is an essential component of phospholipids, it
serves as a methyl donor in transmethylation reactions, and it is a precursor of acetylcholine. Disturbances of phospholipid metabolism in
cells lead to structural and functional alterations of cell membranes,
and such alterations are considered as one of the basic pathogenetic

mechanisms of choline deficiency-induced fatty liver and, perhaps, hemorrhagic kidneys. Little is known about the functional disturbances resulting from the structural alterations of the membranes resulting from choline deficiency. One aspect of considerable interest for future studies is to investigate the integrity of different types of membrane receptors. Disturbances of receptor-mediated cellular functions caused by structural changes of the membrane could lead to a variety of pathological conditions, including altered responses to carcinogens. Further studies along this line may elucidate what specific cell functions are altered by a dietary deficiency of choline.

Choline and its related lipotropes participate in transmethylation reactions to maintain the proper structures and functions of cellular macromolecules, such as nucleic acids, proteins, and phospholipids. Dietary choline deficiency may create a modified methylation state of important cellular constituents, such as DNA, leading to abnormal cell functions and gene regulations. A possible role of cellular oncogenes in cancer etiology has been under intensive investigation. Hepatocyte proliferation induced by different means has been shown to activate certain types of cellular oncogenes. Studies aimed at clarifying the significance of these observations in relation to choline deficiency-induced liver cell proliferation may provide newer insights into the underlying mechanism of the modifying effects of a choline-deficient diet on liver carcinogenesis.

Recently, considerable interest has been generated about the possible beneficial effects of choline consumption on certain types of neurological disorders related to cholinergic dysfunctions. There is, however, controversy regarding the effects of choline deficiency or supplementation on acetylcholine synthesis and turnover in the brain. To provide a rational basis for the use of choline as a therapeutic regimen, animal models of choline deficiency or supplementation should be exploited further to analyze regulatory mechanisms of acetylcholine turnover in the brain.

Knowledge relating to the basic pathogenetic mechanisms of choline deficiency has steadily increased both at the cellular and molecular levels. Therefore the experimental model of choline deficiency remains as an attractive tool to investigate the mechanisms of cell injuries caused by defined biochemical and molecular defects.

ACKNOWLEDGMENT

The authors wish to thank Ms. L. A. Witkowski and L. Barilaro for their assistance in preparing the manuscript and Mrs. D. Pronio for the typing. Supported in part by grants from the NIH, CA26556 and HL17199 and HL28193.

REFERENCES

1. C. H. Best, J. M. Hershey, and M. E. Huntsman, The effect of lecithine of fat deposition in the liver of the normal rat. *J. Physiol.* 75:56-66 (1932).
2. C. H. Best and M. E. Huntsman, The effects of the components of lecithine upon deposition of fat in the liver. *J. Physiol.* 75:405-412 (1932).
3. C. H. Best and J. H. Ridout, Choline and fatty liver produced by feeding cholesterol. *J. Physiol.* 84:7p-8p (1935).
4. C. H. Best, M. E. Huntsman, and J. H. Ridout, The lipotropic effect of proteins. *Nature* 735:821-822 (1935).
5. W. H. Griffith and N. J. Wade, Some effects of low choline diets. *Proc. Soc. Exp. Biol. Med.* 41:188-190 (1939).
6. W. H. Griffith and N. J. Wade, Choline metabolism. 1. The occurrence and prevention of hemorrhagic degeneration in young rats on a low choline diet. *J. Biol. Chem.* 131:567-577 (1939).
7. C. C. Lucas and J. H. Ridout, *Progress in the Chemistry of Fats and Other Lipids*, Vol. 10. Pergamon, Oxford, 1967, pp. 1-150.
8. V. du Vigneaud, M. Cohn, J. P. Chandler, J. R. Schenck, and S. Simmonds, The utilization of the methyl group of methionine in the biological synthesis of choline and creatine. *J. Biol. Chem.* 140:625-641 (1941).
9. S. Simmonds, M. Cohn, J. P. Chandler, and V. du Vigneaud, The utilization of the methyl groups of choline in the biological synthesis of choline. *J. Biol. Chem.* 149:519-525 (1943).
10. V. du Vigneaud, C. Ressler, and J. R. Rachele, The biological synthesis of "labile methyl groups." *Science* 112:267-271 (1950).
11. C. H. Best and J. H. Ridout, Choline as a dietary factor. *Annu. Rev. Biochem.* 8:349-370 (1939).
12. W. H. Sebrell, Jr. and R. S. Harris (eds.), Choline, in *The Vitamins, Chemistry, Physiology, Pathology, Methods*, Vol. 3. Academic Press, New York, pp. 2-150.
13. R. E. Olson, Scientific contributions of Wendell H. Griffith to our understanding of the function of choline. *Fed. Proc.* 30:131-138 (1971).
14. A. Kuksis and S. Mookerjea, Choline. *Nutr. Rev.* 36:201-207 (1978).
15. S. H. Zeisel, Dietary choline: Biochemistry, physiology and pharmacology. *Annu. Rev. Nutr.* 1:95-121 (1981).
16. E. P. Kennedy and S. B. Weiss, The function of cytidine coenzymes in the biosynthesis of phospholipids. *J. Biol. Chem.* 222:193-214 (1956).

17. W. Stern, C. Kovac, and P. A. Weinhold, Activity and properties of CTP:choline-phosphate cytidylyltransferase in adult and fetal rat lung. *Biochim. Biophys. Acta 441*:280-293 (1976).

18. K. Ishidate, M. Tsuruoka, and Y. Nakazawa, Induction of choline kinase by polycyclic aromatic hydrocarbon carcinogens in rat liver. *Biochem. Biophys. Res. Commun. 96*:946-952 (1980).

19. J. A. Bremer and D. M. Greenberg, Methyl transfering enzyme system of microsomes in the biosynthesis of lecithin (phosphatidylcholine). *Biochim. Biophys. Acta 46*:205-216 (1961).

20. R. Mozzi and G. Porcellati, Conversion of phosphatidylethanolamine to phosphatidylcholine in rat brain by the methylation pathway. *FEBS Lett. 100*:363-366 (1979).

21. F. T. Crew, F. Hirata, and J. Axelrod, Identification and properties of methyltransferases that synthesize phosphatidylcholine in rat brain synaptosomes. *J. Neurochem. 34*:1491-1498 (1980).

22. F. Hirata, O. H. Viveros, E. J. Diliberto, and J. Axelrod, Identification and properties of two methyltransferases in conversion of phosphatidylethanolamine to phosphatidylcholine. *Proc. Nat. Acad. Sci. USA 75*:1718-1721 (1978).

23. B. Lombardi, P. Pani, F. F. Schlunk, and S. H. Chen, Labeling of liver and plasma lecithins after injection of 1-2-[^{14}C]-2 dimethylaminoethanol and [^{14}C]-L-methionine-methyl to choline deficient rats. *Lipids 4*:67-75 (1969).

24. W. J. Schneider and D. E. Vance, Effect of choline deficiency on the enzymes that synthesize phosphatidylcholine and phosphatidylethanolamine in rat livers. *Eur. J. Biochem. 85*:181-187 (1978).

25. P. A. Weinhold and R. Sanders, The oxidation of choline by liver slices and mitochondria during liver development in the rat. *Life Sci. 13*:621-629 (1973).

26. L. Jaenicke and H. Rudiger, Formation of methionine methyl groups. *Fed. Proc. 30*:160-166 (1971).

27. C. H. Best, C. C. Lucas, and J. H. Ridout, The lipotropic factors. *Ann. N.Y. Acad. Sci. 57*:646-653 (1954).

28. J. Post, J. G. Benton, R. Breakstone, and J. Hoffman. The effects of diet and choline on fatty infiltration of the human liver. *Gastroenterology 20*:403-410 (1952).

29. C. M. Leevy, M. R. Zinke, T. J. White, and A. M. Gnassi, Clinical observations on the fatty liver. *Arch. Intern. Med. 92*:527-541 (1953).

30. P. Handler, Response of guinea pigs to diet deficient in choline. *Proc. Soc. Exp. Biol. Med. 70*:70-73 (1949).

31. W. G. B. Casselman and G. R. Williams, Choline deficiency in the guinea pig. *Nature 173*:210-211 (1954).
32. T. H. Jukes, Prevention of perosis by choline. *J. Biol. Chem. 134*:789-790 (1940).
33. T. H. Jukes, Effects of choline, gelatin and creatine on perosis in chicks. *Proc. Soc. Exp. Biol. Med. 46*:155-157 (1941).
34. A. E. Schaefer, D. H. Copeland, and W. D. Salmon, Duodenal ulcers, liver damage, anemia and edema of chronic choline deficiency in dogs. *J. Nutr. 43*:201-222 (1951).
35. W. D. Salmon, D. H. Copeland, and M. J. Burns, Hepatomas in choline deficiency. *J. Nat. Cancer Inst. 15*:1549-1565 (1955).
36. R. J. Young, C. C. Lucas, J. M. Patterson, and C. H. Best, Lipotropic dose-response studies in rats: Comparison of choline, betaine, and methionine. *Can. J. Biochem. Physiol. 34*: 713-720 (1956).
37. K. Christensen and W. H. Griffith, Involution and regeneration of thymus in rats fed choline-deficient diets. *Endocrinology 30*:574-580 (1942).
38. P. M. Newberne, R. B. Wilson, and G. Williams, Effects of severe and marginal lipotropic deficiency on responses of postnatal rats to infection. *Br. J. Exp. Pathol. 51*:231-235 (1970).
39. B. M. Gebhardt and P. M. Newberne, Nutritional and immunological responses, T-cell function in offspring of lipotrope and protein deficient rats. *Immunology 26*:489-495 (1974).
40. E. A. J. Williams, B. M. Gebhardt, H. Yee, and P. M. Newberne, Immunological consequences of choline-vitamin B_{12} deprivation. *Rep. Int. 17*:279-292 (1978).
41. E. A. J. Williams, B. N. Gebhardt, B. Morton, and P. M. Newberne, Effects of early marginal methionine-choline deprivation on the development of the immune system in the rat. *Am. J. Clin. Nutr. 32*:1214-1223 (1979).
42. P. M. Newberne, K. M. Nauss, and J. L. V. deCamargo, Lipotropes, immunocompetence and cancer. *Cancer Res. 43*:2426s-2434s (1983).
43. W. S. Hartroft, J. H. Ridout, E. A. Sellers, and C. H. Best, Atheromatous changes in aorta, carotid and coronary arteries of choline-deficient rats. *Proc. Soc. Exp. Biol. Med. 31*:384-393 (1952).
44. G. F. Wilgram, W. S. Hartroft, and C. H. Best, Abnormal lipid in coronary arteries and aortic sclerosis in young rats fed a choline deficient diet. *Science 119*:842-843 (1954).
45. C. F. Wilgram and W. S. Hartroft, Pathogenesis of fatty and schlerotic lesions in the cardiovascular system of choline deficient rats. *Br. J. Exp. Pathol. 36*:298-305 (1955).
46. W. D. Salmon and P. M. Newberne, Cardiovascular disease in the choline deficient rat. *J. Nutr. 76*:483-491 (1962).

47. R. W. Engel and W. D. Salmon, Improved diets for nutritional and pathologic studies of choline deficiency in young rats. *J. Nutr.* *22*:109-121 (1941).
48. G. F. Wilgram, W. S. Hartroft, and C. H. Best, Dietary choline and the maintenance of the cardiovascular system in rats. *Br. Med. J.* *2*:1-5 (1954).
49. D. H. Copeland and W. D. Salmon, The occurrence of neoplasms in the liver, lungs, and other tissues of rats as a result of prolonged choline deficiency. *Am. J. Pathol.* *22*:1059-1081 (1946).
50. G. A. Jervis, Occurrence of brain hemorrhages in choline-deficient rats. *Proc. Soc. Exp. Biol. Med.* *51*:193-195 (1942).
51. J. G. Bellows and H. Chinn, Intraocular hemorrhages in choline deficiency. *Arch. Ophthalmol.* *30*:105-109 (1943).
52. J. L. Burns and S. W. Hartroft, Intraocular hemorrhages in young rats on choline-deficient diets. *Am. J. Opthalmol.* *32*:79-91 (1949).
53. R. A. Bear and W. S. Hartroft, Intraocular hemorrhages in weanling rats. *Am. J. Ophthalmol.* *59*:902-909f (1965).
54. P. J. A. O'Doherty, G. Kakis, and A. Kuksis, Role of luminal lecithin in intestinal fat absorption. *Lipids* *8*:249-255 (1973).
55. I. M. Yousef, P. J. A. O'Doherty, E. F. Whitter, and A. Kaksis, Ribosome structure and chylomicron formation in rat intestinal mucosa. *Lab. Invest.* *34*:256-262 (1976).
56. P. Tso, J. Lam, and W. J. Simmonds, The importance of the lysophosphatidylcholine and choline moiety of bile phosphatidyl choline in lymphatic transport of fat. *Biochim. Biophys. Acta* *528*:364-372 (1978).
57. B. Lombardi and K. N. Rao, Acute hemorrhagic pancreatic necrosis in mice. Influence of age, sex of the animals and of dietary ethionine, choline, methionine and adenine sulfate. *Am. J. Pathol.* *81*:87-100 (1975).
58. B. Lombardi, L. W. Estes, and D. S. Longnecker, Acute hemorrhagic pancreatitis (massive necrosis) with fat necrosis induced in mice by dl-ethionine fed with a choline deficient diet. *Am. J. Pathol.* *79*:465-476 (1975).
59. R. E. Olson and H. W. Deane, A physiological and cytochemical study of the kidney and the adrenal cortex during acute choline deficiency in weanling rats. *J. Nutr.* *39*:31-49 (1949).
60. K. J. Isselbacher and D. H. Alpers, Fatty liver, biochemical and clinical aspects, in *Disease of Liver*, 3rd ed. (L. Schiff, ed.). Lippincott, Philadelphia, 1969, pp. 672-688.
61. A. M. Hoyumpa, H. L. Greene, G. D. Dunn, and S. Schenker, Fatty liver: Biochemical and clinical considerations. *Dig. Dis.* *20*:1142-1170 (1975).

62. D. L. MacLean and C.H. Best, Choline and liver fat. *Br. J. Exp. Pathol.* 15:193-199 (1934).
63. W. S. Hartroft, Accumulation of fat in liver cells and in lipo-diastaemata preceding experimental dietary cirrhosis. *Anat. Rec.* 106:61-87 (1950).
64. W. S. Hartroft, The sequence of pathologic events in the development of experimental fatty liver and cirrhosis. *Ann. N.Y. Acad. Sci.* 57:633-641 (1954).
65. C. H. Best and M. E. Huntsman, The effect of choline on the liver fat of rats in various states of nutrition. *J. Physiol.* 83:255-274 (1935).
66. W. H. Griffith, Choline metabolism. III. The effect of cystine, fat, and cholesterol on hemorrhagic degeneration in young rats. *J. Biol. Chem.* 132:639-644 (1940).
67. H. J. Channon, S. W. F. Hanson, and P. A. Loizides, The effect of variations of diet fat on dietary fatty livers in rats. *Biochem. J.* 36:214-220 (1942).
68. D. A. Benton, A. E. Harper, and C. A. Elvehjem, The effect of different dietary fats on liver fat deposition. *J. Biol. Chem.* 218:693-700 (1956).
69. D. H. Benton, H. E. Spivey, F. Quiros-Perez, A. E. Harper, and C. A. Elvehjem, Effect of different dietary fats on choline requirement of rats. *Proc. Soc. Exp. Biol. Med.* 94:100-103 (1957).
70. D. Stetten, Jr. and J. Salcedo, Jr., The effect of chain length of the dietary fatty acid upon the fatty liver of choline deficiency. *J. Nutr.* 29:167-170 (1945).
71. H. D. Kesten, J. Salcedo, Jr., and D. Stetten, Jr., Fatal myocarditis in choline deficient rats fed ethyl laurate. *J. Nutr.* 29:171-177 (1945).
72. C. Carroll and L. Williams, Influence of dietary fat on fatty liver of choline-deficient rats. *J. Nutr.* 107:1263-1268 (1977).
73. B. Lombardi, Considerations on the pathogenesis of fatty liver. *Lab. Invest.* 15:1-20 (1966).
74. S. H. Chen, L. W. Estes, and B. Lombardi, Lecithin depletion in hepatic microsomal membranes of rats fed on a choline-deficient diet. *Exp. Mol. Pathol.* 17:176-186 (1972).
75. D. E. Leelavathi, S. L. Katyal, and B. Lombardi, Lecithin depletion in liver mitochondria of rats fed a choline-deficient diet. Effect on β-hydroxybutyrate dehydrogenase. *Life Sci.* 14:1203-1210 (1974).
76. R. Pascale, L. Pirisi, L. Diano, S. Zanetti, A. Satta, E. Bartoli, and F. Feo, Role of phosphatidylethanolamine methylation in the synthesis of phosphatidylcholine by hepatocytes isolated from choline-deficient rats. *FEBS Lett.* 145:293-297 (1982).

77. D. R. Hoffman, E. O. Uthus, and W. E. Cornatzer, Effect of diet on choline phosphotransferase, phosphatidylethanolamine methyltransferase and phosphatidyldimethylethanolamine methyltransferase in liver microsomes. *Lipids 15*:439-446 (1979).

78. A. Chalvardjian, Mode of action of choline. V. Sequential changes in hepatic and serum lipids of choline-deficient rats. *Can. J. Biochem. 48*:1234-1240 (1970).

79. I. C. Wells and J. M. Buckley, Lipid components from choline deficient rats. *Proc. Soc. Exp. Biol. Med. 119*:242-243 (1965).

80. J. A. Balint, D. A. Beeler, D. H. Treble, and H. L. Spitzer, Studies in the biosynthesis of hepatic and biliary lecithin. *J. Lipid Res. 8*:486-493 (1967).

81. S. J. Robins and M. J. Armstrong, Biliary lecithin secretion. II. Effect of dietary choline and biliary lecithin synthesis. *Gastroenterology 70*:397-402 (1976).

82. B. Lombardi and A. Oler, Choline deficiency fatty liver. *Lab. Invest. 17*:308-321 (1967).

83. O. Stein and Y. Stein, Lipid synthesis, intracellular transport and secretion. I. Electron microscopic radioautographic study of liver after injection of tritiated palmitate or glycerol in fasted and ethanol treated rats. *J. Cell Biol. 33*:319-339 (1967).

84. L. Estes and B. Lombardi, Effect of choline deficiency on the Golgi apparatus of rat hepatocytes. *Lab. Invest. 21*:374-385 (1969).

85. P. Gyorgy and H. Goldblatt, Choline as a member of the vitamin B_2 complex. *J. Exp. Med. 72*:1-10 (1940).

86. W. H. Griffith and N. J. Wade, Choline metabolism. II. The interrelationship of choline, cystine and methionine in the occurrence and prevention of hemorrhagic degeneration in young rats. *J. Biol. Chem. 132*:627-637 (1940).

87. R. W. Engel and W. D. Salmon, Improved diets for nutritional and pathologic studies of choline deficiency in young rats. *J. Nutr. 22*:109-120 (1941).

88. J. M. Patterson and E. W. McHenry, Choline and the prevention of hemorrhagic kidneys in the rat. *J. Biol. Chem. 145*: 207-211 (1942).

89. J. M. Patterson and E. W. McHenry, Choline and the prevention of hemorrhagic kidneys in the rat. *J. Biol. Chem. 156*: 265-269 (1944).

90. H. C. Moore, The acute renal lesions produced by choline deficiency in the male weanling rat. *J. Pathol. Bacteriol. 74*: 171-184 (1957).

91. W. H. Griffith, The renal lesions in choline deficiency. *Am. J. Clin. Nutr. 6*:263-273 (1958).

92. I. C. Wells, Hemorrhagic kidney degeneration in choline deficiency. *Fed. Proc. 30*:151-154 (1971).

93. M. O. Keith and L. Tryphonas, Choline deficiency and the reversibility of renal lesions in rats. *J. Nutr.* *108*:434-446 (1978).

94. W. H. Griffith, Choline metabolism. IV. The relation of the age, weight and sex of young rats to the occurrence of hemorrhagic degeneration on a low choline diet. *J. Nutr.* *19*:437-448 (1940).

95. D. J. Mulford and W. H. Griffith, Choline metabolism. VIII. The relation of cystine and methionine to the requirement of choline in young rats. *J. Nutr.* *23*:91-100 (1942).

96. S. S. Sobin and E. M. Landis, Blood pressure of the rat during acute and chronic choline deficiency. *Am. J. Physiol.* *148*:557-562 (1947).

97. W. S. Hartroft and C. H. Best, Hypertension of renal origin in rats following less than one week of choline deficiency in early life. *Br. Med. J.* *1*:423-426 (1949).

98. C. C. Kratzing and J. J. Perry, Hypertension in young rats following choline deficiency in maternal diets. *J. Nutr.* *101*:1657-1661 (1971).

99. U. F. Michael, S. L. Cookson, R. Chavez, and V. Pardo, Renal function in the choline deficient rat. *Proc. Soc. Exp. Biol. Med.* *150*:672-676 (1975).

100. K. Christensen, Renal changes in the albino rat on low choline and choline-deficient diets. *Arch. Pathol.* *34*:633-646 (1942).

101. W. S. Hartroft, Pathogenesis of renal lesions in weanling and young adult rats fed choline deficient diets. *Br. J. Exp. Pathol.* *24*:483-494 (1948).

102. A. J. Monserrat, E. A. Porta, and W. S. Hartroft, Sequential renal changes in choline deficient weanling rats. *Arch. Pathol.* *85*:419-432 (1968).

103. J. H. Baxter and H. Goodman, Renal and lipid alterations in choline deficiency. *Proc. Soc. Exp. Biol. Med.* *89*:682-687 (1955).

104. J. B. Simon, R. Sheig, and G. Klatskin, Relationship of early lipid changes in kidney and liver to hemorrhagic renal necrosis of choline-deficient rats. *Lab. Invest.* *19*:503-509 (1968).

105. S. B. Wolbach and O. A. Bessey, Tissue change in vitamin deficiency. *Physiol. Rev.* *22*:233-289 (1943).

106. C. T. Ashworth and A. Grollman, Electron microscopy in experimental hypertension (glomerular changes in choline-deficiency-induced hypertension) in the rat. *Arch. Pathol.* *68*:148-153 (1959).

107. F. I. Dessau and J. J. Oleson, Nature of renal changes in acute choline deficiency. *Proc. Soc. Exp. Biol Med.* *64*:278-279 (1947).

108. A. L. Nagler, W. D. Dettbarn, E. Seifter, and S. M. Levenson, Tissue levels of acetycholine and acetyl cholinesterase in weanling rats subjected to acute choline deficiency. *J. Nutr.* 94:13-19 (1968).

109. A. L. Nagler, W. D. Dettbarn, and S. M. Levenson, Tissue levels of acetylcholine and acetylcholinesterase in weanling germ free rats subjected to acute choline deficiency. *J. Nutr.* 95:603-606 (1968).

110. A. L. Nagler, W. D. Bettbarn, and S. M. Levenson, Status of the microcirculation during acute choline deficiency. *J. Nutr.* 97:232-236 (1969).

111. M. E. Fewster, J. F. Nyc, and W. H. Griffith, Renal lipid composition of choline-deficient rats. *J. Nutr.* 90:252-258 (1966).

112. M. E. Fewster and M. O. Hall, The renal phospholipid composition of choline-deficient rats. *Lipids* 2:239-243 (1967).

113. A. J. Monserrat, E. A. Porta, A. K. Goshal, and S. B. Hartman, Sequential renal lipid changes in weanling rats fed a choline-deficient diet. *J. Nutr.* 104:1496-1502 (1974).

114. P. F. Park and R. C. Smith, Chemical composition of kidneys from choline-supplemented and choline-deficient weanling rats. *J. Nutr.* 96:263-268 (1968).

115. J. L. Farber, Biology of disease. Membrane injury and calcium homostasis in the pathogenesis of coagulative necrosis. *Lab. Inves.* 47:114-123 (1982).

116. G. L. Plaa and H. Witschi, Chemicals, drugs and lipid peroxidation. *Annu. Rev. Pharmacol. Toxicol.* 16:125-141 (1976).

117. P. M. Newberne, M. R. Breshnahan, and N. Kula, Effects of two synthetic antioxidants, vitamin E, and ascorbic acid on the choline-deficient rat. *J. Nutr.* 97:219-231 (1969).

118. A. J. Monserrat, A. K. Ghoshal, W. S. Hartroft, and E. A. Porta, Lipoperoxidation in the pathogenesis of renal necrosis in choline-deficient rats. *Am. J. Pathol.* 55:163-190 (1969).

119. I. C. Wells, A blood clotting defect in choline deficient rats. *Biochim Biophys. Acta* 86:339-345 (1964).

120. M. C. E. Gwee and M. K. Sim, Free choline concentration and cephalin-N-methyltransferase activity in the maternal and foetal liver and placenta of pregnant rats. *Clin. Exp. Pharmacol. Physiol.* 5:649-653 (1978).

121. S. L. Katyal and B. Lombardi, Effects of dietary choline and N,N-dimethylaminoethanol on lung phospholipid and surfactant of newborn rats. *Pediatr. Res.* 12:952-955 (1978).

122. R. J. King and J. A. Clements, Surface active materials from dog lung. II. Composition and physiological correlations. *Am. J. Physiol.* 223:715-726 (1972).

123. L. M. G. VanGolde, Metabolism of phospholipids in the lung. *Am. Rev. Respir. Dis.* 114:977-1000 (1976).

124. R. W. Yost, A. Chander, and A. B. Fisher, Stimulation of methylation pathway for phosphatidylcholine biosynthesis in rat lung by choline deficiency. *Fed. Proc.* *42*:1268 (1983).

125. K. Geiger, M. L. Gallagher, and J. Hedlye-White, Cellular distribution and clearance of aerosolized dispalmitoyl lecithin. *J. Appl. Physiol.* *39*:759-766 (1975).

126. E. L. Cohen and R. J. Wurtman, Brain acetylcholine: Increase after systemic choline administration. *Life Sci.* *16*:1095-1102 (1975).

127. E. L. Cohen and R. J. Wurtman, Brain acetylcholine; control by dietary choline. *Science* *191*:561-562 (1976).

128. D. R. Haubrich and A. B. Pflueger, Choline administration: Central effect mediated by stimulation of acetylcholine synthesis. *Life Sci.* *24*:1083-1090 (1979).

129. J. H. Growdon and R. J. Wurtman, Dietary influences in the synthesis of neurotransmitters in the brain. *Nutr. Rev.* *37*: 129-136 (1979).

130. J. K. Blusztajn and R. J. Wurtman, Choline and cholinergic neurons. *Science* *221*:614-620 (1983).

131. B. J. Morley, G. R. Robinson, G. B. Brown, G. E. Kemp, and R. J. Bradley, Effects of dietary choline on nicotinic acetylcholine receptors in brain. *Nature* *266*:848-850 (1977).

132. G. Pepeu, D. Freedman, and N. J. Giarman, Biochemical and pharmacological studies of dimethylaminoethanol (DEANOL). *J. Pharmacol. Exp. Ther.* *129*:291-295 (1960).

133. F. Pedata, A. Wieraszko, and G. Pepeu, Effects of choline, phosphorylcholine and dimethylaminoethanol on brain acetylcholine level in the rat. *Pharmacol. Res. Commun.* *9*:755-761 (1977).

134. N. R. Zahniser, S. L. Katyal, T. M. Shih, I. Hanin, J. Moossy, A. J. Martinez, and B. Lombardi, Effects of *N*-methylaminoethanol, and *N*,*N*-dimethylaminoethanol in the diet of pregnant rats on neonatal rat brain cholinergic and phospholipid profile. *J. Neurochem.* *30*:1245-1252 (1978).

135. D. R. Haubrich, P. F. L. Wang, T. Chippendale, and E. Proctor, Choline and acetylcholine in rats: Effects of dietary choline. *J. Neurochem.* *27*:1305-1313 (1976).

136. G. E. Crane, Tardive dyskinesia in patients treated with major neuroleptics: A review of literature. *Am. J. Psychiatr.* *124*: 40-54 (1968).

137. J. H. Growdon, A. J. Gelenberg, J. Doller, M. J. Hirsch, and R. J. Wurtman, Lecithin can suppress tardive dyskinesia. *N. Engl. J. Med.* *297*:1029-1030 (1978).

138. B. M. Cohen, A. L. Miller, and J. F. Lipinski, Lecithin in mania: A preliminary report. *Am. J. Psychiatr.* *137*:242-243 (1980).

139.   A. Barbeau, Emerging treatments: Replacement therapy with choline or lecithin in neurological diseases. *Can. J. Neurol. Sci.* 5:157-160 (1978).

140.   J. H. Growdon, E. L. Cohen, and R. J. Wurtman, Huntington's disease: Clinical and chemical effects of choline administration. *Ann. Neurol.* 1:418-422 (1977).

141.   A. Barbeau, Lecithin in neurologic disorders. *N. Engl. J. Med.* 299:200 (1978).

142.   A. Barbeau, Cholinergic treatment in the Tourette syndrome. *N. Engl. J. Med.* 302:1310-1311 (1980).

143.   D. A. Drachman and B. J. Sahakian, Effects of cholinergic agents on human learning and memory, in *Nutrition and the Brain*, Vol. 5 (A. Barbeau, J. Growdon, R. J. Wurtman, eds.). Raven Press, New York, 1979, pp. 351-366.

144.   R. T. Bartus, R. L. Dean, A. J. Goas, and A. S. Lippas, Age related changes in passive avoidance retention: Modulation with dietary choline. *Science* 209:301-303 (1980).

145.   D. M. Bowen, C. B. Smith, P. White, and A. N. Davison, Neurotransmitter related enzymes and indices of hypoxia in senile dementia and other abiotrophies. *Brain* 99:459-496 (1976).

146.   K. L. Davis, L. E. Hollister, and P. A. Berger, Studies on choline chloride in neuropsychiatric disease: Human and animal data. *Psychopharmacol. Bull.* 14:56-58 (1978).

147.   R. W. Engel, D. H. Copeland, and W. D. Salmon, Carcinogenic effects associated with diets deficient in choline and related nutrients. *Ann. N.Y. Acad. Sci.* 49:49-67 (1947).

148.   W. D. Salmon and D. H. Copeland, Liver carcinoma and related lesions in chronic choline deficiency. *Ann. N.Y. Acad. Sci.* 57:664-677 (1954).

149.   W. D. Salmon, D. H. Copeland, and M. J. Burns, Hepatomas in choline deficiency. *J. Nat. Cancer Inst.* 15:1549-1565 (1955).

150.   A. E. Schaefer, D. H. Copeland, W. D. Salmon, and O. M. Hale, The influence of riboflavin, pyridoxine, inosotal, and protein depletion-repletion upon the induction of neoplasms by choline deficiency. *Cancer Res.* 10:786-792 (1950).

151.   P. M. Newberne, W. W. Carlton, and G. N. Wogan, Hepatomas in rats and hepatorenal injury induced by peanut meal or *Aspergillus flavus* extract. *Pathol. Vet.* 1:105-132 (1964).

152.   W. D. Salmon and P. M. Newberne, Occurrence of hepatomas in rats fed diets containing peanut meal as a major source of protein. *Cancer Res.* 23:571-575 (1963).

153.   A. K. Ghoshal and E. Farber, Induction of liver cancer by a diet deficient in choline and methionine. *Proc. Am. Assoc. Cancer Res.* 24:98 (1983).

154. L. A. Poirier, Y. B. Mikol, K. Hoover, and D. Creasia, Liver tumor formation in rats fed methyl-deficient, amino acid defined diets with and without diethylnitrosamine. *Proc. Am. Assoc. Cancer Res.* *24*:97 (1983).

155. J. M. Ward, Morphology of foci of altered hepatocytes and naturally occurring hepatocellular tumors in F344 rats. *Virchows Arch. Pathol. Anat.* *390*:339-345 (1981).

156. J. M. Ward and G. Reznik, Refinements of rodent pathology and the pathologist's contribution to evaluation of carcinogenesis bioassays. *Prog. Exp. Tumor Res.* *26*:266-291 (1983).

157. D. B. Clayson, Nutrition and experimental carcinogenesis: A review. *Cancer Res.* *35*:3293-3300 (1975).

158. A. E. Rogers and P. M. Newberne, Dietary effects on chemical carcinogenesis in animal models for colon liver tumors. *Cancer Res.* *35*:3427-3431 (1975).

159. A. E. Rogers and P. M. Newberne, Lipotrope deficiency in exppperimental carcinogenesis. *Nutr. Cancer* *2*:104-112 (1982).

160. A. E. Rogers, Variable effects of a lipotrope-deficient, high fat diet on chemical carcinogenesis in rats. *Cancer Res.* *35*: 2469-2474 (1975).

161. A. E. Rogers, G. Lenhart, and G. Morrison, Influence of dietary content of lipotropes and lipid on aflatoxin B_1,N-2- fluorenylacetamide, and 1,2-dimethylhydrazine carcinogenesis in rats. *Cancer Res.* *40*:2802-2807 (1980).

162. A. E. Rogers, and P. M. Newberne, Diet and aflatoxin B_1 toxicity in rats. *Toxicol. Appl. Pharmacol.* *20*:113-121 (1971).

163. A. E. Rogers, O. Sanchez, and F. M. Feinsod, Dietary enhancement of nitrosamine carcinogenesis. *Cancer Res.* *34*:96-99 (1974).

164. A. E. Rogers, J. S. Wishnock, and M. C. Archer, Effect of diet on DEN clearance and carcinogenesis in rats. *Br. J. Cancer* *31*:693-695 (1977).

165. A. E. Rogers, Reduction of N-nitrosodiethylamine carcinogenesis in rats by lipotrope or amino acid supplementation of a marginally deficient diet. *Cancer Res.* *37*:194-199 (1977).

166. P. M. Newberne and A. E. Rogers, Nutritional modulation of carcinogenesis, in *Fundamentals in Cancer Prevention* (P. N. Magee, ed.). Univ. Tokyo Press, Tokyo, 1976, pp. 15-40.

167. A. E. Rogers, Influence of dietary content of lipids and lipotropic nutrients on chemical carcinogenesis in rats. *Cancer Res.* *43*:2477s-2484s (1983).

168. A. E. Rogers and P. M. Newberne, Dietary enhancement of intestinal carcinogenesis by dimethylhydrazine in rats. *Nature* *246*:491-492 (1973).

169. H. Shinozuka, B. Lombardi, S. Sell, and R. M. Iammarino, Enhancement of ethionine liver carcinogenesis in rats fed a choline-deficient diet. *J. Nat. Cancer Inst.* 22:36-39 (1978).

170. B. Lombardi and H. Shinozuka, Enhancement of 2-acetylamino-fluorene liver carcinogenesis in rats fed a choline devoid diet. *Int. J. Cancer* 23:565-570 (1979).

171. H. Shinozuka, S. L. Katyal, and B. Lombardi, Azaserine carcinogenesis: Organ susceptibility change in rats fed a diet devoid of choline. *Int. J. Cancer* 22:36-39 (1978).

172. T. C. Campbell and J. R. Hayes, Role of nutrition in the drug metabolizing enzyme system. *Pharmacol. Rev.* 26:171-197 (1974).

173. R. Saito, L. W. Estes, and B. Lombardi, Reduced response to phenobarbital by the liver of rats fed a choline-deficient diet. *Biochim. Biophys. Acta* 381:185-194 (1975).

174. J. L. Suit, A. E. Rogers, and M. E. R. Jetten, Effects of diet on conversion of aflatoxin B_1 to bacterial mutagens by rats in vivo and by rat hepatic microsomes in vitro. *Mutat. Res.* 46:313-323 (1977).

175. T. V. Reddy, R. Ramanathan, H. Shinozuka, and B. Lombardi, Effects of dietary choline deficiency on the mutagenic activation of chemical carcinogens by rat liver fractions. *Cancer Lett.* 18:41-48 (1983).

176. D. S. R. Sarma, S. Rajalakshimi, and E. Farber, Chemical carcinogenesis: Interactions of carcinogens with nucleic acids, in *Cancer*, Vol. 1, 2nd ed. (F. F. Becker, ed.). Plenum Press, New York, 1981, pp. 335-409.

177. L. A. Poirier, P. H. Grantham, and A. E. Rogers. The effects of a marginally lipotrope-deficient diet on the hepatic levels of S-adenosylmethionine and on the urinary metabolites of 2-acetyl-aminofluorene in rats. *Cancer Res.* 37:744-748 (1977).

178. N. Shivapurkar and L. A. Poirier, Tissue levels of S-adenosylmethionine and S-adenosylhomocysteine in rats fed methionine-deficient, amino acid-defined diets for one to five weeks. *Carcinogenesis* 4:1051-1057 (1983).

179. C. L. Hyde and L. A. Poirier, Hepatic levels of S-adenosyl-ethionine and S-adenosylmethionine in rats and hamsters during chronic feeding of DL-ethionine. *Carcinogenesis* 3:309-312 (1982).

180. Y. B. Mikol and L. A. Poirier, An inverse correlation between hepatic ornithine decarboxylase and S-adenosylmethionine in rats. *Cancer Lett.* 13:196-201 (1981).

181. N. M. Shivapurkar and L. A. Poirier, Decreased levels of S-adenosylmethionine in the liver of rats fed phenobarbital and DDT. *Carcinogenesis* 3:589-591 (1982).

182. A. Razin and A. D. Riggs, DNA methylation and gene function. *Science 210*:604-610 (1980).

183. P. A. Jones and S. M. Taylor, Cellular differentiation, cytidine analogs and DNA methylation. *Cell 20*:85-93 (1980).

184. M. Ehrlich and R. Y. H. Wang, 5-methylcytosine in eukaryotic DNA. *Science 212*:1350-1357 (1981).

185. M. W. Liberman, M. B. Kastan, and F. G. Barr, Methylation of deoxycytidine incorporated by repair synthesis following damage with ultraviolet radiation and chemical carcinogens. *Proc. Am. Assoc. Cancer Res. 24*:335 (1983).

186. R. Cox, DNA methylase inhibition in vitro by N-methyl-N-nitro-N-nitrosoguanidine. *Cancer Res. 40*:61-63 (1980).

187. V. L. Wilson and P. A. Jones, Inhibition of DNA methylation by chemical carcinogens in vitro. *Cell 32*:239-246 (1983).

188. A. E. Rogers and R. A. MacDonald, Hepatic vasculature and cell proliferation in experimental cirrhosis. *Lab. Invest. 14*: 1710-1726 (1965).

189. J. H. Boss, E. Rosenmann, and G. Zajicek, Alphafetoprotein and liver cell proliferation in rats fed choline deficient diets. *Z. Ernahrungswiss. 15*:211-216 (1976).

190. S. E. Abanobi, B. Lombardi, and H. Shinozuka, Stimulation of DNA synthesis and cell proliferation in the liver of rats fed a choline-devoid diet and their suppression by phenobarbital. *Cancer Res. 42*:412-415 (1982).

191. N. R. L. Bucher, U. Patel, and S. Cohen, Hormonal factors and liver growth. *Adv. Enzyme Regul. 16*:205-213 (1978).

192. G. Michalopoulos, H. D. Cianciulli, A. R. Novotny, A. D. Kligerman, and R. L. Jirtle, Liver regeneration studies with rat hepatocytes in primary culture. *Cancer Res. 42*:4673-4682 (1982).

193. L. I. Giambarresi, S. L. Katyal, and B. Lombardi, Promotion of liver carcinogenesis in the rat by a choline-devoid diet. Role of liver necrosis and regeneration. *Br. J. Cancer 46*: 825-829 (1982).

194. I. Berenblum and P. Shubik, The role of croton oil applications associated with a single painting of carcinogen in tumor induction of the mouse's skin. *Br. J. Cancer 1*:379-391 (1947).

195. H. C. Pitot and A. E. Sirica, The stages of initiation and promotion in hepatocarcinogenesis. *Biochim. Biophys. Acta 605*:191-215 (1980).

196. T. J. Slaga, A. Sivak, and R. K. Boutwell, eds., *Carcinogenesis, Mechanisms of Tumor Promotion and Cocarcinogenesis*, Vol. 2. Raven Press, New York, 1978.

197. E. Hecker, N. E. Fusenig, F. Marks, H. W. Kunz, and H. W. Theilmann, eds., *Carcinogenesis, Cocarcinogenesis, and the*

Biological Effects of Tumor Promoters, Vol. 7, Raven Press, New York, 1982.

198. D. B. Solt and E. Farber, A new principle for the sequential analysis of chemical carcinogenesis including a quantitative assay for initiation in liver. Nature 263:701-703 (1976).

199. H. Shinozuka, M. A. Sells, S. L. Katyal, S. Sell, and B. Lombardi, Effects of a choline-devoid diet on the emergence of γ-glutamyltranspeptidase positive foci in the liver of carcinogen-treated rats. Cancer Res. 39:2515-2521 (1979).

200. H. Shinozuka and B. Lombardi, Synergistic effect of a choline-devoid diet and phenobarbital in promoting the emergence of foci of γ-glutamyltranspeptidase positive hepatocytes in the liver of carcinogen-treated rats. Cancer Res. 40:3846-3849 (1980).

201. S. Takahashi, B. Lombardi, and H. Shinozuka, Progression of carcinogen-induced foci of γ-glutamyltranspeptidase positive hepatocytes to hepatomas in rats fed a choline-devoid diet. Int. J. Cancer 29:445-450 (1982).

202. A. Columbano, S. Rajalakshmi, and D. S. R. Sarma, Requirement of cell proliferation for the initiation of liver carcinogenesis as assayed by three different procedures. Cancer Res. 41:2079-2083 (1981).

203. T. B. Leonard, J. D. Dent, M. E. Graichen, O. Lyght, and J. A. Popp, Comparison of hepatic carcinogen initiation-promotion systems. Carcinogenesis 3:851-856 (1982).

204. E. Cayama, H. Tsuda, D. R. S. Sarma, and E. Farber, Initiation of chemical carcinogenesis requires cell proliferation. Nature 275:60-62 (1978).

205. A. Columbano, S. Rajalakshmi, and D. S. R. Sarma, Requirement of cell proliferation for the induction of presumptive preneoplastic lesions in rat liver by a single dose of 1,2-dimethylhydrazine. Chem. Biol. Interact. 32:347-351 (1980).

206. T. S. Ying, D. S. R. Sarma, and E. Farber, Role of acute hepatic necrosis in the induction of early steps in liver carcinogenesis by diethylnitrosamine. Cancer Res. 41:2098-2102 (1981).

207. A. K. Ghoshal and E. Farber, The induction of resistant hepatocytes during initiation of liver carcinogenesis with chemicals in rats fed a choline deficient methionine low diet. Carcinogenesis 4:801-804 (1983).

208. A. J. Demetris, Functional, morphological and biological analyses of acute stage methapyrilene liver carcinogenesis. Proc. Am. Assoc. Cancer Res. 24:54 (1983).

209. R. K. Boutwell, Some biological aspects of skin carcinogenesis. Prog. Exp. Tumor Res. 4:207-250 (1964).

210. N. H. Colburn, E. J. Wendel, and G. Abruzzo, Dissociation of mitogenesis and late-stage promotion of tumor cell phenotype by phorbol esters: Mitogen resistant variants are sensi-

tive to promotion. *Proc. Nat. Acad. Sci. USA* 78:6912-6916 (1981).

211. T. J. Slaga, A. J. P. Klein-Szanto, S. M. Fischer, C. E. Weeks, K. Nelson, and S. Major, Studies on mechanism of action of antitumor-promoting agents: Their specificity in two stage promotion. *Proc. Nat. Acad. Sci. USA* 77:2251-2254 (1980).

212. T. J. Slaga, S. M. Fisher, K. Nelson, and G. L. Gleason, Studies on the mechanism of skin tumor promotion: Evidence for several stages in promotion. *Proc. Nat. Acad. Sci. USA* 77:3659-3663 (1980).

213. G. Furstenberger, D. L. Berry, B. Sorg, and F. Marks, Skin tumor promotion by phorbol esters is a two stage process. *Proc. Nat. Acad. Sci. USA* 78:7722-7726 (1981).

214. T. J. Slaga and A. J. P. Klein-Szanto, Initiation-promotion versus complete skin carcinogenesis in mice: Importance of dark basal keratinocytes (stem cells). *Cancer Invest.* 1:425-436 (1983).

215. G. Ohde, J. Schuppler, R. Schulte-Hermann, and H. Keiger, Proliferation of rat liver cells in preneoplastic nodules after stimulation of liver growth by xenobiotic inducers. *Arch Toxicol. Suppl.* 2:451-455 (1979).

216. R. Schulte-Hermann, G. Ohde, J. Schuppler, and I. Timmermann-Trosiener, Enhanced proliferation of putative preneoplastic cells in rat liver following treatment with the tumor promoters phenobarbital, hexachlorocyclohexane, steroid compounds and nafenopin. *Cancer Res.* 41:2556-2562 (1981).

217. E. Farber, Sequential events in chemical carcinogenesis in cancer: A comprehensive treatise, in *Cancer*, Vol. 1, 2nd ed. (F. F. Becker, ed.). Plenum Press, New York, 1982, pp. 485-506.

218. H. Shinozuka, B. Lombardi, and S. E. Abanobi, A comparative study of the efficacy of four barbiturates as promoters of the development of γ-glutamultranspeptidase-positive foci in the liver of carcinogen-treated rats. *Carcinogenesis* 3:1017-1021 (1982).

219. J. M. Bishop, Cellular oncogenes and retroviruses. *Annu. Rev. Biochem.* 52:301-354 (1983).

220. G. M. Cooper, Cellular transforming genes. *Science* 218:801-806 (1982).

221. M. Gayette, C. J. Petropoulos, P. R. Shank, and N. Faust, Expression of cellular oncogenes during liver regeneration. *Science* 219:510-512 (1983).

222. N. Fausto and P. R. Shank, Oncogene expression in liver regeneration and hepatocarcinogenesis. *Hepatology* 3:1016-1023 (1983).

223. G. M. Williams, Liver carcinogenesis: The role for some chemicals of an epigenetic mechanism of liver-tumor promotion

involving modification of the cell membrane. *Food Cosmet. Toxicol. 19*:577-583 (1981).

224. G. M. Williams, S. Telang, and C. Tong, Inhibition of intercellular communication between liver cells by the liver tumor promoter 1,1,1-tricholoro-2,2-bios-(p-chlorophenyl)ethane (DDT). *Cancer Lett. 11*:399-403 (1981).

225. G. M. Williams, Epigenetic effects of liver tumor promoters and implications for health effects. *Environ. Health Perspect. 50*: 177-183 (1983).

226. H. Shinozuka, S. E. Abanobi, and B. Lombardi, Modulation of tumor promotion in liver carcinogenesis. *Environ. Health Perspect. 50*:163-168 (1983).

8

Role of Nutritional Status in Drug Metabolism and Toxicity

Rosemary E. McDanell and André E. M. McLean

University College London
London, England

I. INTRODUCTION

It is well known that nutritional status may have a marked effect on the way drugs are metabolized by the body (1-3). This awareness is of particular significance because of the enormous variation in food intake in different areas of the world. A report (4) from the World Health Organization (WHO) estimates that 500 million people throughout the world are underfed. Primary malnutrition caused by poverty and food shortage is a frequent, usually intermittent, feature of the less-developed countries (LDCs), whereas in more affluent societies, secondary malnutrition may result from a number of disease states.

Concerning food intake, the WHO estimates (4) that protein usually forms about 11% of dietary energy intake. Although there is only a threefold variation in total dietary protein content in different parts of the world, the type of protein varies 20-fold, with a much larger proportion of vegetable protein being consumed in the poorer countries, while in the so-called developed countries the diet is very much richer in animal proteins. For example, the median consumption for meat protein varies from 3.6 g/person a day in Rwanda, to 71.8 g/person a day in Uruguay. While amino acid intake does not show such great variation, there are many social groups suffering from local protein shortage, which is of particular relevance in growing children where the protein requirements are greater.

There are also enormous differences in total fat intake, varying from about 10% of total energy in individuals and countries where traditional food patterns based on cereals or starches have been retained, to 40 to 50% in developed communities. Other major dietary variants are sugar, where the amount consumed is again very much higher in the more-developed countries, and fiber where consumption drops as fat and sugar intake rises.

In this review we use the term drug metabolism in its widest sense, to include the enzyme systems that metabolize numerous foreign compounds (5-7), examples of which are shown in Table 1. These systems were first noted as being concerned with the elimination of drugs.

Man is continuously exposed to compounds that are absorbed into the body but can be used neither as energy sources nor for synthesis of body components. These compounds are collectively termed xenobiotics (Gk. *xenos*:strange, foreign; *bios*: life), and may enter the body by ingestion with food, inhalation via the lungs or absorption through the skin. They include many natural compounds of plant and animal origin and also drugs, pesticides, food additives, and other man-made environmental contaminants, including both natural and synthetic carcinogens. Current estimates from the United States Environmental Protection Agency (EPA) and the Food and Drug Administration (FDA) indicate that there are about 63,000 chemicals in common use including 4000 active ingredients in drugs, 1500 in pesticides, and 5500 used as food additives and preservatives (5).

Xenobiotic metabolism occurs in two different phases (8,9). In phase 1 the compound undergoes reactions, which can be classified as oxidations, reductions, and hydrolyses, during which it acquires hydroxyl, carboxyl, amino, or sulfhydryl groups. The oxidizing system is a mixed-function oxidase (MFO) system requiring the reduced coenzyme nicotinamide adenine dinucleotide phosphate (NADPH) and molecular oxygen.

Table 1 Examples of Compounds Which Act as Substrates for the Phase-1 and Phase-2 Drug Metabolizing Enzymes

Group	Examples
Drugs	Benzodiazepines, warfarin, barbiturates, and many others
Carcinogens/mutagens	Nitrosamines, aflatoxin, acetylamino-fluorene, benzo[a]pyrene
Pesticides	Dieldrin, parathion
Food additives	Antioxidants (BHA, BHT), terpenes
Environmental pollutants	PCBs
Endogenous	Steroids, thyroxine, fatty acids

The system catalyzes the insertion of one atom of oxygen into the foreign molecule with the second atom giving rise to water. This enzyme system is bound to the membrane of the smooth endoplasmic reticulum and consists essentially of an electron transport chain from NADPH through the flavoprotein cytochrome P450 reductase, and on to the group of hemoprotein cytochrome P450s which act as the terminal oxidases (10). Phosphatidylcholine, or a similar phospholipid is essential for optimal drug metabolizing activity.

The second phase (phase 2) consists essentially of synthetic reactions or conjugations which require a source of energy in the form of ATP. Conjugation involves the coupling of a compound or its phase-1 metabolite with a conjugating agent. In man there are eight major conjugation reactions utilizing glucuronic acid, glycine, cysteine, glutamine, glutathione, methionine, sulfate, and acetyl radicals (8). Conjugations are characterized by the formation of an active intermediate, which in most cases is a nucleotide. There are two kinds of conjugation reactions depending upon whether the conjugating agent or the xenobiotic is activated.

The liver has long been considered the major organ of xenobiotic metabolism, although other tissues such as lung, gut, and skin are now thought to play a significant role, particularly as they are often direct portals of entry for xenobiotics into the body (11). This whole enzyme system is remarkable for the diversity of substances that can serve as substrates (Table 1), and for its inducibility, often by these same substrates (12,13). Thus although the basal levels of drug metabolizing enzyme activity may be low, particularly in extrahepatic tissues, the enzyme systems in these tissues are highly inducible by the many foreign chemicals to which they are commonly exposed.

The oxidizing enzyme system was originally thought to be entirely one of the detoxification whereby substances were inactivated and made more water-soluble for excretion; however it also converts a number of relatively inert compounds to highly reactive metabolites (6,14). Most of these activating processes probably occur through phase 1 oxidations, although, more recently, some instances of activation by conjugation have been demonstrated. Examples of these are shown in Table 2.

II. MEASUREMENT OF DRUG-METABOLIZING ACTIVITY

Several methods have been employed for measuring drug-metabolizing activity in man and experimental animals. In man the method of choice is usually some pharmacokinetic measurement of the drug under investigation or of a model drug such as antipyrine. These measurements include plasma half-life, elimination rate constant or its equivalent, apparent volume of distribution, and clearance. Exhalation of isotopic

Table 2   Examples of Compounds Activated to Toxic Metabolites by the Drug-Metabolizing Enzyme System

Phase 1		
Carbon tetrachloride	→ free radical	→ hepatic necrosis
Paracetamol	→ quinone imine	→ hepatic necrosis
Benzo [a]pyrene	→ diol epoxide	→ carcinogen
Phase 2		
N-hydroxy-2-acetyl-aminofluorine	→ sulfate and glucuronide conjugate	→ carcinogen
Dichloroethane	→ glutathione conjugate	→ carcinogen ?
Isoniazid	→ acetylation and hydrolysis	→ liver damage

carbon dioxide is also used, employing various substrates such as amino-pyrine and caffeine that can distinguish between different subspecies of cytochrome hemoproteins.

In experimental animals, direct in vivo measurements, such as carbon dioxide exhalation, sleeping time induced by hexobarbital, or zoxazol-amine paralysis time, have been employed. More usually, in vitro tech-niques are used, whereby the metabolism of a model substrate is meas-ured either in whole homogenates of tissue or in the microsomal prepara-tion from these homogenates. Almost all of these studies have investi-gated activity in the liver, although more recently extrahepatic tissues also have been used. Another recent trend has been the use of tissue culture, particularly hepatocytes isolated from liver perfusion.

III. CONSEQUENCES OF DIETARY MANIPULATION

In considering dietary effects on drug metabolism, it is important to realize that in altering the amount of one component of the diet, one can either maintain the levels of intake of the other dietary components and thus alter the caloric content of the dietary input, or maintain the caloric content and inversely change the level of one of the other dietary components.

In relation to the former option Nakajima and his co-workers (15) have demonstrated that in rats subjected to caloric restriction, MFO activity increased in a linear fashion with decreased caloric intake. Concerning the second option, in many of these dietary studies main-tenance of an isocaloric diet has been achieved by altering the protein/carbohydrate ratio so that, for example, a low-protein diet is equivalent to a high-carbohydrate diet. In this instance care is required in in-terpreting whether the observed effect is due to an individual or to a combined effect of the two dietary components. In human studies (16) on dietary carbohydrate and protein, it has been suggested that the magnitude of effect produced by the two components individually is less than observed when changes in dietary carbohydrate and protein are relative to each other.

A second consideration in interpreting these studies is that apart from specific dietary alterations in the drug-metabolizing enzyme system, there are also histological and biochemical alterations caused by long-term, severe dietary deficiencies, which may have indirect effects on drug metabolism (17,18). Examples are found in the severe fatty liver and reduced plasma proteins in kwashiorkor (17) or in the fatty liver of experimental choline deficiency (18).

IV. PROTEIN

Protein is probably the most widely studied dietary constituent in terms of its effect on drug metabolism. These studies have varied a great deal in terms of what is thought to be a normal level of dietary protein. Generally, researchers have assumed that the level of dietary protein considered normal or adequate for control groups is between 18 to 30% of the diet, while low-protein diets range from 0 to 10%.

In considering the amount of protein required for growth, it has been shown that in rats fed 10% dietary protein, maximum growth is achieved, while at 6% protein, half maximum growth takes place, and at 3% growth ceases although the animals may stay alive for many weeks or months (19). In man protein usually forms about 11% of energy intake (4). Uncomplicated protein deficiency is relatively uncommon in man and usually involves caloric deficiency, in which case it is referred to as protein-calorie malnutrition (PCM) or protein-energy malnutrition (PEM). The effects of general nutrient deficiency as opposed to protein deficiency will be discussed in the section on undernutrition.

A large number of observations of the effect of protein deficiency on drug metabolism have been made in experimental animals and more recently in human volunteers. In general, a dietary reduction in either the quantity of protein or the quality of protein in terms of essential amino acids causes a depression of MFO activity. Glutathione and sulfate required for conjugation reactions are also depleted in protein deficiency, and this effect is discussed in Section X, Phase-2 Metabolism. The subsequent effect on toxicity of compounds metabolized by the drug-metabolizing enzyme system will therefore depend on whether the compounds are activated or inactivated by this system (Table 2).

In considering the effects of altered dietary protein on drug metabolism, it is appropriate to consider first the effect on the enzyme system itself and second the resultant toxicity of compounds metabolized by this system.

A. Effect of Protein on the Drug-Metabolizing Enzymes

1. *Experimental Animals*

In animal studies investigating the effect of dietary protein on drug metabolism, results have varied enormously according to the sex, strain, and species of the experimental animal and to the enzyme substrate employed. This variation indicates the need for great caution in extrapolating results to man. While such studies have been extensively reviewed (1-3), various more recent and illustrative examples will be discussed.

Mgbodile and Campbell (20) showed a depression of hepatic microsomal MFO activity in weanling rats fed a 5%-casein semipurified diet for 2 weeks, compared with control animals fed 20% casein. They demonstrated a 64 to 66% decrease in V_{max} per milligram of microsomal protein for ethylmorphine and aniline, and an equivalent reduction in the content of cytochrome P450 and in the activities of cytochrome P450 and cytochrome c reductase. Three-quarters of this depression have been attributed to a specific effect on the enzyme system. After 2 weeks of feeding a low-protein diet an approximate 50% increase in cell size for most liver cells was observed, largely accounted for by a three-fold increase in lipid and glycogen content. Liver microsomal protein was rapidly depressed, primarily because of losses per cell, despite a compensatory increase in tissue size in these animals. These workers indicated that the portion (25%) of the enzyme activity that was lost because of the depressed rate of cell proliferation and decreased microsomal protein content, could be readily recovered if protein depletion was not too severe. Probably the more substantial effect of protein deficiency is on the enzyme system itself. The same group of workers subsequently demonstrated (21,22) that the MFO components most affected by protein deficiency were cytochrome P450 and cytochrome P450 reductase, and they suggested that the interaction between these two components is the primary mechanism responsible for the dietary protein effect, as opposed to an effect on the specific activity of either component alone.

Recent studies have shown that depression of MFO activity associated with dietary protein deficiency occurs within 24 hr (22). When the effects of protein deficiency were studied over a 2-month period, Kuwano and Hiraga (23) demonstrated that the cytochrome P450 and b_5 content in liver microsomes decreased in a biphasic fashion, with a rapid decrease in the first 4 days followed by a more gradual decrease. Aminopyrine demethylase, nitroanisole demethylase, and aniline hydroxylase also decreased in a biphasic fashion. Although MFO activity decreases within 24 hr of feeding a low-protein diet, recovery of ethylmorphine demethylase and aminopyrine hydroxylase activity on returning to a control diet is stated to take 2 to 4 weeks (21). However since only a day 15 measurement is given, it is not clear if there is a biphasic recovery from a low-protein diet with most of the effect taking place in the first 4 days. A long lag time would be suggestive of pathological and structural changes and alteration of cell populations in the liver, in addition to the adaptive changes in protein synthesis in otherwise normal cells. The latter are usually complete after a few days of dietary switch.

Decreased activity of the MFO system with low-protein diet has been demonstrated for a range of substrates both in vivo and in vitro (24-33). One of the first demonstrations of this effect was by Kato et al. (24) who established that a low-protein diet decreased the microsomal metabolism of hexobarbitone, strychnine, and meprobamate in female rats. Kato also showed (25) that the microsomal metabolism of aminopyrine, n-methylaniline, hexobarbital, aniline, and p-nitrobenzoic acid in male and female rats was increased by high-protein diet and decreased by low-protein diet. Moreover, the demethylation of aminopyrine, hydroxylation of hexobarbital, and reduction of p-nitrobenzoic acid was increased more markedly in male rats by high-protein diet and decreased more markedly in male than female rats by low-protein diet. They proposed that protein deficiency may impair the ability of androgens to increase drug-metabolizing activity.

McLean and McLean (26) have also demonstrated an 80% decrease in hepatic aminopyrene (Pyramidon) demethylation and benzo[a]pyrene hydroxylation in rats fed a no-protein or a 3%-casein diet for 4 days. In more recent work (27) rats fed a low-protein diet had decreased arylhydrocarbon hydroxylase (AHH) activity but increased activity of epoxide hydrase and UDP-glucuronyltransferase. This is one of the few examples of an effect on conjugation. In our laboratory rats fed a low-protein diet or starved overnight had decreased activity of hepatic glutathione transferase. When considering the effect of diet on toxicity, it is important to consider both phases of drug metabolism. These phases will be further discussed later in Section X, Phase-2 Metabolism.

In accord with the in vitro results, Kato et al. (28) established that a decreased level of dietary protein intake was directly correlated both with delayed plasma clearance and prolonged anesthesia in rats given pentobarbital and also with the toxicity of strychnine, pentobarbitone, and zoxazolamine. This effect was more marked in male than in female rats. Similarly, Weatherholtz et al. (29) showed increased heptachlor levels and decreased conversion of the pesticide to its epoxide metabolite in protein-deficient rats.

Several workers have demonstrated an effect of protein deficiency on the inducibility of the drug-metabolizing enzyme system. Generally in protein deficiency the system is still responsive to induction, but to a lesser extent than in control animals. Kato et al. (24) showed that

phenobarbitone increased hexobarbital hydroxylase and strychnine
oxidase in female rats fed a low-protein or protein-deficient diet by a
similar factor to that in control rats; however in the protein-deficient
rats the final enzyme activity after phenobarbitone was less than that
in phenobarbitone-treated controls. Marshall and McLean (30) reported
a similar effect for hepatic cytochrome P450. Hayes and his co-workers
(31-33) have carried out a series of studies on the effect of a low-
protein diet on the inducibility of the rat hepatic microsomal enzyme
system by phenobarbitone and 3-methylcholanthrene. These workers
measured microsomal protein, phosphatidylcholine, and cytochrome
P450 (31), as well as kinetics and interaction of ethylmorphine and
aniline with cytochrome P450 (32). As with other workers, they
showed that even though the microsomal enzyme system of protein-
deprived animals could be induced, the induced activity never reached
that of induced protein-sufficient controls.

2. Humans

In studies on man, both normal and malnourished individuals have been
investigated. Kappas and his co-workers (16,34) demonstrated that
when healthy volunteers were switched from their customary home diets
to a high-protein (44%) diet, the average plasma half-life of aminopyrine
and theophylline decreased 41 and 36%, respectively. When subjects
were switched from the high- to a low-protein (10%) diet the average
antipyrine and theophylline half-lives increased 63 and 46%, respectively. In further exploration of the influence of protein on drug metabolism, they fed protein supplements to subjects on a calculated well-
balanced diet (15% protein). Addition of 100 g of sodium caseinate after
2 weeks on the balanced diet resulted in increased removal rates of
antipyrine and theophylline (34).
    In a similar study Balabaud et al. (35) looked at phenytoin metabolism
in 10 healthy volunteers given the same diet as that described by Kappas
et al. (34). However they found no significant differences in peak
plasma levels or half-life of phenytoin between the high- and low-protein diets. They suggest that the large intraindividual differences encountered in their subjects may have masked any dietary effect. In
other studies with asthmatic children on long-term theophylline therapy
Feldman et al. (36) showed that the metabolism of the drug was greater
during consumption of a high-protein (20%) compared with a low-protein
(3%) diet.
    Mucklow et al. (37) have compared antipyrine metabolism in Indo-
Pakistani lactovegetarians and nonvegetarians. Clearance of the drug
was significantly slower in 16 lactovegetarians than in the subjects who
ate meat regularly. The absence of meat from the diet was associated
with a significantly smaller intake of dietary protein which was abnor-
mally low by western standards. They suggest that it is this contrast

in protein intake between the dietary subgroups that is responsible
for differences observed in antipyrine metabolism. In a similar study
Brodie et al. (38) found that in white subjects native to Britain, a
vegetarian diet is not characterized by a low-protein intake, and mean
antipyrine clearance values in vegetarians and nonvegetarians do not
significantly differ.

Further studies in humans have looked at drug metabolism in malnourished individuals and these effects will be discussed in Section V,
Undernutrition.

B. Dietary Protein and Toxicity

In recent years it has been shown that the drug-metabolizing enzyme
system, whose primary function is detoxification, can also bring about
the reverse effect and convert biologically inert compounds to active
metabolites and potentially toxic entities (6,14). For example, many
chemical carcinogens require activation to yield the high reactive species
that bind covalently to DNA giving rise to mutations and carcinogenesis.
Because a reduction in dietary protein reduces the activity of the MFO
system, the ultimate toxicity of a compound will depend upon whether
the compound is activated or inactivated by this system and upon the
rate of subsequent removal by conjugation. The presence of alternative
pathways of metabolism will determine the total amount of reactive metabolite that reaches the target organ, and its rate of delivery.

Many workers in this field have looked at the effect of protein deficiency on the metabolism of only one or two substrates in vitro. It
is essential to consider several factors before attempting extrapolation
to the whole animal. If the compound is metabolized primarily by an
MFO-catalyzed reaction to less-reactive products, then protein deficiency should prolong the compound's clearance and duration of action.
However for those compounds that are activated by the MFO system,
the ultimate toxicity will depend upon the amount of active metabolite
reacting with the target organ and the subsequent metabolic routes to
which the reactive intermediate is subjected.

McLean and McLean (26) demonstrated that rats fed a protein-free or
3%-protein diet for 4 days were resistant to the lethal effects of carbon
tetrachloride. The LD_{50} in control rats was 6.4 ml/kg compared with
14.7 ml/kg in rats fed the protein-free diet. It is suggested that carbon tetrachloride is metabolized by the MFO system to a free radical,
which then catalyzes chains of autooxidation in the structural lipids of
the cell, resulting in fatty accumulation and cellular necrosis. In simple
terms, rats fed a low-protein diet have decreased MFO activity and
therefore a decreased rate of conversion of carbon tetrachloride to its
active metabolite. In the meantime the "safe" pathway of carbon tetrachloride exhalation via the lung goes on unaltered. Similarly, starving
rats overnight leads to increased carbon tetrachloride toxicity without
any change in the amount of P450 per cell. The increase in toxicity

may be due to removal of glycogen, which forms a ready source of
glucose and perhaps acts as an alternative target for toxic chemicals.
In addition, ketosis will lead to an increased rate of carbon tetrachloride metabolism.

Weatherholtz et al. (29) showed that rats fed a 5%-protein diet for
10 days were much less susceptible to the toxic effects of the pesticide
heptachlor than rats fed 20 or 40% protein. This was due to a much
slower rate of conversion to the toxic epoxide metabolite. Similarly,
Kato et al. (28) found that the acute toxicity of octamethylpyrophos-
phamide (OPMA) is decreased in protein-deficient rats. In contrast,
increased toxicity in protein deficiency has been demonstrated for a
number of compounds that are inactivated by the MFO system. Kato
et al. (28) reported an increase in the acute toxicity of strychnine,
pentobarbital, and zoxazolamine in protein-deficient rats.

One of the earlier studies by Boyd (39) undertaken at the request of
WHO, looked at pesticide toxicity in weanling rats to establish which of
these pesticides would be least toxic for use in countries where the diet
was deficient in protein. Of 15 pesticides investigated, the toxicity of
five was not significantly altered, whereas others showed greatly increased toxicity, particularly captan which is detoxified by the MFO
system and also reacts with glutathione in a detoxification reaction.
Loss of glutathione is one of the major effects of protein deficiency.
Boyd et al. (40) similarly showed that phenacetin is another compound
whose toxicity is increased by protein deficiency, and Webb et al. (41)
observed that the organophosphate pesticides malathion and parathion,
as well as their more toxic metabolites maloxon and paraoxon, have
increased toxicity in protein deficiency.

McLean and Day (42) demonstrated that a low-protein diet causes a
marked increase in the toxicity of paracetamol (acetaminophen). It would
be expected that as paracetamol is activated by the MFO system, but
removed by sulfate and glucuronide conjugation which are more resistant to the effect of low-protein diet, that a low-protein diet would decrease paracetamol toxicity. However because of the concurrent depletion of glutathione in protein-deficient rats, paracetamol toxicity is increased as the second detoxifying pathway, via glutathione, is removed.

Several workers have also looked at the effect of a low-protein diet on
mutagenicity and carcinogenicity. The mutagenicity of certain chemical
carcinogens in vitro is different when the microsomes used to activate
the carcinogens are derived from protein-deficient animals. Dimethylnitrosamine (DMN), a carcinogen requiring activation, was less mutagenic to bacteria when the microsomes were isolated from protein-deficient mice. Conversely, the mutagenicity of the directly reactive
carcinogen N-methyl-N-nitrosoguanidine was increased in protein deficiency (43).

The original work on low-protein diet and tumor production (44)
showed a dramatic decrease, from 98 to 0% in the prevalence of

spontaneous mammary tumors in mice fed low-cystine diets compared with those fed a high-cystine or stock diet. Silverstone and Tannenbaum (45) went on to indicate that the frequency of benign hepatomas in mice was strikingly lower in those animals fed 9% casein compared with 18% casein. They concluded that it was the specific amino acid content of the diet that was important since supplementing the 9%-casein diet with methionine and cystine increased the frequency of hepatomas to that of the 18%-casein group.

Other workers have looked at the effect of dietary protein on chemically induced neoplasia. Walters and Roe (46) showed that mice injected within 24 hours of birth with DMBA, developed significantly more lung tumors than mice similarly injected but later fed a low-protein diet. It has similarly been reported (47) that in rats fed a low-protein diet during the administration of aflatoxin B_1, the number of precancerous lesions and liver tumors is greatly reduced and the covalent binding to DNA decreased by 70%, presumably as a result of a reduced rate of microsomal enzyme activity. In considering the effects of diet on carcinogenesis it is important to note that this is a multistage process. Low-protein diets usually reduce tumor-promoting effects (45) so that unless tumor initiation and promotion are clearly separated, dietary effects may seem confused and contradictory. Similarly, in considering tumor initiation, a clear distinction must be drawn between the rate of activation of a carcinogen, and the total quantity of activated carcinogen that reaches the target molecules.

Clinton et al. (48) looked at the effect of dietary protein changes on DMBA-induced mammary tumors in rats. The role of protein in the initiation phase was examined, and it was demonstrated that in rats fed varying protein concentrations before dosing, the prevalence of tumors increased as the protein content of the diet decreased. When the promotion phase of DMBA carcinogenesis was examined by maintaining rats on the experimental diets for 25 weeks after dosing, no effect on tumor incidence was observed. More recently Appleton and Campbell (49) have also investigated effects of dietary protein on initiation and promotion. They showed that the emergence of aflatoxin-induced liver lesions in rats is dependent upon both the level of dietary casein and the time when the protein is consumed. Feeding a low-protein diet during the dosing period enhanced the development of lesions thought to be characteristic of aflatoxin-induced acute toxicity. However feeding a low-protein diet during the postdosing period resulted in a marked decrease in the number of liver foci, which probably represents a lower tendency for neoplastic development. Thus a low-protein diet results in distinctly different histological effects depending upon the time of feeding. This suggests that different mechanisms may be involved in the dietary effects on the two stages of carcinogenesis.

C. Protein Quality

It has been observed that protein quality, in terms of essential amino acid content, can affect drug-metabolizing enzyme activity. Miranda and Webb (50) compared the effects of feeding diets containing 18% of either casein or gluten as the sole source of protein, to male rats for 10 days. They showed that body weight gains, liver weight, microsomal protein, cytochrome P450, and drug metabolism in vitro and in vivo, were all relatively lower in the gluten-fed animals. They suggest that this effect is due to the imbalance and/or deficiency of amino acids in the gluten diet. The acute toxicity of heptachlor, which undergoes metabolic activation to the toxic metabolite, was also lower in gluten-fed rats.

More recently Kato and his co-workers (51) fed male rats eight different types of protein at 10% in the diet. In general, the relative biological values of the proteins, as shown by gain in body weight, were positively correlated with the activity of the microsomal enzyme system, including aminopyrine-N-demethylase, aniline hydroxylase, cytochrome P450, and NADPH cytochrome c reductase.

Protein quality, in terms of amino acid content, also is likely to affect phase-2 drug metabolism, particularly with the sulfur-containing amino acids, which are the main source of sulfate for sulfate conjugation. This aspect is further discussed under Section X, Phase-2 Metabolism.

V. UNDERNUTRITION

Most experiments investigating the effect of diet on drug metabolism have attempted to examine the effect of a single dietary component. This is obviously of importance in elucidating the mechanisms by which these effects occur. However in humans the more common situation is multinutrient deficiency. Malnutrition caused by lack of protein or calories, or both, is an enormous problem. As a rough generalization one can say that protein plus caloric deficiency is clinically expressed as kwashiorkor, whereas a predominantly caloric deficiency leads to the state of marasmus. The more common situation is some gradation between these two extremes and is referred to as protein-calorie malnutrition (PCM) or protein-energy malnutrition (PEM). In addition to energy and protein deficiency the other major public health problems of nutritional origin are deficiencies of vitamin A, iron, and vitamin B complex, and trace metal inadequacies. The effects of undernutrition on drug metabolism have been extensively studied in animals and to a lesser extent in man.

1. *Experimental Animals*

The effect of starvation on the drug-metabolizing system was first shown by Dixon et al. (52), who demonstrated a depression of drug metabolism both in vivo and in vitro when male mice were starved for 36 hr. Effects were seen on oxidation of hexobarbitone and chloramphenicol, pyramidon dealkylation, and acetanilide hydroxylation. Kato and Takanata (53) later confirmed that the biological half-lives of pentobarbital and carisoprodol were markedly prolonged in fasted male rats but slightly shortened in fasted female rats. Similarly, the duration of hexobarbital anesthesia was increased in fasted male rats but decreased in fasted female rats. The duration of zoxazolamine paralysis decreased in both male and female rats. Kato and Gillette (54) suggested that these apparently contrasting results may be explained by sex differences and suggested that starvation may cause impairment by interfering with the stimulating effects of androgenic steroids.

It has been observed (55) that there is no sex difference in response to starvation in mice or rabbits, and it is likely that the effects of starvation on drug metabolism will vary widely with sex, strain, and species of experimental animal. Recently, Nakajima and Sato (56) have shown that hepatic microsomal metabolism of aromatic and chlorinated hydrocarbons is enhanced in 1-day fasted rats of both sexes, even though fasting produced no significant increase in microsomal protein or cytochrome P450 content. A sex difference was noted in the metabolism of these hydrocarbons both in fed and 1-day fasted rats. However food deprivation for 3 days decreased the extent of this sex difference.

Apart from altering hepatic activity, starvation also effects drug-metabolizing enzyme activity in the intestine. Marselos and Laitinen (57) looked at drug hydroxylation and glucuronidation enzyme levels in the liver and small-intestinal mucosa of male rats after 3 days of starvation. They reported that liver microsomal P450 content and NADPH cytochrome c reductase activity were unaffected, while p-nitro anisole demethylase activity was increased. In liver, the specific activity of UDP-glucose dehydrogenase, total beta-glucuronidase, and 3-hydroxy-acid-dehydrogenase were increased, while UDP-glucuronyltransferase was unaffected. In small-intestinal mucosa, specific enzyme activities were lower in starved animals, with the exception of UDP-glucuronyltransferase, which was unchanged. Phenobarbitone treatment proved more effective in inducing several microsomal enzymes in the starved rats than in those fed ad libitum. Wattenberg et al. (58) showed almost total loss of benzo[a]pyrene hydroxylase activity in the small-intestinal mucosa of male rats starved for 48 hr. Similarly, from the same laboratory, workers established that 3-methyl-4-methylamino-azobenzene-N-demethylase activity in the small intestine was decreased to less than 5% of controls by 48-hr starvation (59). In our laboratory we have showed decreased activity of 7-ethoxyresorufin (7ERR) deethylation in rat small intestine when the diet is restricted for 5 days. Although the liver weights are reduced in the animals on a restricted diet, 7ERR deethylation and cytochrome P450, calculated for the whole liver,

are similar whether the diet is restricted or fed ad libitum (Table 3).
It is thought that basal levels of drug-metabolizing enzymes are minimal in the intestine until induced by exogenous compounds in the diet.
This point is clearly illustrated in Table 3, where small-intestinal mixed
function oxidase activity is barely detectable in rats fed a purified diet
compared with those fed the normal stock diet.

2. Humans

Krishnaswamy (17,60,61,62) in India, has studied extensively the effect
of undernutrition on drug metabolism in both adults and children. She
concludes that in mild or moderate undernutrition, particularly in adults,
the rate of drug metabolism is either normal or slightly increased. Only
in severely malnourished adults with nutritional edema is drug metabolism impaired and similar to that observed in children with marasmus or
kwashiorkor.

In considering the effects of undernutrition on drug metabolism, it
is important to be aware of the pathophysiological alterations that occur
in such situations. These include changes in the absorptive functions
of the gastrointestinal tract, fatty liver, major changes in body fluids,
loss of plasma proteins, and many metabolic and hormonal changes, including alterations in cardiac and renal physiology (17,61). The effects
of these changes on drug metabolism have been demonstrated in several
studies (61-67) and are particularly significant in drug metabolism
studies when the volume of distribution is altered.

Krishnaswamy and Naidu (62) in their study on antipyrine metabolism
in adult subjects with nutritional edema, showed that the plasma half-
life of the drug varied widely, but that in one-third of the malnourished
subjects it was increased 44 to 88% compared with controls. Unfortunately, apparent distribution volumes and metabolic clearance rates for
antipyrine were not calculated in this study, and the subjects were not
restudied after nutritional rehabilitation. Mehta et al. (63,64) looked
at the metabolism of chloramphenicol, acetaminophen, sulfadiazine, and
antipyrine in children suffering from protein-energy malnutrition (PEM).
The plasma half-lives of all four drugs were significantly increased in
the PEM group compared with age-matched controls. There was a concomitant decrease in the plasma elimination rate constant. They obtained
similar results when using the same children after nutritional rehabilitation as their control group. A urinary excretion pattern of chloram0pheni-
col, before and after 4 to 8 weeks of rehabilitation, was also available
for five children with PEM. In PEM only 35 to 55% of the drug excreted
was in the conjugated form, and urinary drug concentration was 100
μg/ml at 30 hr. After rehabilitation, most of the drug was excreted by
24 hr, with 55 to 65% in the conjugated form. Similarly, with sulfadiazine the PEM children excreted less drug per body weight and a
decreased proportion of acetylated drug. These observations of altered

Table 3 Mixed Function Oxidase Activity in Homogenates of Liver, Small Intestine, and Large Intestine, From Rats Fed 41B Stock Diet Ad Libitum or 41B Stock Diet Restricted

| Diet | 7-Ethoxyresorufin deethylation (pmol/mg protein/min) | | | Cytochrome P450 (nmol/g wet weight) | Liver weight (% body wt) |
	Small intestine	Large intestine	Liver	Liver	
Purified	7.1 (1.8)	11.3 (6.2)	50.5 (6.3)	46.6 (4.0)	5.1 (0.4)
41B ad libitum	206[a] (80)	9.4 (1.2)	131[a] (19)	43.4 (4.1)	4.9 (0.3)
41B restricted	98 (126)	17.3 (10)	143[a] (65)	54.7[a] (9.0)	3.4[a] (0.2)

[a]Significantly altered from controls (P < .05).
Figures are for mean (± standard deviation) for groups of four rats.
Diets fed for 5 days.
41B restricted so that rats maintained their original body weight. Weight gain (g) in ad libitum fed rats was 47 ± 3.
(For methods see R. E. McDanell and A. E. M. McLean, *Biochem. Pharmacol.* 33(12), 1977 (1984).)

drug kinetics are of particular importance with drugs having a narrow
therapeutic index of which sulfadiazine is one.

Monckberg et al. (65) looked at the kinetics of salicylate in infants
with marasmus. The half-life of free salicylates was 7.5 hr in the malnourished group compared with 1.1 hr in controls. Similarly, the elimination rate constant was very much lower in children with marasmus.
Studies on children with kwashiorkor (66) showed that they excreted
higher amounts of unmetabolized chloroquine than was found after dietary rehabilitation. Balmer et al. (67) demonstrated that tetrachloroethylene, used in the treatment of hookworm infestation, is less toxic
in malnutrition, perhaps because of decreased formation of toxic metabolites.

VI. CARBOHYDRATE

Only a few studies have examined the effect of carbohydrates on drug
metabolism. The earlier experiments looked at the effects of feeding
carbohydrate as the sole dietary source. This, however, is an un-
realistic situation in terms of normal dietary consumption, and therefore
later studies took account of the effect of altering the level of carbohydrate in the diet.

As discussed previously, in studies on carbohydrate and protein there
is usually an inverse variation between the two components so that, for
example, a low-protein diet could also be considered a high-carbohydrate
diet. It recently has been suggested that carbohydrate and protein
have an opposite effect on drug metabolism, and therefore the magnitude
of response is greatest when these components are altered relative to
each other, rather than when they are examined individually (16). This
will be discussed later under human studies.

A. Experimental Animals

Kato (68) has shown that feeding sucrose instead of the normal stock diet
for 72 hr decreases hepatic aminopyrine demethylation, pentobarbital
hydroxylation, NADPH dehydrogenase, and cytochrome P450. He went
on to establish (69) the same effect for a range of NADPH-dependent
enzymes in male and female rats fed a sucrose diet, and he demonstrated
that phenobarbitone induction still occurred in sucrose-fed animals.

In accordance with the in vitro results, Kato (3) found a decrease in
the in vivo metabolism of the muscle relaxant carisoprodol and enhancement of carisoprodol paralysis and strychnine mortality in sucrose-fed
rats. Peters and Strother (70) similarly have shown prolonged hexobarbitone sleeping time associated with a decrease in the in vitro metabolism of hexobarbitone and benzphetamine in rats whose normal diet was
supplemented with 30% glucose in the drinking water. There was also

a significant decrease in NADPH oxidase, NADPH cytochrome c reductase, cytochrome P450 and cytochrome P450 reductase. A similar in vitro response has been observed in mice (71). Comparing glucose, sucrose, and fructose, these workers found that glucose was the most effective in decreasing drug-metabolizing activity.

Venkatesan et al. (72) showed a marked decrease in DMN demethylase activity in hepatic microsomes from rats fed glucose for 24 hr before sacrifice, and Dickerson et al. (73) have reported that rats fed sucrose, or glucose plus fructose, have lower levels of hepatic cytochrome P450. In terms of toxicity Boyd et al. (74) reported that high-sucrose diets, compared with starch, potentiated the lethal reactions to benzylpenicillin in rats. However in these studies there was a very high mortality in the control groups.

Sato, Nakajima, and their colleagues have published recent papers (15,75) investigating the effect of dietary carbohydrate on the metabolism of eight volatile hydrocarbons in rats. They reported that a diet deficient in carbohydrate significantly enhanced the metabolism of these compounds irrespective of the protein or fat content of the diet. They conclude that it is dietary carbohydrate, and not protein as was previously believed, that regulated hydrocarbon metabolism. However their measured rates of carbon tetrachloride metabolism were very low. Measured as clearance of carbon tetrachloride from the head space of a sealed vial, this group measured rates from 0.07-0.24 μmol/g of liver per hour, depending upon the diet. In contrast, Seawright and McLean (76) found a 10-times higher rate of metabolism of carbon tetrachloride to carbon dioxide, in the region of 1-2 μmol/g of liver per hour. In addition, the liquid diets used by Nakajima et al. (15) were mostly fed in amounts that were partly caloric restricted to ensure complete consumption. Some of their diets contained 50%, or more, of the energy as fat and may have caused ketosis and a stimulation of drug metabolism by ketone bodies (77).

B. Humans

Less work has been done on the specific effects of carbohydrate in humans. Once again the problem arises that in altering the proportion of one component of the diet, the proportion of others must vary in a reciprocal manner. Kappas and his co-workers (16,36) looked at the effect of varying carbohydrate and protein on the metabolism of antipyrine and theophylline. In six healthy volunteers, the isocaloric change from a high-protein to a high-carbohydrate diet resulted in a substantial increase in the plasma half-lives of both drugs. They also demonstrated that supplementing a well-balanced control diet with an extra 200 g of sucrose daily for 2 weeks led to a decrease in the removal rates of the drugs from plasma, whereas supplementing the diet with protein increased removal rates. They conclude that when fed as supplementary calories to the normal diet, protein and carbohydrate have opposite influences on drug oxidations, but that the magnitude of effect is less than when carbohydrate and protein are varied relative to each other. This conclusion illustrates again that care must be observed in interpreting the effect of altering ratios of dietary components. The effects seen may as easily be due to a decrease in one component of the diet as to an increase in another, or even to some secondary effect such as ketosis, fat mobilization, or glycogen deposition.

VII. LIPIDS

Approximately 30 to 55% of the dry weight of the endoplasmic reticulum is lipid, mostly in the form of phospholipid. The phospholipid, phosphatidylcholine, is an integral component of the MFO system and is essential for optimal drug-metabolizing activity (78). The relative amount of phospholipids in the liver endoplasmic reticulum, as well as their fatty acid compositions, are greatly dependent upon the diet (79). In animals the amount of dietary lipid and the amount of saturated and unsaturated fatty acids in the diet can influence the activity and inducibility of the MFO system (80). So far however this effect has not been demonstrated in man.

A. Experimental Animals

As with other dietary components, results of animal studies on dietary lipid are varied. Most investigators agree, however, that MFO activity is less in animals fed low levels of dietary fat compared with controls. Feeding a lipid-free diet causes an inadequate synthesis of microsomal hydroxylating enzymes and cytochrome P450. Caster et al. (81) fed male rats diets containing 0.2 to 10% corn oil for 21 days and showed that drug metabolism both in vitro and in vivo was decreased at low levels of dietary corn oil. Wade et al. (80-82) similarly demonstrated that the apparent V_{max} for aniline hydroxylase, hexobarbital oxidase, and ethylmorphine demethylase in male rats, was enhanced 56, 53, and 47%, respectively, by diets containing 3% corn oil compared with fat-free controls. The apparent K_ms were statistically similar regardless of the fat content of the diet. Associated with the increased drug metabolism in corn oil-fed rats were increases in the concentration of cytochrome P450, decreased hexobarbitone sleep times, and decreased glucose-6-phosphate dehydrogenase activity (82). The microsomal contents of linoleic and arachidonic acids also increased as the corn oil content of the diet increased. Lang et al. (83) observed that rats fed a fat-free diet for 5 weeks exhibited significant decreases in hepatic O-demethylation of p-nitroanisol and 3-hydroxylation of benzo[a]pyrene.

Several reports have indicated that microsomes from animals fed diets containing polyunsaturated or essential fatty acids, have greater MFO activity than microsomes from rats fed fat-free diets or diets containing saturated fats. Century (84) reported that rats fed 7% beef fat (saturated), metabolized hexobarbital and aminopyrine more slowly than rats fed unsaturated fatty acids. Hietenan et al. (85) showed that rats fed cocoa butter or cocoa butter plus cholesterol in concentrations such that the total fat content of the diet was 24 to 34% by weight (i.e., 60% of the energy content) resulted in hepatic microsomes having only 10% of the activity of those from rats fed a diet containing an equivalent concentration of cholesterol.

Lambert and Wills (86) reported experiments in which depression of benzo[a]pyrene hydroxylase activity resulted from fat deficiency and was not recovered by the addition of 10% herring oil. On addition of 10% corn oil to the diet benzo[a]pyrene-hydroxylating activity was significantly increased. They suggest that the effect was due to the linoleic acid content of the diet. Work in our laboratory has shown that enzyme activity can be restored with herring oil as well as linoleic acid (87).

Dietary lipid has been shown to affect the inducibility as well as the basal activity of the MFO system. Marshall and McLean (87) demonstrated that in rats fed adequate (20%) dietary protein, the response to phenobarbitone is largely determined by the dietary fat content. They showed that to permit maximum induction of cytochrome P450 synthesis after administration of phenobarbitone, the dietary addition of either herring oil, linoleic acid, or 0.1% oxidized sitosterol was required. Since coconut oil, 5% olive oil, or saturated fatty acids were not effective, they suggested that some aspect of the polyunsaturation was responsible for the "permissive" effect. Century and Horwitt (88) similarly demonstrated that the ability of phenobarbitone treatment to stimulate hexobarbitone hydroxylase and aminopyrine demethylase was significantly increased by dietary inclusion of polyunsaturated fat in the form of menhaden or linseed oil. In our laboratory it has also been shown that feeding rats a purified, fat-free diet, depresses phenobarbitone induction, but that reducing dietary intake removes this inhibitory effect (Table 4).

B. Humans

Despite the many observations of the dietary lipid effect on drug metabolism in experimental animals, similar results have not been demonstrated in man. Anderson et al. (89) looked at the effects on drug metabolism rates in man when fat was substituted for carbohydrate or protein in the diet. Antipyrine and theophylline kinetics were studied in six healthy males during the sequential feeding of high-carbohydrate, high-fat, and high-protein diets over a 6-week period. Significant,

Table 4 Effect of Fat Free Diet and Diet Restriction on Induction
of Cytochrome P450 by Phenobarbitone (PB)

Diet	Feeding	Cytochrome P450/g liver
Stock pellets + PB	ad libitum	124 ± 12
20% casein/no fat + PB	ad libitum	62 ± 6
20% casein/no fat + PB	14 g/day	86 ± 12
20% casein/no fat + PB	10 g/day	101 ± 8
20% casein/no fat + PB	5 g/day	157 ± 25

Results are expressed as mean ± ISD for groups of four rats.
Rats fed ad libitum consume approximately 16 g/day
PB given at 1 mg/ml in drinking water for 6 days. Restricted diet
fed for 3 days.
Source: Ref. 42 and A. E. M. McLean and P. A. Day, unpublished
work.

though minor changes in plasma half-life and clearance were produced.
The average plasma half-life was shorter during the high-protein period
than during the high-fat or high-carbohydrate period. Changes in
half-life were accompanied by reciprocal changes in calculated metabolic
clearances, and there were no significant alterations in apparent distribution volumes. The authors suggest that alterations in drug metabolism are due to reduced protein, and that substituting fat for carbohydrate has little or no effect on metabolic rates.

In another study by the same group (89) the effects of saturated and
unsaturated fats were investigated. It was found that substituting
either corn oil (polyunsaturated) or butter (saturated) in place of carbohydrate in the diet had no influence on the metabolism of antipyrine
or theophylline. This effect was despite a concurrent alteration in
plasma lipids. They concluded that diet can alter the metabolic systems
that regulate plasma lipid levels without significantly affecting cytochrome P450-mediated drug metabolism.

Mucklow and his co-workers (90) studied antipyrine and debrisoquine
metabolism in normal volunteers, switching between saturated and unsaturated dietary fat. For 2 weeks dietary fat was provided as 95%
animal fat and 5% vegetable fat, and for the other 2 weeks as 33% animal
fat and 67% vegetable fat. They showed that short-term changes in the
proportion of animal to vegetable fat do not influence antipyrine clearance or debrisoquine hydroxylation.

VIII. VITAMINS

A. Vitamin A

Becking (91) showed that in weanling male rats fed a vitamin A-deficient diet for 20 to 25 days, the rate of metabolism of aminopyrine and
aniline was significantly lower both in vitro and in vivo. There was
no alteration in the in vivo metabolism of nitrobenzoic acid. Microsomal
protein concentration and NADPH cytochrome c reductase activity were
unaffected, but cytochrome P450 was significantly depressed.

Interest in vitamin A arose from observations that it could influence
tumor frequency induced by chemical carcinogens in experimental animals (92-94). This subject has been recently reviewed by Peto et al.
(95) with regard to beta-carotene and human cancer. In this respect
it would seem appropriate to look at the metabolism of the carcinogen
under study and the effect of vitamin A on MFO activity.

B. Vitamin B

1. *Thiamine* (B_1)

Unlike most other dietary components, a deficiency of thiamine has
been found to increase the metabolism both of drugs and a carcinogen
(96,97). Conversely, a high dietary level of thiamine (20 mg/g diet)
depresses the metabolism of many drugs and decreases cytochrome
P450 content and cytochrome c reductase activity (98-101). Wade and
his co-workers (98,99) have shown that dietary administration of high
levels of thiamine hydrochloride depresses the metabolism of aniline,
heptachlor, zoxazolamine, and aminopyrine in male rats without significantly altering hexobarbitone oxidation. Associated with these decreased rates of metabolism was a decrease in the content of cytochrome
P450, a decrease in activity of cytochrome c reductase, an increase in
liver weight, increased zoxazolamine paralysis time, and increased
glucose-6-phosphate dehydrogenase activity.

More recently, after using pair-feeding experiments, Wade et al. (100)
suggested that the depression of aniline hydroxylase, cytochrome c
reductase, and ethylmorphine demethylase is due to thiamine, but that
depression of cytochrome P450 and b_5 may be primarily due to the increased amount of carbohydrate ingested by rats fed the thiamine-
enriched diet. When starch was substituted for sucrose cytochrome
P450 was not lowered in the high-thiamine group. The concluded that
since pair-feeding experiments suggest that drug metabolism is depressed by thiamine ingestion and that the concentration of P450
is depressed by sucrose ingestion, the site of thiamine interaction
may not involve synthesis or maintenance of total cytochrome P450.

In more recent studies Ruchiwarat et al. (96,97) have shown that thiamine deficiency in rats increases the metabolism of dimethylnitrosamine and enhances plasma disappearance of paracetamol with increased production of the glucuronide and sulfate metabolites in the rat.

2. *Riboflavin and Niacin*

Since flavoproteins are components of the microsomal electron transport system, it would be expected that dietary deficiency of these vitamins would affect microsomal metabolism of drugs. Studies on the effects of these vitamins are few and conflicting depending upon the animal species and drug substrate used. These effects have been extensively reviewed by Kato (3).

C. Vitamin C

Vitamin C-deficiency studies are limited to those species (man, monkey, and guinea pig) whose requirements must be met via the diet. Most studies with vitamin C have shown decreased metabolism of a variety of pharmacological agents in deficient animals. However there is little information to date about the underlying biochemical basis for the action of the vitamin.

Richards et al. (102) originally demonstrated in 1941 that prolonged phenobarbitone sleep times in scorbutic guinea pigs could be reversed by the administration of ascorbic acid. Other workers subsequently showed an increased half-life for compounds such as acetanilide, aniline, antipyrine, zoxazolamine, and coumarin in scorbutic animals, all of which could be reversed by vitamin C administration (103-105).

Kato et al. (105) looked at the microsomal metabolism of a range of substrates in guinea pigs maintained on a vitamin C-deficient diet for 12 days. They observed decreased metabolism of aniline, hexobarbitone, and zoxazolamine but no effect on various other drugs, or on electron transport components. They conclude that the effect of vitamin C is specific for hydroxylation reactions. Conversely, Zannoni et al. (106,107) have shown an in vitro decrease in both hydroxylation and demethylation, as well as a significant decrease in cytochrome P450 and NADPH cytochrome P450 reductase. However this was only when microsomal ascorbic acid levels reached 30% of normal values. These decreased enzyme activities in vitamin C-deficient animals can be restored to normal levels within a few days of vitamin C repletion (108).

Vitamin C effects on drug metabolism have been reviewed by Zannoni et al. (108).

In studies on humans, Hollaway et al. (109) found that subclinical ascorbic acid deficiency of short duration, in five normal male volunteers, had no effect on antipyrine metabolism. Ascorbic acid deficiency however may occur in some patients with liver disease and be associated

with a prolonged plasma half-life of antipyrine (110). With large doses
of vitamin C there appears to be no effect on the kinetics of antipyrine
or diphenylhydantoin (111). However inhibition of sulfate conjugation
of drugs by high doses of vitamin C has been reported (112) probably
as a result of competition for available sulfate.

D. Vitamin E

Dietary deficiency of vitamin E has been reported to reduce rat liver
MFO activity (113-115). An explanation offered for this is that there
may be an impairment of heme synthesis, as it has been reported that
vitamin E deficiency in rats leads to decreased activities of bone mar-
row aminolevulinic acid synthetase and hepatic delta-aminolevulinic
hydratase (116).

An alternative view of how vitamin E may function in hepatic drug
metabolism is related to the vitamin's ability to scavenge free radicals
and inhibit lipid peroxidation. However Carpenter (113) has shown
that feeding N,N-diphenyl-p-phenylenediamine (DPDD) to vitamin E-
deficient rats for 48 hr, completely eliminated microsomal lipid peroxidation but did not affect the codeine and aminopyrine demethylation
activities in these animals. Another possible role for vitamin E was
proposed by Diplock et al. (117,118), who suggested that it may function as an inhibitor of the oxidation of selenide-containing rpoteins.
However other workers have not found depressed MFO activity in vitamin E deficiency (119). We, in our laboratory, studied rats fed yeast
diets severely deficient in both vitamin E and selenium and did not find
any defect in drug metabolism either in vitro or in vivo (unpublished).

IX. MINERALS

It is important to study the effects of dietary minerals on drug metabolism. Apart from gross deficiencies in malnourished individuals, marginal
mineral deficiencies may occur in many countries. Both essential and
nonessential minerals have been shown to affect drug metabolism, although results of these studies are extremely variable.

A. Calcium and Magnesium

Dingell et al. (120) reported decreased metabolism of hexobarbitone,
aminopyrine, and p-nitrobenzoic acid in rats fed calcium-deficient diets
for 40 days. Becking (121) later demonstrated a much more rapid effect of magnesium deficiency with significantly decreased microsomal
aminopyrine metabolism, cytochrome P450, and cytochrome c reductase
occurring within 12 days of feeding a magnesium-deficient diet. However oxidation of pentobarbitone, phentobarbitone sleep times, and
p-nitrobenzoic acid metabolism were unaffected.

More recent studies (122) have suggested that the magnesium effect on drug metabolism is mediated in some way by thyroid hormone levels because thyroxine levels in magnesium-depleted rats were substantially lower than in controls. It also has been suggested that there may be an effect on the phosphatidylcholine component of the MFO system since a marked decrease in lysophosphatidylcholine levels was noted in rats after 12 days on a magnesium-deficient diet (122).

B. Iron

There is a need for studies of the effect of both iron deficiency and increased iron levels on drug metabolism since both of these conditions occur commonly, even in developed countries, caused by dietary deficiency and bleeding and self-medication, respectively.

It was assumed that iron deficiency would result in a decrease in cytochrome P450 because it is a hemoprotein, requiring iron in its bio-

synthesis. However, Becking (121) demonstrated that the in vitro microsomal metabolism of aminopyrine and aniline was markedly increased during iron depletion but was unaltered after 35 days on a diet containing 1.5 times the normal iron content. Pentobarbitone metabolism and nitroreductase activity were unaltered by iron-deficient or high-iron diets. Conversely, Catz et al. (123) working with mice showed an increased hepatic metabolism of hexobarbitone and aminopyrine but no change in the rate of aniline metabolism. Hoensch et al. (124) demonstrated a marked decrease in intestinal drug metabolism in iron-deficient rats within 2 days. This could be of great significance in man where intestinal metabolism may have a protective effect against toxic compounds ingested in the diet.

Because of widespread marginal iron deficiency, it is important to establish the clinical significance in humans. O'Malley and Stevenson (125) showed that the antipyrine plasma half-life was unaltered in eight patients with severe iron deficiency. Although not all of these cases were due to dietary iron deficiency, it would seem to indicate that there may not be an effect of this deficiency, as such, on drug metabolism.

Other trace metals are likely to have an effect on the drug-metabolizing enzyme system, although these are less well studied and results are variable. These include copper, zinc, selenium, chromium, manganese, nickel, cadmium, cobalt, and lead. All of the studies to date have looked at effects of single minerals, and the interrelationships of different minerals need to be further investigated.

X. PHASE-2 METABOLISM

Most studies on diet and drug metabolism have looked at phase-1 oxidation reactions. In terms of both metabolism and toxicity it is important also to look at subsequent phase-2 conjugations. These are mainly detoxification reactions, although recently a few examples of activation by conjugation have been reported (14).

Most studies on diet and phase-2 metabolism have looked at sulfation and glucuronidation. Sulfation requires a sufficient supply of inorganic sulfate to synthesize the cosubstrate 3'-phosphoadenosine-5'-sulfatophosphate (PAPS), while glucuronidation uses UDPG-glucuronate (UDPGA) as cosubstrate. Decreases in sulfate or UDPGA availability have been shown to decrease the rates of sulfation and glucuronidation (126,127).

1. *Experimental Animals*

Woodcock and Wood (128) demonstrated that male rats fed a protein-free diet for 7 days had higher microsomal UDP-glucuronyltransferase activity than control rats when the substrates p-nitrophenol and o-aminophenol were used. No significant differences in the sulfotransferases were seen with these two substrates. Magdalou et al. (129) similarly showed a large increase in microsomal UDP-glucuronyltransferase and cytochrome P450 in male rats fed a protein-free diet. Repletion with normal protein diet restored control levels of UDP-glucuronyltransferase and cytochrome P450 within 5 days. However repletion with a diet deficient in sulfur containing amino acids maintained the high levels seen in protein-deficient animals. They suggest that a diet deficient in methionine and cysteine leads to a lack of sulfur-containing precursors and therefore a decrease in sulfoconjugation. Because this diet provides enough saccharide in the form of starch to ensure sufficient quantities of UDP-glucuronic acid, it seems that glucuronidation may be increased to compensate for the deficiency in sulfoconjugation.

A few studies have looked at dietary effects on conjugation in vivo. Glazenberg et al. (130) have studied the effect of feeding a methionine-deficient diet on the concentrations of the cofactors for drug conjugation. He showed that in rats fed the methionine-deficient diet, hepatic glutathione decreased to 20% of control values, and the hepatic concentration of active sulfate, PAPS, was also decreased. Sulfate conjugation of paracetamol decreased to 50% of control values, although formation of the glutathione conjugate remained unaffected despite the dramatic desuffering from PEM excreted 74% of a dose of benzoic acid as the glycine conjugate and the remainder as the glucuronide conjugate. This is compared with a 99% excretion of the glycine conjugate in controls. They suggest that this is due to a decrease in availability of glycine in PEM rats and an associated increase in UDP-glucuronide transferase activity with subsequently increased glucuronide formation.

The effect of fasting, as opposed to protein deficiency, has also been studied. Mulder et al. (132) examined conjugation in rats fasted for 72 hr but were unable to show any effect on sulfation or glucuronidation

of harmol and suggest that these processes are not easily disturbed by
dietary manipulation. Alvin and Dixit (133) showed that glucuronide
conjugation of chloramphenicol was decreased in vivo by 18 to 24-hr
fasting in male rats. However other workers (134) have shown that
in rats fasted for 72 hr glucuronidation of p-nitrophenol was not significantly altered. In our laboratory we have shown a 75% decrease
in hepatic glucuronyl transferase in rats that had been fasted overnight.

2. Humans

Few studies have looked at these effects in man. Mehta et al. (135)
have shown that children with PEM excreted 35 to 55% of a dose of
chloramphenicol as the glucuronide conjugate, whereas after dietary
rehabilitation 55 to 65% of the dose was excreted as the glucuronide.
Krishaswamy (61) looked at the metabolism of sulfadiazine in undernourished subjects and showed that a significantly higher proportion
of the acetylated metabolite was excreted in the urine in these subjects
compared with well-nourished individuals. She suggests that this was
caused by the significantly lower plasma protein binding of the drug
in the undernourished group and therefore greater availability of the
free drug for acetylation.

XI. CONCLUSION

Man is continuously exposed, both by choice and circumstance, to a
range of foreign compounds in his environment. Many of these compounds are metabolized by a series of enzymes, which were first associated with drug detoxification. Most foreign compounds are inactivated
by this enzyme system, although some compounds, e.g., carcinogens
such as aflatoxin and benzo[a]pyrene, are now known to be converted
to highly reactive metabolites.
   Many studies in experimental animals and man have shown that the
activity of the drug-metabolizing enzymes is markedly influenced by
diet. This is of obvious importance in the large numbers of people
throughout the world who are malnourished to varying degrees. Malnutrition almost invariably leads to a reduced capacity for metabolism,
with a consequent alteration in the toxicity of the substrate, depending upon whether it is activated or inactivated.
   One feature of a review in this field is the enormous variation in experimental results from different workers. There are a number of reasons for these disparities. First, methods are often poorly standardized,
particularly with regard to kinetics under the extreme physiological
and pathological states likely to be encountered in dietary deficiency.
Similarly, a whole series of experimental models are employed, from
in vivo studies in the whole animal to studies in isolated systems such

as tissue homogenates, microsomes, or cell cultures. It often is difficult to extrapolate with any validity from one system to another. Finally, there are enormous genetic variations in drug-metabolizing activity, both within and between species.

In any study of dietary effects on drug metabolism it is important to recognize that in altering the component of the diet under investigation, the relative proportions of the other components will change. This makes exact interpretation of the causal factor difficult. Similarly, in long-term nutrient deficiency a range of other systems are altered that may produce a subsequent effect on xenobiotic metabolism.

REFERENCES

1. T. K. Basu and J. W. T. Dickerson, Inter-relationship of nutrition and the metabolism of drugs. *Chem. Biol. Interact.* *8*:193 (1974).
2. T. C. Campbell and J. R. Hayes, Role of nutrition in the drug metabolizing enzyme system. *Pharmacol. Rev. 26*(3):171 (1974).
3. R. Kato, Drug metabolism under pathological and abnormal physiological states in animals and man. *Xenobiotica* 7:25 (1977).
4. Energy and protein requirements. *WHO Tech. Rep. Ser. 522* (1973).
5. W. E. Blumberg, Enzymic modification of environmental intoxicants. *Q. Rev. Biophys. 11*:482 (1978).
6. R. M. Welch, Toxicological implications of drug metabolism. *Pharmacol. Rev. 30*:457 (1979).
7. H. Autrup, Carcinogen metabolism in human tissues and cells. *Drug Metab. Rev. 13*:603 (1982).
8. R. T. Williams, Nutrients in drug detoxication reactions, in *Nutrition and Drug Interrelations* (J. H. Hathcock and J. Coon, eds.). Academic Press, New York, 1978, p. 303.
9. A. Kappas and A. P. Alvares, How the liver metabolizes foreign substances. *Sci. Am. 232*:22 (1975).
10. M. J. Coon, T. A. van der Hoeven, D. A. Haugen, F. P. Guengen, J. L. Vermilon, and D. P. Ballou, Biochemical characterization of highly purified cytochrome P450 and other components of P450 system of liver microsomal membranes, in *Cytochrome P450 and b$_5$* (D. Y. Cooper, ed.). Plenum Press, New York, 1975, p. 303.
11. T. E. Gram, ed., in *Extrahepatic Metabolism of Drugs and other Foreign Compounds*. MTP Press Ltd., Lancaster, U.K., 1980.
12. A. H. Conney, Pharmacological implications of microsomal enzyme induction. *Pharmacol. Rev. 19*:317 (1967).

13. A. H. Conney, Induction of microsomal enzymes by foreign chemicals and carcinogenesis by polycyclic aromatic hydrocarbons. *Cancer Res. 42*:4875 (1982).

14. G. J. Mulder, Detoxification or toxification? Modification of the toxicity of foreign compounds by conjugation in the liver. *Trends in the Biochemical Sciences 6*:512 (1979).

15. T. Nakajima, Y. Koyama, and A. Sato, Dietary modification of metabolism and toxicity of chemical substances with special reference to carbohydrate. *Biochem. Pharmacol. 31*:1005 (1982).

16. K. E. Anderson, A. H. Conney, and A. Kappas, Nutritional influences on chemical biotransformations in humans. *Nutr. Rev. 40*:161 (1982).

17. K. Krishnaswamy, Drug metabolism and pharmacokinetics in malnutrition. *Clin. Pharmacokinet. 3*:216 (1978).

18. A. E. Rogers, Nutrition, in *The Laboratory Rat* Vol. 1 (H. J. Baker, J. R. Lindsey, and S. H. Weisbroth, eds.). Academic Press, New York, 1979, p. 123.

19. L. R. Njaa, Weight maintenance and protein intake of the young rat. *Br. J. Nutr. 19*:433 (1965).

20. M. U. K. Mgbodile and T. C. Campbell, Effect of protein deprivation in male weanling rats on the kinetics of hepatic microsomal enzyme activity. *J. Nutr. 102*:53 (1972).

21. J. R. Hayes, M. U. K. Mgbodile, A. H. Merril, L. S. Neurukar, and T. C. Campbell, The effect of dietary protein depletion and repletion on rat hepatic mixed function oxidase activities. *J. Nutr. 108*:1788 (1978).

22. T. C. Campbell, J. R. Hayes, A. H. Merrill, M. Maso, and M. Goetchius, Nutritional status and drug metabolism. *Drug Metab. Rev. 9(2)*:173 (1979).

23. S. Kuwano and K. Hiraga, Effect of dietary protein deficiency on the rat hepatic drug metabolizing system. *Jpn. J. Pharmacol. 30*:75 (1980).

24. R. Kato, E. Chiesara, and P. Vassanelli, Factors influencing induction of hepatic microsomal drug metabolizing enzymes. *Biochem. Pharmacol. 11*:211 (1962).

25. R. Kato, Sex differences in the activities of microsomal drug metabolizing enzyme systems in relation to dietary protein. *Jpn. J. Pharmacol. 16*:221 (1966).

26. A. E. M. McLean and E. K. McLean, The effect of diet and DDT on microsomal hydroxylating enzymes and on sensitivity of rats to carbon tetrachloride poisoning, *Biochem. J. 100*:564 (1966).

27. E. Hietanen, Modifications of hepatic drug metabolizing enzyme activities and their induction by dietary protein. *Gen. Pharmacol. 11*:443 (1980).

28. R. Kato, T. Oshima, and S. Tomizawa, Toxicity and metabolism of drugs in relation to dietary protein. *Jpn. J. Pharmacol. 18*:356 (1968).

29. W. M. Weatherholtz, T. C. Campbell, and R. E. Webb, Effect of dietary protein levels on the toxicity and metabolism of heptachlor. *J. Nutr. 98*:90 (1969).

30. W. J. Marshall and A. E. M. McLean, The effect of oral phenobarbitone on hepatic microsomal drug metabolizing enzymes. *Biochem. Pharmacol. 18*:153 (1969).

31. J. R. Hayes, M. U. K. Mgbodile, and T. C. Campbell, Effect of protein deficiency on the inducibility of the hepatic microsomal drug metabolizing system I. *Biochem. Pharmacol. 22*:1005 (1973).

32. M. U. K. Mgbodile, J. R. Hayes, and T. C. Campbell, Effect of protein deficiency on the inducibility of the hepatic microsomal drug metabolizing system II. *Biochem. Pharmacol. 11*:1125 (1973).

33. J. R. Hayes and T. C. Campbell, Effect of low protein diet on the hepatic microsomal drug metabolizing enzyme system III. *Biochem. Pharmacol.* 23:1721 (1974).

34. A. Kappas, K. E. Anderson, A. H. Conney, and A. P. Alvares, Influence of dietary protein and carbohydrate on theophylline metabolism in man. *Proc. Nat. Acad. Sci. USA 73*:2501 (1976).

35. C. Balabaud, G. Vinon, and J. Paccalin, Influence of dietary protein and carbohydrate on phenytoin metabolism in man. *Br. J. Clin. Pharmacol. 9*:523 (1980).

36. C. H. Feldman, V. E. Hutchinson, C. E. Pippinger, T. Blumenfeld, B. R. Feldman, and W. J. Davis, Effect of dietary protein and carbohydrate on theophylline metabolism in children. *Pediatrics 66*:956 (1980).

37. J. C. Mucklow, M. T. Caraher, D. B. Henderson, P. H. Chapman, D. F. Rioberts, and M. D. Rawlins, The relationship between individual dietary constituents and antipyrine metabolism in Indo-Pakistani immigrants to Britain. *Br. J. Clin. Pharmacol. 13*:481 (1982).

38. M. J. Brodie, A. R. Boobis, E. L. Toverud, W. Ellis, S. Murray, C. T. Dollery, S. Webster, and R. Harrison, Drug metabolism in white vegetarians. *Br. J. Clin. Pharmacol. 9*:523 (1980).

39. E. M. Boyd, Dietary protein and pesticide toxicity in male weanling rats. *Bull. WHO 40*:80 (1969).

40. E. M. Boyd, M. A. Boulanger, and E. S. DeCastro, Phenacitin toxicity and dietary protein. *Pharmacol. Res. Commun. 1*:15 (1969).

41. R. E. Webb, C. C. Bloomer, and C. K. L. Miranda, Effect of casein diets on the toxicity of malathion and parathion and their oxygen analogues. *Bull. Environ. Contam. Toxicol. 9*:102 (1973).

42. A. E. M. McLean and P. A. Day, The effect of diet on the toxicity of paracetamol and the safety of paracetamol-methionine mixtures. *Biochem. Pharmacol. 24*:37 (1975).

43. P. Czygan, H. Greim, A. Garr, F. Schaffner, and H. Popper, The effect of dietary protein deficiency on the ability of isolated hepatic microsomes to alter the mutagenicity of a primary and secondary carcinogen. *Cancer Res. 34*:119 (1974).

44. J. White and H. B Andervant, Effect of a diet relatively low in cystine on the production of spontaneous mammary gland tumors in strain C3H male mice. *J. Nat. Cancer Inst. 3*:449 (1943).

45. H. Silverstone and A. Tannenbaum, Proportion of dietary protein and the formation of spontaneous hepatomas in the mouse. *Cancer Res. 11*:442 (1951).

46. M. A. Walters and F. J C. Roe, The effect of dietary casein on the induction of lung tumors by the injection of dimethylbenzanthracene into new born mice. *Br. J. Cancer 18*:312 (1964).

47. T. V. Madahavan and C. Goplan, The effect of dietary protein on the carcinogenicity of aflatoxin. *Arch. Pathol. 85*:133 (1968).

48. S. K. Clinton, C. R. Truex, and W. J. Visek, Dietary protein, aryl hydrocarbon hydroxylase and chemical carcinogenesis in rats. *J. Nutr. 109*:55 (1979).

49. B. Scott Appleton and T. C. Campbell, Effect of high and low dietary protein on the dosing and post-dosing periods of aflatoxin B_1 induced hepatic preneoplastic lesion development in the rat. *Cancer Res. 43*:2150 (1983).

50. C. L. Miranda and R. E. Webb, Effects of dietary protein quality on drug metabolism in the rat. *J. Nutr. 103*:1425 (1973).

51. N. Kato, T. Tani, and A. Yoshida, Effect of dietary quality of protein on the liver microsomal mixed function oxidase system, plasma cholesterol and urinary ascorbic acid in rats fed PCB. *J. Nutr. 111*:123 (1981).

52. R. L. Dixon, W. Shultice, and J. R. Fouts, Factors affecting drug metabolism by liver microsomes IV. Starvation. *Proc. Soc. Exp. Biol. Med. 103*:333 (1960).

53. R. Kato and A. Takanata, Effect of starvation on the in vivo metabolism and effect of drugs in male and female rats. *Jpn. J. Pharmacol. 17*:208 (1967).

54. R. Kato and J. R. Gillette, Effect of starvation on NADPH dependent enzymes in liver microsomes of male and female rats. *J. Pharmacol. Exp. Ther. 150*:279 (1965).

55. R. Kato, K. Onoda, and A. Takanata, Species differences in drug metabolism by liver microsomes in alloxan diabetic or fasted animals. *Jpn. J. Pharmacol. 20*:546 (1970).

56. T. Nakajima and A. Sato, Enhanced activity of liver drug metabolizing enzymes for aromatic and chlorinated hydrocarbons following food deprivation. *Tox. App. Pharmacol. 50*:549 (1979).

57. M. Marselos and M. Laitinen, Starvation and phenobarbitone treatment effects on drug hydroxylation and glucuronidation in the rat liver and small intestinal mucosa. *Biochem. Pharmacol.* *24*:1529 (1975).

58. L. W. Wattenberg, J. L. Leong, and P. J. Strand, Benzo[a]pyrene hydroxylase activity in the gastrointestinal tract. *Cancer Res.* *22*:1120 (1962).

59. R. Billings and L. W. Wattenberg, The effects of dietary alterations on 3-methyl-4-aminoazobenzene-N-demethylase activity. *Proc. Soc. Exp. Biol. Med.* *139*:965 (1972).

60. K. Krishnaswamy, Nutrition and drug metabolism. *Indian J. Med. Res.* *68*(Suppl):109 (1978).

61. K. Krishnaswamy, Drug metabolism and pharmacokinetics in malnutrition. *Trends Pharmacol. Sci.* *4(7)*:295 (1983).

62. K. Krishnaswamy and A. N. Naidu, Microsomal enzymes and malnutrition as determined by plasma half-life of antipyrine. *Br. Med. J.* *1*:538 (1977).

63. S. Mehta, H. K. Kalsi, S. Jayaraman, and V. S. Mathur, Chloramphenicol metabolism in children with protein calorie malnutrition. *Am. J. Clin. Nutr.* *28*:977 (1975).

64. S. Mehta, C. K. Nain, B. Sharma, and V. S. Mathur, Drug metabolism of malnourished children. *Prog. Clin. Biol. Res.* *77*: 379 (1980).

65. F. Monckberg, M. Bravo, and O. Gonzalez, Drug metabolism and infantile undernutrition, in *Nutrition and Drug Interrelations* (J. H. Hathcock and J. Coon, eds.). Academic Press, New York, 1978, p. 399.

66. B. A. Wharton and E. W. McChesney, Chloroquine metabolism in kwashiorkor. *J. Trop. Pediatr.* *16*:130 (1970).

67. S. Balmer, G. Howells, and B. A. Wharton, The effects of tetrachloroethylene in children with kwashiorkor and hookworm infestation. *J. Trop. Pediatr.* *16*:20 (1970).

68. R. Kato, Possible role of P450 in the oxidation of drugs in liver microsomes. *J. Biochem.* *59*:574 (1966).

69. R. Kato, Effect of phenobarbitone treatment on the activities of NADPH dependent enzymes of liver microsomes in fasted or sucrose fed rats. *Jpn. J. Pharmacol.* *17*:181 (1967).

70. M. A. Peters and A. Strother, A study of some possible mechanisms by which glucose inhibits drug metabolism in vivo and in vitro. *J. Pharmacol. Exp. Ther.* *180*:151 (1972).

71. A. Strother, J. K. Throckmorton, and C. Herzer, The influence of high sugar consumption by mice on the duration of action of barbiturates and in vitro metabolism of barbiturates, aniline and p-nitroanisole. *J. Pharmacol. Exp. Ther.* *179*:490 (1971).

72. N. Venkatesan, J. C. Arcos, and M. F. Argus, Amino acid induction and carbohydrate repression of dimethylnitrosamine demethylase activity in rat liver. *Cancer Res. 30*:2563 (1970).

73. J. W. T. Dickerson, T. K. Basu, and D. V. Parke, Activity of drug metabolizing enzymes in the liver of growing rats fed on diets high in sucrose, glucose or fructose or equimolar glucose and fructose, *Proc. Nutr. Soc. 30*:27A (1971).

74. E. M. Boyd, I. Dobos, and F. Taylor, Benzylpenicillin toxicity in albino rats fed synthetic high starch versus high sugar diets. *Chemotherapy 15*:1 (1970).

75. A. Sato and T. Nakajima, A vial equilibration method to evaluate the drug metabolizing enzyme activity for volatile hydrocarbons. *Tox. App. Pharmacol. 47*:41 (1979).

76. A. A. Seawright and A. E. M. McLean, The effect of diet on carbon tetrachloride metabolism. *Biochem. Pharmacol. 105*: 1055 (1967).

77. G. J. Traiser and G. L. Plaa, Effect of aminotriazole on iso-propanol and acetone potentiated carbon tetrachloride hepato-toxicity. *Can. J. Physiol. Pharmacol. 51*:291 (1973).

78. H. W. Strobel, A. Y. H. Lu, J. Heideman, and M. J. Coon, Phosphatidylcholine requirement in the enzymatic reduction of hemoprotein P450 and in fatty acid, hydrocarbon and drug hydroxylation. *J. Biol. Chem. 245*:4851 (1970).

79. W. P. Norred and A. E. Wade, Dietary fatty acid induced alterations of hepatic microsomal drug metabolism. *Biochem. Pharmacol. 21*:2887 (1972).

80. A. E. Wade and W. P. Norred, Effect of dietary lipid on drug metabolizing enzymes. *Fed. Proc. 35*:2475 (1976).

81. W. O. Caster, A. E. Wade, F. E. Greene, and J. S. Meadows, Effect of different levels of corn oil in the diet upon the rate of hexobarbital, heptachlor and aniline metabolism in the liver of the male white rat. *Life Sci. 9*:181 (1970)

82. A. E. Wade, W. P. Norred, and J. S. Evans, Lipids in drug detoxication, in *Nutrition and Drug Interrelations* (J. N. Hathcock and M. J. Coon, eds.). Academic Press, New York, 1978, p. 475.

83. M. Lang, E. Hietenan, O. Hanninen, and M. Laitinen, Inducibility of hepatic drug metabolizing enzymes during fat deficiency. *Gen. Pharmacol. 9*:381 (1978).

84. B. Century, Lipids affecting drug metabolism and cellular functions, in *Drugs Affecting Lipid Metabolism* (C. W. Holmes, ed.). Plenum Press, New York, 1969, p. 629.

85. E. Hietenen, M. Laitinen, H. Vanio, and O. Hanninen, Dietary fats and properties of the endoplasmic reticulum. II. Dietary lipid induced changes in activities of drug metabolizing enzymes of liver and duodenum of the rat. *Lipids 10*:467 (1975).

86. L. Lambert and E. D. Wills, The effect of dietary lipids on benzo[a]pyrene metabolism in hepatic endoplasmic reticulum. *Biochem. Pharmacol.* *26*:1423 (1977).

87. W. J. Marshall and A. E. M. McLean, A requirement for dietary lipids for induction of cytochrome P450 by phenobarbitone in rat liver microsomal fraction. *Biochem. J.* *122*:569 (1971).

88. B. Century and M. K. Horwitt, A role of dietary lipid in the ability of phenobarbitone to stimulate hexobarbitone and antipyrene metabolism. *Fed. Proc.* *27*:349 (1968).

89. K. E. Anderson, A. H. Conney, and A. Kappas, Nutrition and oxidative drug metabolism in man. Relative influence of dietary lipids, carbohydrate and protein. *Clin. Pharmacol. Ther.* *26*: 493 (1979).

90. J. C. Mucklow, M. T. Carahar, J. R. Idle, M. D. Rawlins, T. Sloan, R. L. Smith, and P. Wood, The influence of changes in dietary fat on the clearance of antipyrine and the 4-hydroxylation of debrisoquine. *Br. J. Clin. Pharmacol.* *9*:283P (1980).

91. G. C. Becking, Vitamin A status and drug metabolism in the rat. *Can. J. Physiol. Pharmacol.* *51*:6 (1972).

92. E. W. Chu and R. A. Malmgreen, An inhibitory effect of vitamin A on the induction of tumors of the forestomach and cervix in the Syrian hamster by carcinogenic polycyclic hydrocarbons. *Cancer Res.* *25*:884 (1965).

93. U. Saffioti, R. Monteseno, A. R. Sellakum, and S. A. Borg, Experimental cancer of the lung: Inhibition by vitamin A of the induction of tracheobronchial squamous metaplasia and squamous cell tumors. *Cancer Res.* *20*:857 (1967).

94. A. E. Rogers, B. J. Hernden, and P. M. Newbern, Induction by dimethylhydrazine of intestinal carcinoma in normal rats and rats fed high or low levels of vitamin A. *Cancer Res.* *33*:1003 (1973).

95. R. Peto, R. Doll, J. D. Buckley, and M. B. Sporn, Can dietary beta-carotene materially reduce human cancer rates. *Nature*, *290*:201 (1981).

96. M. Ruchiwarat, A. Aramphongphan, V. Tanphaichitr, and W. Bandittanukool, The effect of thiamine deficiency on the metabolism of acetaminophen (paracetamol). *Biochem. Pharmacol.* *30*: 1901 (1981).

97. M. Ruchiwarat, W. Mahathanatrakul, Y. Srhashafanant, and D. Kitikool, Effects of thiamine deficiency on the metabolism and acute toxicity of dimethylnitrosamine in the rat. *Biochem. Pharmacol.* *27*:1783 (1978).

98. A. E. Wade, F. E. Green, R. H. Ciordia, J. S. Meadows, and W. O. Caster, Effects of dietary thiamine intake on hepatic drug metabolism in the male rat. *Biochem. Pharmacol.* *18*:2288 (1969).

99. W. Grosse and A. E. Wade, The effect of thiamine consumption on liver microsomal drug metabolizing pathways. *J. Pharmacol. Exp. Ther.* *176*:758 (1971).

100. A. E. Wade, B. Wu, C. M. Holbrook, and W. O. Caster, Effects of thiamine antagonists on drug hydroxylation and properties of cytochrome P450 in the rat. *Biochem. Pharmacol. 22:* 1573 (1973).

101. A. E. Wade, B. Wu, and J. Lee, Nutritional factors affecting drug metabolizing enzymes of the rat. *Biochem. Pharmacol. 24:* 785 (1975).

102. R. K. Richards, K. Keuter, and T. I. Klatt, Effects of vitamin C deficiency on the action of different types of barbiturates. *Proc. Soc. Exp. Biol. Med. 48:* 403 (1941).

103. J. Axelrod, S. Undenfriend, and B. B. Brodie, Ascorbic acid in aromatic hydroxylation. III. Effects of ascorbic acid on hydroxylation of acetanilid, aniline and antipyrine in vivo. *J. Pharmacol. Exp. Ther. 111:* 176 (1954).

104. A. H. Conney, G. A. Bray, C. Evans, and J. J. Burns, Metabolic interactions between L-ascorbic acid and drugs. *Ann. N.Y. Acad. Sci. 92:* 115 (1961).

105. R. Kato, A. Takanata, and T. Oshima, Effect of vitamin deficiency on the metabolism of drugs and TPNH linked electron transport system in liver microsomes. *Jpn. J. Pharmacol. 19:* 25 (1968).

106. V. G. Zannoni, E. J. Flynn, and M. M. Lynch, Ascorbic acid and drug metabolism. *Biochem. Pharmacol. 21:* 1377 (1972).

107. V. G. Zannoni and P. H. Sato, Effects of ascorbic acid on drug metabolism. *Ann. N.Y. Acad. Sci. 258:* 119 (1975).

108. V. G. Zannoni, P. H. Sato, and L. E. Rikans, Ascorbic acid and drug metabolism, in *Nutrition and Drug Interrelations* (J. H. Hathcock and J. Coon, eds.). Academic Press, New York, 1978, p. 347.

109. D. E. Holloway, S. W. Hutton, F. J. Peterson, and W. C. Duane, Lack of effect of subclinical ascorbic acid deficiency upon antipyrine metabolism. *Fed. Proc. 40:* 915 (1981).

110. A. D. Beattie and S. Sherlock, Ascorbic acid deficiency in liver disease. *Gut 17:* 571 (1976).

111. J. T. Wilson, C. J. VanBoxtel, G. Alvan, and F. J. Sjoqvist, Failure of vitamin C to affect the pharmacokinetic profile of antipyrine in man. *J. Clin. Pharmacol. 16:* 265 (1976).

112. J. B. Houston and G. Levy, Drug biotransformation interactions in man. VI. Acetaminophen and ascorbic acid. *J. Pharm. Sci. 65:* 1218 (1976).

113. M. P. Carpenter, Vitamin E and microsomal drug hydroxylations. *Ann. N.Y. Acad. Sci. 203:* 93 (1972).

114. L. Horn, M. Brin, and M. Barker, Effect of vitamin E deficiency on drug metabolism. *Fed. Proc. 33:* 672 (1974).

115. L. R. Horn, L. J. Machlin, M. O. Barker, and M. Brin, Drug metabolism and hepatic heme proteins in the vitamin E deficient rat. *Arch. Biochem. Biophys. 172:* 270 (1976).

116. P. I. Caasi, J. Haughworth, and P. P. Nair, Biosynthesis of heme in vitamin E deficiency. *J. Biol. Chem. 245*:5498 (1970).
117. A. T. Diplock, H. Baum, and J. A. Lucy, The effect of vitamin E on the oxidation state of selenium in rat liver. *Biochem. J. 123*:721 (1971).
118. A. S. M. Giasuddin, C. P. J. Caygill, A. T. Diplock, and E. Jeffrey, The dependence on vitamin E and selenium of drug demethylation in rat liver microsomes. *Biochem. J. 146*:339 (1975).
119. R. F. Burk and B. S. S. Masters, Some effects of selenium on the hepatic microsomal cytochrome P450 system in the rat. *Arch. Biochem. Biophys. 170*:124 (1975).
120. J. V. Dingell, P. E. Joiner, and L. Hurwitz, Impairment of drug metabolism in calcium deficiency. *Biochem. Pharmacol. 15*:97 (1966).
121. G. C. Becking, Hepatic drug metabolism in iron-, magnesium-, and potassium-deficient rats. *Fed. Proc. Am. Soc. Exp. Biol. 35*:2480 (1976).
122. G. C. Becking, Dietary minerals and drug metabolism, in *Nutrition and Drug Interrelations* (J. H. Hathcock and J. Coon, eds.). Academic Press, New York, 1978, p. 371.
123. C. S. Catz, M. R. Juchall, and S. J. Yaffe, Effects of iron, riboflavin and iodide deficiencies on hepatic drug metabolizing enzyme systems. *J. Pharmacol. Exp. Ther. 174*:197 (1970).
124. H. Hoensch, C. H. Woo, and R. Schmid, Cytochrome P450 and drug metabolism in intestinal and villous and crypt cells of rats; Effect of dietary iron. *Biochem. Biophys. Res. Commun. 65*:399 (1975).
125. K. O'Malley and I. H. Stevenson, Iron deficiency anemia and drug metabolism. *J. Pharm. Pharmacol. 25*:339 (1973).
126. G. J. Mulder and K. Keulemans, Metabolism of inorganic sulfate in the isolated perfused rat liver. *Biochem. J. 176*:959 (1978).
127. P. Moldeus, B. Anderson, and V. Gergely, Regulation of glucuronidation and sulfate conjugation in isolated hepatocytes. *Drug Metab. Dispos. 7*:416 (1979).
128. B. G. Woodcock and G. C. Wood, Effect of protein-free diet on UDP-glucuronyltransferase and sulfotransferase activities in rat liver. *Biochem. Pharmacol. 20*:2703 (1971).
129. J. Magdalou, D. Steimetz, A. M. Batt, B. Poullain, G. Seist, and G. Debry, The effect of dietary sulfur-containing amino acids on the activity of drug metabolizing enzymes in rat liver microsomes. *J. Nutr. 109*:864 (1979).
130. E. J. Glazenberg, I. M. C. Jekel-Halsema, E. Scholtens, A. J. Baars, and G. J. Mulder, Effects of variation in the dietary supply of cysteine and methionine on liver concentrations of

glutathione and "active" sulfate (PAPS) and serum levels of
sulfate, cystine, methionine and taurine in relation to the metabolism of acetaminophen. *J. Nutr.* *113*:1363 (1983).

131. M. I. Thabrew, O. O. Olorunsogo, J. O. Olowookere, and
     E. A. Bababunmi, Possible defect in xenobiotic activation be-
     fore glycine conjugation in protein-energy malnutrition. *Xeno-
     biotica* *12*:849 (1982).

132. G. J. Mulder, T. J. M. Temmink, and H. J. Koster, The effect
     of fasting on sulfation and glucuronidation in the rat in vivo.
     Biochem. Pharmacol. *31*:1941 (1982).

133. J. Alvin and B. N. Dixit, Pharmacological implications of altera-
     tions in the metabolism of chloramphenicol. *Biochem. Pharmacol.*
     23:139 (1974).

134. M. Marselos and M. Laitinen, Starvation and phenobarbitone
     treatment effects on drug hydroxylation and glucuronidation in
     the rat liver and small intestinal mucosa. *Biochem. Pharmacol.*
     24:1529 (1975).

135. S. Mehta, C. K. Nain, B. Sharma, and V. S. Mathur, Drug
     metabolism in malnourished children, in *Nutrition in Health and
     Disease and International Development* (A. E. Harper and G. K.
     Davies, eds.). Alan R. Liss Inc., New York, 1981, p. 739.

9

Pathological Changes Associated with Drug-Induced Malnutrition

Daphne A. Roe

Cornell University
Ithaca, New York

I. INTRODUCTION

Drug-induced malnutrition results from the intake of drugs that reduce the absorption, impair the utilization, or promote the excretion of
nutrients to an extent that nutrient loss exceeds nutrient gain.   The
malnutrition associated with drug intake may be multifactorial, either
because two or more drugs are absorbed having additive or synergistic
effects, or because concurrent inadequacies of diet or disease contribute to nutrient depletion. Drug-induced malnutrition may also be
complex in the sense that the drug has adverse effects on more than
one nutrient.

II. CLASSIFICATIONS OF DRUG-INDUCED MALNUTRITION

A. Nutrient Depletion by Causes Independent of Cellular Injury

Drug-induced nutrient depletion may be due to precipitation, chelation,
complexation, or sequestration of a nutrient by a drug with excretion
of a nonutilizable product via the intestine or urine. Examples include
phosphate depletion, induced by aluminum-containing antacids (1);
zinc depletion or copper depletion induced by penicillamine (2,3); riboflavin depletion, following boric acid absorption (4); pyridoxine

Table 1 Drug-Induced Nutrient Depletion Independent of Biochemical and Cellular Injury

Drug	Nutrient depletion	Mechanism
Sodium bicarbonate	Folate	Luminal pH ↑ in proximal small intestine
Aluminum hydroxide	Phosphate	Precipitation of dietary phosphate
Mineral oil	Beta-carotene	Solubilization
Boric acid	Riboflavin	Complexation
Penicillamine	Zinc; copper	Chelation
Cholestyramine	Fat, fat-soluble vitamins, vitamin B_{12}	Bile acid sequestration
	Folate	Adsorption
Isoniazid	Vitamin B_6	Schiff-base formation and excretion of the product

deficiency, induced by isoniazid, due in part to Schiff-base formation and in part to urinary excretion of the product (5); and folate depletion with chronic intake of the bile acid sequestrant, cholestyramine (6). Nutrient depletion from such causation may be independent on both biochemical and cellular injury. A listing of drug-induced depletion by causes that are independent of cellular injury is given in Table 1.

B. Nutrient Depletion Secondary to Drug-Induced Inhibition of Enzymes Required for Coenzyme Biosynthesis

Vitamin antagonists can cause an acute vitamin deficiency despite continued intake of that particular vitamin. Indeed, by designating a drug as a vitamin antagonist it can be assumed that the drug will either inhibit an enzyme essential to coenzyme biosynthesis or will inhibit an enzyme necessary to nutrient activation. For example, methotrexate induces acute folate deficiency primarily because it inhibits the dihydrofolate reductase enzyme and therefore blocks utilization of dietary folate. Methotrexate has multiple antifolate effects. In the tissues, the drug binds tightly to the dihydrofolate reductase enzyme, and folate displaced from this enzyme by the drug, is hyperexcreted in the urine. Methotrexate is taken up by cells such as hepatocytes, polyglutamates are formed, and the synthesis of folate polyglutamates is impaired. Methotrexate polyglutamates also inhibit dihydrofolate reductase and thymidylate synthetase. These effects of methotrexate result in inhibition of DNA, RNA, and protein synthesis (7).

Coumarin anticoagulants inhibit the reductase system required for the production of the active form of vitamin K from the storage form, vitamin K-2,3-epoxide (8-10).

It is important to consider the types of biochemical lesion that can result from use of specific vitamin antagonists and also to relate such lesions to observed cell injury and the signs of toxicity. For example, methotrexate acutely produces mitotic arrest. Mitotic arrest inhibits cell repair mechanisms, and hence, in tissues with a rapid turnover such as the epithelium of the small intestine, the methotrexate effect prevents the replication of crypt cells, resulting in villous atrophy. The cellular damage and chronic pathology that result from prolonged administration of methotrexate occurs because intracellular formation of methotrexate polyglutamates allows accumulation of the free drug intracellularly, and the drug continues to exert a toxic effect (7,11).

The pathology that ensues from coumarin anticoagulant intoxication is related to the hemorrhagic phenomena that occur when the clotting mechanism is impaired through inadequate synthesis of vitamin K-dependent clotting proteins (12).

The forms of nutrient depletion that are induced by drugs that are vitamin antagonists are listed in Table 2.

C. Drug-Induced Nutrient Depletion Which Is Dependent on Tissue Injury

Nutrient depletion can result from cellular injury caused by the effects of xenobiotics. Cellular injury may result from metabolism of a parent drug to a proximate toxin and adduction of the proximate toxin to cellular macromolecules. Drugs can cause cell damage through denaturation of membranes, which enhances the risk of oxidative injury to the cells. Drug-induced phototoxicity can impair the integrity of cell membranes through lysosomal damage. In discussing the mode of action of toxins in causing alterations in cellular physiology, Smuckler (13) emphasizes that while two modalities have been proposed, i.e., adduction of cellular macromolecules which may alter function and initiation of chain reactions involving free radicals or peroxides, these theories do not specifically answer the question of how toxins actually cause cell pathology (14).

Concerning the present discussion, the major question is not the precise relationship between biochemical lesion and cellular injury, but rather the impact of cellular injury on the intracellular biosynthesis of nutrients; on the transport of nutrients into cells, within cells, and out of cells; and also the utilization of nutrients by cells, that are our present concern. Up until recently, we have been most interested in the recognition of cells that have undergone toxic injury. The fatty

Table 2   Nutrient Depletion by Vitamin Antagonists

Drug	Nutrient depletion (impaired utilization)
Methotrexate	Folate
Pyrimethamine	Folate
Pentamidine isothionate	Folate
Trimethoprim	Folate
Triamterene	Folate
Sulfasalazine	Folate
Isoniazid	Vitamin B_6
Hydralazine	Vitamin B_6
Penicillamine	Vitamin B_6
Nitrous oxide	Vitamin B_{12}
Phenytoin	Vitamin D
Phenobarbital	Vitamin D
Isoniazid	Vitamin D
Coumarin anticoagulants	Vitamin K

infiltration of the hepatocyte, the appearance of the Mallory body in the hepatocyte, and the appearance of the Heinz body in the erythrocyte, all are examples of such histological markers of cell damage. On the other hand, temporary and reversible drug-induced changes in the cell membrane may inhibit uptake of nutrients. For example, anionic drugs that inhibit inorganic anion transport in human erythrocytes also depress the uptake of folate into these cells (15).

A listing of drug-induced nutrient depletion secondary to tissue injury is given in Table 3.

The chemically induced lesions that cause nutrient depletion and the drugs responsible for these lesions will be discussed to clarify relationships between pathology and adverse nutritional outcome.

III.   THE PATHOPHYSIOLOGY OF MALABSORPTION IN DRUG-INDUCED ENTEROPATHIES

Malabsorption results from several different effects of drugs on the gastrointestinal tract. Mechanisms by which drugs can cause malabsorption can be grouped according to their pathophysiological impact.

Functional deficits in nutrient absorption, caused by drugs, may be related to changes in gut motility or function or to abnormalities of epithelial cell transport. With some drugs, more than one mechanism may operate to cause malabsorption. Malabsorption may be secondary to maldigestion or can be secondary to a drug-induced nutritional deficiency. Not infrequently, drug-induced malabsorption is conditioned by a pre- or coexistent disease process. Indeed, there are examples of drugs that only appear to induce significant malabsorption in the presence of a specific disease process.

A. Biguanides

A high prevalence of intestinal side effects is associated with the use of biguanides, phenformin and metformin, in patients with insulin-independent diabetes. These drugs may inhibit gastric emptying and may also cause malabsorption (16).

Anorexia, dyspepsia, and diarrhea have been reported in patients maintained on biguanide therapy. Prolonged use may cause vitamin B_{12} malabsorption. Berchtold et al. (17) and Tomkin et al. (18) showed that long-term therapy with the biguanide metformin may result in impairment of vitamin B_{12} status. Tomkin (19) also reported that malabsorption of vitamin B_{12} may occur with intake of the related

Table 3 Drug-Induced Nutrient Depletion Secondary to Tissue Injury

Drug	Nutrient depletion	Predominant cells affected by toxic injury
Phenformin	Vitamin B_{12}	Enterocytes
Metformin (biguanides)	Glucose	Enterocytes
Methotrexate	Fat	Enterocytes
Aminopterin	Folate	Hepatocytes
Colchicine	Sodium	Enterocytes
	Potassium	Enterocytes
	Nitrogen	Enterocytes
	Fat	Enterocytes
	Disaccharides	Enterocytes
	Vitamin B_{12}	Enterocytes
Alcohol	Riboflavin	Hepatocytes, enterocytes
	Thiamin	Neurons
	Folate	Enterocytes, hepatocytes

biguanide phenformin. With either drug, this selective malabsorption
is reversible when the drug is discontinued. Biguanide-induced malabsorption of vitamin B$_{12}$ may be due to either competitive inhibition

of vitamin absorption in the distal ileum or to drug-induced inactivation of vitamin B$_{12}$. In the rat, phenformin impairs vitamin B$_{12}$ status

only in animals in which chemical diabetes has been induced by alloxan
treatment. However in human subjects biguanide-induced malabsorption of vitamin B$_{12}$ has not been linked to diabetic diarrhea, which is

due to a diabetic autonomic neuropathy, nor has it been related to
pancreatic exocrine dysfunction.

It has been found that 30% of diabetic patients receiving metformin
for 4 to 5 years show vitamin B$_{12}$ malabsorption (18). However hema-

tological evidence of vitamin B$_{12}$ deficiency with megaloblastic anemia

is rare. An isolated case of megaloblastic anemia induced by vitamin
B$_{12}$ deficiency was described by Callaghan et al. (20) in a 58-year-old

woman who had been taking metformin for 8 years.

Relationships between biguanide-induced changes in absorptive capacity and alterations in the ultrastructure of the enterocytes were
investigated by Arvanitakis et al. in 1973 (21). Absorption rates for
glucose, water, and sodium were determined using a triple-lumen tube
technique before and after administration of 100 mg of phenformin in
tablet form. Five patients without evidence of intestinal disease were
studied by this technique, and thereafter in separate experiments,
jejunal mucosal biopsies were obtained in three other volunteers before,
and 75 min after, the administration of 100-mg phenformin. Jejunal
absorption rates for glucose, water, and sodium were markedly reduced in the five study subjects, and these rates returned to normal
approximately 175 min after administration of the drug. There were
no changes in mucosal biopsy structure as shown by light microscopy,
but on electron microscopy, marked changes were seen in the mitochondria of the enterocytes after administration of phenformin. It
was demonstrated that following administration of the drug matrix
granules completely disappeared from the absorptive cell mitochondria
throughout most of the villus. Further, the intermembrane spaces
had an altered electron density. The investigators intimate that the
mitochondrial changes may lead to inhibition of mitochondrial energy
metabolism, which would limit the availability of ATP necessary for
active intestinal absorption of glucose and sodium.

Since it has been shown that the biguanides impair glucose absorption, the suggestion has been made that the action of these drugs in
reducing blood sugar may be related to glucose malabsorption.

The question that remains unanswered is the relationship between
the structural changes induced by the biguanides in the enterocyte

and the defect in vitamin B_{12} absorption, which was identified by previous investigators. The absorption of vitamin B_{12} in the ileum involves a non-energy-requiring attachment of the vitamin intrinsic factor complex to a membrane protein. In experimental animals vitamin B_{12} is transiently attached to mitochondria during its residence time in the ileal enterocytes (22-24). Whether or not the process could be interrupted by mitochondrial injury is presently unknown (22).

B. Folate Antagonists

Arrest of mitosis in the epithelium of the small intestine, either induced by ionizing radiation or by folate antagonists produces morphological changes that resemble those seen in tropical sprue. The morphological changes induced by methotrexate in the human proximal small intestine were first examined by Trier (25). In this study, 48 peroral duodenal and proximal jejunal biopsy specimens were obtained from 14 patients who had no previous evidence of intestinal disease. The samples were obtained before and sequentially after a single 2 to 5-mg/kg dose of methotrexate over a 96-hr period. The villus pattern of the intestinal mucosa was unaltered by this treatment, but the number of mitoses in the crypts first decreased, and then after 6 to 48 hr there was mitotic arrest. These acute effects of methotrexate lasted approximately 96 hr after the dose.

Sprue-like changes were first produced in mice by another folate antagonist, aminopterin (4-aminopteroylglutamic acid) which was administered over 1 to 10 days. The sequence of intestinal changes induced by this folate antagonist consisted in an arrest of mitoses in the crypt, followed by sloughing of the epithelial intestinal cells, and an appearance of atrophic villi (26).

Preliminary electron microscopic examination of the intestinal lesions of aminopterin-treated mice confirmed the development of the sprue-like lesion and also showed excessive numbers of fat droplets in the jejunal mucosal cells of fasted drug-treated mice compared with control animals (27).

Both aminopterin and methotrexate may produce malabsorption with long-term dosage. The effect of brief aminopterin dosage on fat absorption has been studied at various times following administration of the drug. After mitoses have been inhibited by aminopterin, fat absorption remains normal in the rat for at least 32 hr. There is then a 2-day period of severe fat malabsorption. While crypt mitosis begins to recover on the fourth day following drug administration, fat malabsorption continues until about 24 hr later (28).

The relationship between mitotic arrest and the development of malabsorption is not specific for methotrexate or even for the group of drugs that are known to be folate antagonists. Rather, as Creamer

(29) has pointed out, when a drug induces decreases in the rate of
cell turnover and there is a reduction in the number of adult intestinal
epithelial cells, absorptive capacity is reduced.

Repeated oral administration of methotrexate slows the rate of absorption of the drug. Methotrexate can also cause malabsorption of
other drugs such as phenobarbital and isoniazid (30).

Methotrexate can cause nutrient malabsorption both with short- and
long-term dosage. Methotrexate-induced malabsorption has been studied in children under treatment for acute lymphoblastic leukemia. It
has been found that the impact of the drug on the absorptive process,
as reflected by changes in D-xylose absorption, is related to the spacing of drug treatments such that if the spacing is at 7-day intervals,
the effects are greater than if the drug dosage is more widely spaced
(31).

It has been demonstrated that certain drugs (e.g., triparanol) and
certain dietary deficiencies (e.g., niacin deficiency) that cause
marked intestinal mucosal injury, sensitize the mucosa to gluten toxicity. However this effect has not been found in laboratory rats that
received methotrexate. We may assume, therefore, that the malabsorption induced by methotrexate in human subjects is not influenced by
gluten exposure (32).

IV. SPINDLE POISONS

The foregoing discussion has shown that mitochondrial injury and mitochondrial arrest can each be associated with drug-induced malabsorption. It is important, however, to establish the concept that those
drugs that induce the same ultrastructural changes may differ with
respect to their effects on absorptive processes. This concept is well
illustrated in relation to the spindle poisons. Spindle poisons include
the antiinflammatory, anti-gout drug, colchicine; the antifungal
agent, griseofulvin; podophyllotoxin and its synthetic analogues; and
the vinca alkaloids. These drugs cause interference with the microtubular system within cells which results in destruction of the mitotic
spindle. The microtubular system is composed of alpha- and beta-
tubulin. Tubulin has a specific receptor for colchicine, podophyllotoxin, and the vinca alkaloids. Whereas colchicine and podophyllotoxin share a binding site, the vinca alkaloids bind to a different site
on the tubulin molecule (33,34).

When cells are exposed to vinca alkaloids such as vincristine or vinblastine, there is a dissolution of microtubules and the formation of
intracellular crystals which contain the drug bound to tubulin. Exposure of cells undergoing mitosis to either colchicine or the vinca alkaloids results both in the disappearance of the spindle, because of a
block in microtubular assembly, and condensation of the chromosomes,
which leads to inhibition of mitosis (35-37).

A. Vincristine Toxicity

Vincristine toxicity includes gastrointestinal side effects. Clinical signs include constipation, which may progress to adynamic ileus and intestinal ulceration (38,39). Guinea pigs treated with vincristine intraperitoneally or intravenously, and killed 1 to 5 days later, show a sequence of intestinal changes. On the first day after injection, necrotic cells are seen in the crypts, and mitoses are in metaphase. Over the following days, gross abnormalities occur with severe villus atrophy which resembles that seen in the acute gastrointestinal radiation syndrome. The epithelial cells of the mucosa are flattened, poorly stained, and are vacuolated. Occasionally polynucleated cells are seen. Signs of mucosal recovery begin on the fifth day after injection of the drug. Alterations in the intestinal intramural nerve plexus (Auerbach's plexus) are also seen. These changes develop within 24 hr after drug administration, and partial recovery is evidenced by the fifth day postdrug administration. Hobson et al. (40), who made these observations, concluded that vincristine has a twofold effect on the small intestine. It interferes with normal replicative function of the intestinal mucosal cells so that the mucosal renewal is inhibited, which results in mucosal atrophy. Also, the drug is neurotoxic to the intestine, as it is in other tissues. Guinea pigs that are injected with vincristine exhibit rapid weight loss, but growth resumes within 1 week after injection. Diarrhea is present during the period of weight loss.

Vincristine is administered intravenously. Frequently, it is used in combination with prednisone to treat acute lymphocytic leukemia. It is also used to treat patients with advanced Hodgkin's disease and in the chemotherapy of acute myelogenous leukemia and histiocytic lymphoma. Occasionally, it is used to treat solid tumors. The major toxicity is not to the gut but to the peripheral nervous system. Indeed, neurotoxicity is the dose-limiting side effect (41).

The fact that vincristine treatment is intermittent, rather than continuous, may explain the absence of a clinically important malabsorption syndrome. Further, the neurotoxic effects of vincristine to the gut wall lead to constipation and ileus rather than to a diarrheal state that might cause transient malabsorption.

B. Colchicine Toxicity

Colchicine, which has long been used as the drug of choice in acute gout, is a spindle poison that arrests mitosis in metaphase (42). Effects of colchicine on the gastrointestinal mucosa were described 20 years ago and were subsequently likened to the effects of x-rays (43, 44). Whereas histological changes in the intestinal mucosa follow colchicine administration, changes under light microscopy are variable. Malabsorption, which occurs with higher dosages of colchicine, may not parallel the extent of the histological abnormalities (45).

Diarrhea occurring in individuals receiving colchicine has been associated with drug-induced malabsorption. Race et al. (46) studied the effects of colchicine on intestinal function in human subjects. The drug was shown to produce increases in the fecal excretion of sodium, potassium, nitrogen, and fat. Colchicine also impairs the absorption of vitamin B_{12}. Disaccharidase activity is impaired by the drug, which explains secondary lactose intolerance (45,46).

Involvement of the microtubular system in intracellular transport is indicated by several studies. In particular, evidence has been presented that supports a role for microtubules in the transport of lipid across the intestinal epithelial cells. Pavelka and Gangl (47) studied the effects of colchicine on the intestinal mucosa and on the intestinal transport of lipid in fasted rats. After colchicine treatment, electromicrographs of intestinal epithelial cells from the tips of the jejunal villi showed a reduced number of microtubules, and lipid particles had accumulated in the cytoplasm of these cells. The Golgi apparatus, which normally has a supranuclear location, was displaced to lie in an apical position. Membranes containing the lipid particles were considered to be derived from the Golgi apparatus and from the smooth endoplasmic reticulum. The triglyceride content of the intestinal mucosa was increased in colchicine-treated animals, while serum triglyceride levels were reduced. By employing radiotracer methods, these investigators showed that while colchicine did not affect the uptake of fatty acids by the intestinal mucosa, there were changes in fatty acid esterification.

The most interesting aspect of this study is that it links the colchicine-dependent ultrastructural changes in the enterocyte with effects of the drug on intracellular lipid transport. Experiments have also been carried out by Stopa et al. (48) to investigate effects of colchicine on vitamin B_{12} uptake. These investigators use guinea pigs in their studies. In vitro assays of solubilized intrinsic factor-vitamin B_{12} receptors from the intestine of these animals and from the intestine of control and cascara-treated animals were carried out. Receptor activity of the solubilized extract was quantitated. Receptor activity was reduced by colchicine treatment and the activity increased again 2.5 days after colchicine administration was discontinued. It has therefore been concluded that colchicine reduces vitamin B_{12} absorption because of the drug's effects on the intrinsic factor-vitamin B_{12} receptor.

V. THE NEUROTOXICITY OF VITAMIN ANTAGONISTS

Vitamin antagonists that impair utilization of vitamin B_6, niacin, vitamin B_{12}, folate, or thiamin produce characteristic neuropathies and encephalopathies. The clinical signs and pathology related to the neurotoxicity of vitamin antagonists may closely resemble the neural signs of vitamin deficiency or may differ in that the signs and pathology are related to the toxicity of the agent on the central or peripheral nervous system.

A. Neurotoxicity of Isoniazid (Isonicotinic Acid Hydrazide)

Isoniazid (INH) neuropathy in human subjects is a purely sensory neuropathy, whereas in certain experimental animals such as rats, the neuropathy involves both sensory and motor nerves. Acute overdosage with isoniazid results in an encephalopathy with edema and demyelination of the cerebral white matter (49). Neurotoxic side effects associated with isoniazid include not only the chronic sensory polyneuropathy and the encephalopathy, but also acute encephalopathy, which follows massive overdosage with the drug.

The neurotoxic effects of isoniazid are explained by a drug-dependent vitamin B_6 deficiency. Indeed, 30 years ago Jones and Jones (50) suggested that vitamin B_6 deficiency might be involved in the etiology of isoniazid neuropathy, and at about the same time, Riley and his associates (51) showed that vitamin B_6 blocked the acute convulsant effects of isoniazid, both in experimental animals and in humans.

Biehl and Vilter (52) commented on the similarity between the neuropathy induced by the vitamin B_6 antagonist deoxypyridoxine and by isoniazid. Vitamin B_6 protects patients from the peripheral neuropathy even when high dosages (20 mg/kg) of isoniazid are administered (53).

Risk factors for isoniazid-induced neurotoxicity are both genetic and acquired; slow acetylators of the drug show a higher susceptibility. Acquired risk factors include alcoholism, diabetes mellitus, and malnutrition (54).

Isoniazid is an example of a hydrazide drug that combines with pyridoxal phosphate to form pyridoxal-INH hydrazone complexes which inactivate the coenzyme action of vitamin B_6. The acetyl-INH metabolite of isoniazid cannot form hydrazones with pyridoxal phosphate, and therefore the N-acetylation reaction of isoniazid to form acetyl-INH represents the major metabolic mechanism for the protection of the body against the drug-induced neurotoxicity (55).

Hydrazines and hydrazides produce neurotoxicity by the same mechanism, and with both isoniazid and monomethylhydrazine (MMH), a component of missile and rocket fuel systems, the neurotoxicity is prevented by rapid acetylation of the compound. Coadministration of pyridoxine with either INH or MMH was shown to protect against neurotoxicity of these compounds (56,57).

Whereas it has been established that chronic isoniazid neurotoxicity can be prevented by coadministration of vitamin B_6 in dosages between 10 and 25 mg daily, it has also been shown that a single high dose of vitamin B_6 can be used as an antidote to isoniazid overdosage. In the latter instance, however, it is necessary to give a single dose of pyridoxine hydrochloride equivalent to the gram amount of isoniazid that has been ingested. This intervention causes resolution of seizures and of the metabolic acidosis that accompanies acute isoniazid poisoning (58).

Pyridoxine is also protective against MMH intoxication, although the protection afforded by the vitamin, according to animal studies, is not as complete as with isoniazid (59).

Whereas it has been emphasized that vitamin B_6 may be used in megadosages to counteract the neurotoxicity of isoniazid and hydrazine overdosage, continued administration of massive doses of vitamin B_6 imposes a risk of development of a pyridoxine neuropathy which has been recently described in patients receiving continued megadoses of this vitamin (60).

1. Isoniazid-Induced Pellagra

Pellagra is precipitated in individuals on a marginal intake of niacin under circumstances when the conversion of tryptophan to niacin is also impaired (61).

Pyridoxal phosphate is a necessary coenzyme in the metabolism of tryptophan and its conversion to niacin; therefore in individuals who are receiving isoniazid without coadministration of pyridoxine, conversion of tryptophan to niacin may be inhibited. Pellagra, a rare complication of isoniazid therapy, was described by Pegum (62) and by DiLorenzo (63). Shanker (64) described 50 cases of pellagra among whom 10% had coexisting tuberculosis for which they were receiving isoniazid. This investigator assumed that the drug could have contributed to the development of pellagra in these individuals. Both cutaneous and neurological signs of pellagra may occur when the disease is precipitated by isoniazid.

B. Subacute Combined Degeneration Induced by Nitrous Oxide

Nitrous oxide oxidizes some forms of vitamin B_{12}, and it has been shown that when vitamin B_{12} has been oxidized in this manner, it no longer functions as a coenzyme. Nitrous oxide-induced vitamin B_{12} deficiency may produce a neuropathy as well as hematological effects that are manifest as megaloblastosis and development of a megaloblastic anemia (65-67). Oxidized vitamin B_{12} cannot activate methionine synthetase, and therefore normal levels of methionine can no longer be maintained (68).

In monkeys subacute combined degeneration of the spinal cord can be induced by exposure of the animals to an atmosphere of nitrous oxide. Spinal cord changes in these animals included spongy degeneration in the posterior columns, and in the lateral corticospinal, the spinocerebellar, and the anterior corticospinal tracts. Monkeys showing these histological changes have been ataxic and have also shown an inability to hold objects steady. Monkeys that received methionine supplements did not exhibit such neurotoxicity or development of subacute combined degeneration after nitrous oxide exposure (69).

C. Methotrexate Encephalopathy

High-dose regimens with methotrexate are advocated for the prevention or treatment of meningeal leukemia, lymphoma, or carcinoma. The optimum dosage, the duration of infusion, the route of infusion, and the frequency of administration vary between protocols. In high-dose therapy, methotrexate may be introduced directly into the intraventricular fluid via subcutaneous Ommaya reservoirs (7). This mode of therapy is known to prevent meningeal spread of lymphoreticular neoplasms that may be resistant to the systemic delivery of methotrexate.

Intracranial introduction of methotrexate has distinct therapeutic advantages. Adverse outcomes include brain damage. Focal brain damage, due to methotrexate, has been described in patients who have received this form of therapy for the treatment of a central nervous system tumor that is in relapse. If the catheter is displaced so that methotrexate is delivered directly into the brain substance, necrotizing cerebritis may develop. Focal damage to the brain is followed by dense glial scarring (70).

Methotrexate neurotoxicity is not only an outcome of penetration of the drug directly into the brain substance, but it may also occur if intrathecal methotrexate is given in combination with cranial irradiation. With this combination therapy, cortical thinning, intracerebral calcification, and mild dementia may ensue (71).

In the present context, it is important to evaluate the extent to which methotrexate neurotoxicity within the brain represents an antivitaminic effect of the drug. Other factors may contribute to brain damage including trauma, ionizing radiation, and the presence of intracerebral tumor tissue. As far as we are aware, the precise relationship between the antifolate effects of methotrexate and the neurotoxic consequences of intrathecal administration of the drug have not been well delineated, nor has there been research on the effects of long-term methotrexate neurotoxicity on the transport or utilization of nutrients within the brain. Since clinical experience of severe methotrexate neurotoxicity indicates that older patients are more at risk and that there is an enhanced risk in the presence of meningeal leukemia, it is unlikely that further animal studies of intracerebral neurotoxicity induced by this drug would necessarily clarify post hoc propter hoc relationships between this vitamin antagonist and pathological outcome (72).

D. Wernicke-Korsakoff Syndrome

The Wernicke-Korsakoff syndrome consists of an organic brain syndrome which occurs most frequently in alcoholics. The acute phase of this disease, Wernicke's encephalopathy, is manifested clinically by an acute confusional state accompanied by ocular abnormalities including nystagmus, paralysis of the ocular muscles including the lateral rectus, weakness of congugate gaze, ataxia, and commonly a polyneuropathy with weakness of the legs. There is a variable depression of the state of consciousness with common presentation as a stupor from which the patient can be roused. The disease occurs mainly in alcoholics who have abstained from food or who have eaten irregularly for long periods. Occasionally Wernicke's encephalopathy may occur in persons without a history of alcohol abuse who suffer either from gross malabsorption syndromes, metastatic malignancies, or have received prolonged parenteral hyperalimentation. Untreated Wernicke's encephalopathy commonly progresses into the chronic phase of the disease, which is known as Korsakoff's psychosis.

Korsakoff's psychosis is characterized by a gross memory deficit and loss of mental function. Brain pathology found at autopsy includes focal necrosis of brain cells within the midbrain, thalamus, and pons, with symmetrical loss of brain cells and myelin degeneration within nuclei in these areas of the brain and, more particularly, in relation to the ventricles. The condition is recognized as a thiamin deficiency or dependency disease in that in the acute phase (Wernicke's encephalopathy) the patients respond to the injection of thiamin in dosages of approximately 200 mg, intravenously, twice daily. If the disease is not treated in its early phases, there is progression to Korsakoff's psychosis in which response to thiamin is no longer evident (73).

The association of an encephalopathy with thiamin deficiency has been demonstrated in small animal species. A polio encephalomalacia has been found in dogs, cats, and foxes in association with chronic thiamin deficiency (74-76). Signs of the encephalopathy that have been described in dogs have been reversed by administration of megadoses of thiamin hydrochloride.

Whereas we are now able to describe and define the cause of drug-induced and alcohol-induced neurological disorders of the central nervous system, relationships between mechanisms and the localization of lesions are as yet not well defined.

VI. HEPATOTOXICITY OF VITAMIN ANTAGONISTS

A. Methotrexate

Hepatotoxicity is recognized as a chronic side effect of methotrexate administration. The hepatic pathology includes fatty infiltration, toxic hepatitis, and fibrosis or cirrhosis (77).

The prevalence of hepatic fibrosis in patients who have received methotrexate varies in different reports. Factors influencing the prevalence of fibrosis include drug dose, frequency of drug administration, the total dose of the drug and preexistent alcoholic liver disease. Dahl et al. (78), on the basis of a study of 44 patients with postmethotrexate hepatic disease, in which dosage schedules were either with prolonged small-dose protocol or by intermittent larger doses, concluded that the prolonged drug administration was more likely to be associated with hepatotoxicity.

A high total dose of methotrexate carries a higher risk of hepatic fibrosis (79,80). All studies indicate that the frequency of cirrhosis is low when the total dose of methotrexate is less than 1.5 g. Longer exposure to the drug was shown by Zachariae and Nyfors (81) to lead to a higher frequency of cirrhosis. Reports have stressed the association of chronic methotrexate hepatotoxicity and prior alcohol abuse (82,83).

The biochemical basis of methotrexate hepatotoxicity is known. Methotrexate is taken up by the hepatocytes through an active transport system, which is the same transport system used by physiological forms of circulating folate. In high concentrations of the drug it may also diffuse into the cell, and it is believed that with high-dose therapy this mode of access of the drug to the hepatocyte is more important than the active transport system. After the methotrexate is inside the hepatocyte, polyglutamates are formed. It has been well established that methotrexate polyglutamate synthesis increases not only with an increased drug concentration but also with the duration of exposure. The methotrexate polyglutamate remains for an extended period within the hepatocyte, where it is a direct inhibitor both of thymidylate synthetase and also of another enzyme, aminoamidazole-carboxamide ribonucleotide transformylase. This latter enzyme is involved in the de novo purine synthesis. Inhibitory functions of the drug are linked to the degree of cytotoxicity (84).

B. Isoniazid

The hepatotoxic metabolites of isoniazid are acetyl-isoniazid and acetyl-hydrazine. The hydrazine yields a chemically reactive acetylating agent which is bound covalently to liver cell protein (85). Formation of acetyl-isoniazid and acetyl-hydrazine from isoniazid is greater in people who are fast acetylators, and production of these metabolites is also enhanced when phenobarbital is administered (86). Isoniazid injury to the liver is usually manifest as an acute hepatocellular injury. Hepatocellular degeneration as well as cholestasis may be manifested. Chronic hepatic injury from isoniazid also has been seen with a histological picture of both macro- and micronodular fibrosis.

Pyridoxal does not form Schiff bases with the acetyl derivatives of isoniazid, and therefore administration of vitamin B_6 has no protective function after the formation of these metabolites. While it may be possible to protect the liver against isoniazid toxicity by giving massive dosages of vitamin B_6, such intervention may be unjustifiable because it may reduce the therapeutic efficacy of the drug (87).

VII. NUTRIENT DEPLETION SECONDARY TO DRUG-INDUCED TISSUE INJURY

It is well known that nutrient depletion may follow tissue injury. However the concept of leaky cells is too simplistic to explain observed effects. While the mode of action of a few drugs in inducing nutritional deficiency is known, in most cases such mechanisms are still unclear. Further, the proposed mechanisms are largely conjectural. The imperfect state of the art is pointed out by comments by Leo and Lieber (88) about existing theories relative to the production of hepatic vitamin A depletion in alcoholics. They have emphasized that malnutrition of dietary origin could contribute to depletion of hepatic vitamin A in ethanol-fed animals. However when ethanol is given to experimental animals that are maintained on a normal or even a high-vitamin A diet, there is a depression of hepatic vitamin A that cannot then be attributed to an inadequate intake. They have also noted that while enhanced degradation of vitamin A in the liver, in association with chronic alcoholic liver disease or enhanced mobilization of vitamin A, could produce depletion of the vitamin from the liver; neither of these mechanisms has been properly investigated.

We have observed that in ethanol-fed hamsters the flavin depletion occurs at levels of riboflavin intake that are adequate in control animals. In ethanol-fed hamsters concentrations of the flavin coenzymes flavin adenine dinucleotide (FAD) and flavin mononucleotide (FMN) are reduced. However we are still unable to explain why ethanolic damage to the liver or liver damage by acetaldehyde impairs synthesis of flavin coenzymes.

In the present state of the art, knowledge of interrelationships between tissue damage and subsequent or consequent nutrient depletion is meager and research in this area is greatly needed.

VIII. CONCLUSIONS

In this review, mechanisms responsible for drug-induced malnutrition have been examined. Present knowledge indicates that nutrient depletion can arise as the outcome of physicochemical interaction between drug and nutrient, as a result of drug-induced impairment in the absorption or metabolism of a nutrient, or as a result of tissue injury. To further understand the antinutrient effects of drugs, biochemists should collaborate with pathologists to design studies which will not only identify specific effects of drugs on nutrient metabolism but will also include ultrastructural investigations of subcellular components in which the drug-nutrient interaction is located. There is a further need to develop a nutritional profile to be used with studies of organ function in animal testing and clinical trials of new drugs to permit prospective assessment of the drugs' potential for causing nutritional deficiencies.

REFERENCES

1. M. Lotz, E. Zisman, and F. C. Bartter, Evidence of a phosphorus depletion syndrome in man. *N. Engl. J. Med.* 278:409-415 (1968).
2. J. M. Walshe, Penicillamine: A new oral therapy for Wilson's disease. *Am. J. Med.* 21:487-495 (1956).
3. Multicentre Trial Group, 1973. Controlled trial of D(-)penicillamine in severe rheumatoid arthritis. *Lancet 1*:275-280 (1973).
4. D. A. Roe, D. B. McCormick, and R-T. Lin, Effects of riboflavin on boric acid toxicity. *J. Pharm. Sci.* 61:1081-1085 (1972).
5. R. W. Vilter, The vitamin B_6-hydrazide relationship, in *Vitamins and Hormones* (R. S. Harris, J. A. Loraine, and I. G. Wool, eds.). Academic Press, New York, 1964.
6. R. J. West and J. K. Lloyd, The effect of cholestyramine on intestinal absorption. *Gut 16*:93 (1975).
7. J. Jolivet, K. H. Cowan, G. A. Curt, N. J. Glendeninn, and B. A. Chabner, The pharmacology and clinical use of methotrexate. *N. Engl. J. Med.* 309:1094-1104 (1983).
8. A. K. Willingham and J. T. Matschiner, Changes in phylloquinone epoxidase activity related to prothrombin synthesis and microsomal clotting activity in the rat. *Biochem. J. 40*:435-441 (1974).

9. G. L. Nelsestuen and J. W. Suttie, The mode of action of vitamin K. Isolation of a peptide containing the vitamin K-dependent portion of prothrombin. *Proc. Nat. Acad. Sci. USA 70*: 3366-3370 (1973).

10. R. G. Bell, J. A. Sadowski, and J. T. Matschiner, Mechanism of action of warfarin. Warfarin and metabolism of vitamin K-1. *Biochemistry 11*:1959-1961 (1972).

11. N. J. Clendennin, K. H. Cowan, B. T. Kaukfmann, M. V. Nadkarni, and B. A. Chabner, Dihydrofolate reductase from a methotrexate-resistant human breast cancer cell line: Purification, properties and binding of methotrexate and polyglutamates. *Proc. Am. Assoc. Cancer Res. 24*:276 (abs.) (1983).

12. D. A. Roe, *Drug-Induced Nutritional Deficiencies*. AVI Publishing Co., Westport, Conn., 1976, pp. 36-38.

13. E. A. Smuckler, Iatrogenic disease, drug metabolism, and cell injury: Lethal synthesis in man. *Fed. Proc. 36*:1708-1714 (1977).

14. J. R. Mitchell and D. J. Jollows, Metabolic activation of drugs to toxic substances. *Gastroenterology 68*:392 (1975).

15. R. F. Branda, Interactions of drugs and folate compounds at the cellular level, in *Nutrition in Health and Disease and International Development*, Symp. XII Internat. Congr. Nutr., Alan R. Liss, Inc., New York, 1981, pp. 773-781.

16. Editorial. Diabetes and the gut. *Br. Med. J. 1*:1742-1744 (1979).

17. P. Berchtold, P. Bolli, U. Arbenz, and G. Keiser, Intestinal malabsorption as a result of treatment with metformin. A question of the mode of action of the biguanide. *Diabetalogica 5*: 405-412 (1969).

18. G. H. Tomkin, D. R. Hadden, J. A. Weaver, and D. A. D. Montgomery, Vitamin B-12 status of patients on long-term metformin therapy. *Br. Med. J. 2*:685-687 (1971).

19. G. H. Tomkin, Malabsorption of vitamin B-12 in diabetic patients treated with fenformin: A comparison with metformin. *Br. Med. J. 3*:673-675 (1973).

20. T. S. Callaghan, D. R. Hadden, and G. H. Tomkin, Megaloblastic anaemia due to vitamin B_{12} malabsorption associated with long-term metformin treatment. *Br. Med. J. 280*:1214 (1980).

21. C. Arvanitakis, V. Lorenzsonn, and W. A. Olsen, Phenformin-induced alterations of small intestinal function and mitochondrial structure in man. *J. Lab. Clin. Med. 82*:195-200 (1973).

22. J. Lindenbaum, Malabsorption of vitamin B-12 and folate, in *Nutrition and Gastroenterology* (M. Winick, ed.) John Wiley & Sons, New York, 1980, pp. 105-123.

23. R. H. Donaldson, Mechanisms of malabsorption of cobalamin, in *Cobalamin* (B. M. Babior, ed.). John Wiley & Sons, New York, 1975, p. 335.

24. T. J. Peters and A. V. Hoffbrand, Absorption of vitamin B_{12} by the guinea pig. I. Subcellular localization of vitamin B_{12} in the ileal enterocyte during absorption. *Br. J. Hematol. 19*: 369 (1970).
25. J. S. Trier, Morphological alterations induced by methotrexate in the mucosa of human proximal intestine. I. Serial observations by light microscopy. *Gastroenterology 42*:295-305 (1962).
26. P. F. Millington, J. B. Finean, O. C. Forbes, and A. C. Frasier, Studies of the effects of aminopterin on the small intestine of rats. I. The morphological changes following a single dose of aminopterin. *Exp. Cell Res. 28*:162-178 (1962).
27. B. J. Rybak, Electromicroscopic changes of intestinal lesions. I. Aminopterin-induced lesions in mice. *Gastroenterology 42*: 306-318 (1962).
28. T. G. Redgrave and W. J. Simmons, Effect of aminopterin on the absorption of fat into the lymph of unanestetized rats. *Gastroenterology 52*:54-66 (1967).
29. B. Creamer, The turnover of the epithelium of the small intestine. *Br. Med. Bull. 23*:226-230 (1967).
30. M. Freeman-Narrod, in *Methotrexate in the Treatment of Cancer* (R. Potter and E. Wiltshore, eds.). Williams & Wilkins, Baltimore, 1962, pp. 17-21.
31. A. W. Kraft, H. E. M. Kay, D. N. Lawson, and T. J. McElwain, Methotrexate-induced malabsorption in children with acute lymphoblastic leukemia. *Br. Med. J. 2*:1511-1512 (1977).
32. J. S. Sandu and D. R. Fraser, Effect of dietary cereals on intestinal permeability in experimental enteropathy in rats. *Gut 24*:825-830 (1983).
33. N. L. R. Bucher, Microtubules. *N. Engl. J. Med. 287*:195-196 (1972).
34. T. Hokfelt and A. Dahlstrom, Effects of two mitosis inhibitors (colchicine and vinblastine) on the distribution and axonal transport of noradrenalin storage particles, studied by fluorescence and electron microscopy. *Z. Zellforsch. Mikrosk. Anat. 119*:460-482 (1971).
35. L. Wilson, Properties of colchicine binding protein from chick embryo brain. Interactions with vinca alkaloids and podophyllotoxine. *Biochemistry 9*:4999 (1970).
36. K. G. Bensoh and S. E. Malawista, Microtubular crystals in mammalian cells. *J. Cell Biol. 40*:95 (1969).
37. J. Bryan, Definition of three classes of binding sites in isolated microtubule crystals. *Biochemistry 11*:2611 (1972).
38. S. Lowenbraun, V. T. DeVita, and A. A. Serpick, Combination chemotherapy with nitrogen mustard, vincristine, procarbazine and prednisone in lymphosarcoma and reticulum cell sarcoma. *Cancer 25*:1018-1025 (1970).

39. S. G. Sandler, W. Tobin, and E. S. Henderson, Vincristine-
     induced neuropathy. *Neurology 19*:367-374 (1969).
40. R. W. Hobson, H. R. Jervis, R. L. Kingry, and J. R. Wallace,
     Small bowel changes associated with vincristine sulfate treat-
     ment: An experimental study in the guinea pig. *Cancer 34*:
     1888-1896 (1974).
41. W. B. Pratt and R. W. Ruddon, *The Anticancer Drugs*. Uni-
     versity Press, Oxford, New York, 1979, pp. 221-233.
42. E. W. Taylor, The mechanism of colchicine inhibition of mitosis.
     I. Kinetics of inhibition and the binding of H_3-colchicine.
     J. Cell Biol. 25:145-160 (1965).
43. P. Dustin, Jr., New aspects of the pharmacology of antimitotic
     agents. *Pharmacol. Rev. 15*:449-480 (1963).
44. J. C. Hampton, A comparison of the effects of x-radiation and
     colchicine on the intestinal mucosa of the mouse. *Radiat. Res.
     28*:37-59 (1966).
45. D. I. Webb, R. B. Chodos, C. Q. Mahar, and W. W. Faloon,
     Mechanisms of vitamin B_{12} malabsorption in patients receiving
     colchicine. *N. Engl. J. Med. 279*:845-850 (1968).
46. T. F. Race, I. C. Paes, and W. W. Faloon, Intestinal malab-
     sorption induced by oral colchicine. Comparison with neomycin
     and cathartic agents. *Am. J. Med. Sci. 259*:32-41 (1970).
47. M. Pavelka and A. Gangl, Effect of colchicine on the intestinal
     transport of endogenous lipid. Ultrastructural, biochemical and
     radiochemical studies in fasting rats. *Gastroenterology 84*:544-
     553 (1983).
48. E. G. Stopa, R. O'Brien, and M. Katz, Effect of colchicine on
     guinea pig intrinsic factor-vitamin B_{12} receptor. *Gastroenter-
     ology 76*:310-314 (1979).
49. J. B. Gavanagh, Toxic substances and the nervous system.
     Br. Med. Bull. 25:268-273 (1969).
50. W. A. Jones and G. P. Jones, Peripheral neuropathy due to
     isoniazid. *Lancet 1*:1073-1074 (1953).
51. R. H. Riley, K. F. Killam, E. H. Jenny, W. H. Marshall,
     T. Tausig, N. S. Apter, and C. C. Pfeiffer, Convulsant ef-
     fects of isoniazid. *J. Am. Med. Assoc. 152*:1317-1321 (1953).
52. J. P. Biehl and R. W. Vilter, Effects of isoniazid on pyridox-
     ine metabolism. *J. Am. Med. Assoc. 156*:1549-1552 (1954).
53. H. B. Carlson, E. M. Anthony, W. F. Russell, Jr., and
     G. Middlebrook, Prophylaxis of isoniazid neuropathy with pyri-
     doxine. *N. Engl. J. Med. 255*:118-122 (1956).
54. K. M. Citron, Drugs used in the treatment of tuberculosis and
     leprosy, in *Side Effects of Drugs. A Survey of Unwanted Ef-
     fects of Drugs Reported in 1968-1971*, Vol. 7 (L. Meyler, and
     A. Herxheimer, eds.). Exerpta Medica, Amsterdam, 1972,
     p. 424.

55. R. D. O'Brien, M. Kirkpatrick, and P. S. Miller, Poisoning of the rat by hydrazine and alkyl hydrazines. *Toxicol. Appl. Pharmacol.* 6:317-377 (1964).

56. D. W. Hein, *The Biochemical Basis of the N-acetylation Polymorphism and Its Toxicological Consequences.* Ph.D. Thesis, Univ. Michigan, Ann Arbor, 1982.

57. D. W. Hein, Isoniazid toxicity as related to acetylator status. *Fed. Proc.* 42:3087-3091 (1983).

58. S. Wason, P. G. Lacoutere, and F. H. Lovejoy, Single high-dose pyridoxine treatment for isoniazid overdose. *J. Am. Med. Assoc.* 246:1102-1104 (1981).

59. M. E. George, M. K. Pinkerton, and K. C. Back, Therapeutics of monomethylhydrazine intoxication. *Toxicol. Appl. Pharmacol.* 63:201-208 (1982).

60. H. Schaumburg, J. Kaplan, A. Windebank, N. Vick, S. Rasmus, D. Pleasure, and M. J. Brown, Sensory neuropathy from pyridoxine abuse: A new megavitamin syndrome. *N. Engl. J. Med.* 309:445-448 (1983).

61. R. W. Vilter, J. F. Mueller, and W. B. Bean, The therapeutic effect of tryptophane in human pellagra. *J. Lab. Clin. Med.* 34:409-413 (1949).

62. J. S. Pegum, Nicotinic acid and burning feet. *Lancet* 2:536 (1952).

63. P. A. DiLorenzo, Pellagra-like syndrome associated with isoniazid therapy. *Acta Dermato.-Venereol.* 47:318-322 (1967).

64. P. S. Shankar, Pellagra in Gulbarga. *J. Indian Med. Soc.* 54: 73-75 (1955).

65. I. Chanarin, Cobalamins and nitrous oxide: A review. *J. Clin. Pathol.* 33:909-916 (1980).

66. R. B. Layzer, Myeloneuropathy after prolonged exposure to nitrous oxide. *Lancet* 2:1227-1230 (1978).

67. J. A. L. Amess, J. F. Burman, G. M. Rees, D. G. Nance-kievill, and D. L. Molen, Megaloblastic haemopoiesis in patients receiving nitrous oxide. *Lancet* 2:339-342 (1978).

68. R. Deacon, M. Lumb, J. Perry, I. Chanarin, B. Minty, M. J. Halsey, and J. F. Nunn, Selected inactivation of vitamin B_{12} in rats by nitrous oxide. *Lancet* 2:1023-1024 (1978).

69. J. M. Scott, P. Wilson, J. J. Dinn, and D. G. Weir, Pathogenesis of subacute combined degeneration: A result of methyl group deficiency. *Lancet* 2:334-337 (1981).

70. R. J. Packer, R. A. Zimmerman, J. Rosenstock, L. B. Rorke, D. G. Norris, and P. H. Berman, Focal encephalopathy following methotrexate therapy. Administration via misplaced intraventricular catheter. *Arch. Neurol.* 38:450-452 (1981).

71. R. W. Nelson and J. T. Frank Intrathecal methotrexate-induced neurotoxicities. *Am. J. Hosp. Pharm.* 38:65-68 (1981).

72. W. A. Bleyer, J. C. Drake, and B. A. Chabner, Neurotoxicity and elevated cerebrospinal fluid methotrexate concentration in meningeal leukemia. *N. Engl. J. Med.* *289*:770-773 (1973).

73. M. Victor, R. D. Adams, and G. H. Collins, *The Wernicke-Korsakoff Syndrome. A Clinical and Pathological Study of 245 Patients, 82 with Post Mortem Examination.* F. A. Davis Co., Philadelphia, 1971.

74. D. H. Reed, R. D. Jolly, and M. R. Alley, Polio encephalomalacia with dogs with thiamine deficiency. *Vet. Pathol. 14*: 103-112 (1977).

75. K. V. Jubb, L. Z. Saunders, and H. V. Coates, Thiamine deficiency encephalopathy in cats. *J. Comp. Pathol. 66*:217-227 (1956).

76. C. A. Evans, W. E. Carlson, and R. G. Green, The pathology of Chastek paralysis in foxes. *Am. J. Pathol. 18*:79-92 (1942).

77. H. H. Roenigk, W. F. Bergfeld, R. St.Jacques, F. J. Owens, and W. A. Hawk, Hepatotoxicity of methotrexate in the treatment of psoriasis. *Arch. Dermatol. 103*:251-261 (1971).

78. M. G. C. Dahl, M. M. Gregory, and P. J. Scheuer, Methotrexate hepatotoxicity in psoriasis: Comparison of different dosage regimens. *Br. Med. J. 1*:654-656 (1972).

79. H. A. Shapiro, J. O. Trowbridge, J. C. Lee, and H. I. Maibach, Liver disease in psoriatics—an effect of methotrexate therapy. *Arch. Dermatol. 110*:547-551 (1974).

80. B. J. Podurgiel, D. B. McGill, J. Ludwig, W. F. Taylor, and S. A. Muller, Liver injury associated with methotrexate therapy for psoriasis. *Mayo Clin. Proc. 48*:787-792 (1973).

81. H. Zachariae and A. Nyfors, Liver biopsies from psoriatics related to methotrexate therapy. 3. Findings in post-methotrexate liver biopsies from 160 psoriatics. *Acta Pathol. Microbiol. Scand. 85*:511-518 (1977).

82. H. Zachariae and T. Schiodt, Liver biopsy in methotrexate treatment. *Acta Dermato.-Vernereol. 51*:215-220 (1971).

83. H. Zachariae, K. Kragbulle, and H. Sogaard, Methotrexate induced liver cirrhosis. *Br. J. Dermatol. 102*:407-412 (1980).

84. R. G. Moran, M. Mulkins, and C. Heidelberger, Role of thymidylate synthetase activity in development of methotrexate cytotoxicity. *Proc. Nat. Acad. Sci. USA 76*:5924-5928 (1979).

85. J. A. Timbrell, J. R. Mitchell, W. R. Snodgrass, and S. D. Nelson, Isoniazid hepatotoxicity: The relationship between covalent binding and metabolism in vivo. *J. Pharmacol. Exp. Ther. 213*:364-369 (1980).

86. J. R. Mitchell, H. J. Zimmerman, K. G. Ishak, U. P. Thorgeirsson, R. G. Moran, M. Mulkins, and C. Heidelberger, Role of thymidylate synthetase activity in development of methotrexate cytotoxicity. *Proc. Nat. Acad. Sci. USA 76*:5924-5928 (1979).

87. M. Black, J. R. Mitchell, H. J. Zimmerman, K. G. Ishak, and G. R. Epler, Isoniazid-associated hepatitis in 114 patients. *Gastroenterology* 69:289-301 (1975).

88. M. A. Leo and C. S. Leiber, Hepatic vitamin A depletion in alcoholic liver injury. *N. Engl. J. Med.* 307:597-601 (1982).

INDEX

A

Abortion, 36
Acetaldehyde, 372
Acetaminophen, 330, 334
Acetanilide, 333, 342
Acetate, 217, 233
Acetylaminofluorene, 296, 299, 302
Acetylators, 367
Acetylcholine, 181, 280, 281, 290, 291, 293, 302
  receptor, 294
Acetylcholinesterase, 29, 181
Acetyl-CoA carboxylase, 11
Acetylhydrazine, 372
Acetyl-isoniazid, 372
N-Acetylneuraminic acid, 167, 173
Achievement tests, 191
Acid detergent fiber, 254
Acoustic startle reflex, 8
Actinomycin D, 13-15, 21, 26-29, 70, 73
Activation, 329
Adaptive responses to dietary fiber, 227-233
Adenosine triphosphate, 11, 67, 68, 323, 362
S-Adenosylmethionine, 281, 299

Adrenal, 10
  adrenalectomy, 14, 22, 28-30, 85
  cortex, 30
  steroids, 30
Adrenocorticotropic hormone (ACTH), 30
Adynamic ileus, 365
Aflatoxin B_1, 27, 28, 295, 296, 299, 331, 346
Agar, 144, 252, 254
Aggression, 7
Aggressive behavior, 8
Aging, 16, 36
Alanine, 93-95
Albumin, 6, 12, 14, 22-24, 30, 66, 85-87, 95, 97
  mRNA, 84
  synthesis, 84, 90, 91
Alcohol, 129
Alcoholism, 97, 118, 120, 121, 367, 370, 372
Aldosterone, 30, 31
Alfalfa, 144, 215, 228, 252
  seed, 245
  sprouts, 245
Alkaline phosphatase, 30, 229
Alloxan, 362
Aluminum, 215
Alzheimer's disease, 294

Milton Keynes UK
Ingram Content Group UK Ltd.
UKHW020320111024
449327UK00040B/1461